ADVENTURES IN THEORETICS:
"Towards a Unified Motion-Force Theory of Universal Space-Time Curvature-Continuum!"

ADVENTURES IN THEORETICS:
"Towards a Unified Motion-Force Theory of Universal Space-Time Curvature-Continuum!"

Leo Emmanuel Lochard

ADVENTURES IN THEORETICS:
"Towards a Unified Motion-Force Theory of Universal Space-Time Curvature-Continuum!"

© Leo Emmanuel Lochard 2015

Published by
Lighthouse Christian Publishing
SAN 257-4330
5531 Dufferin Drive
Savage, Minnesota, 55378
United States of America

www.lighthousechristianpublishing.com

Introduction: Where We Stand Mathematically

We are living in the post-Relativity era. Over a hundred years have elapsed since Dr. Einstein pronounced his scientific discovery of The Special Theory of Relativity and then of The General Theory of Relativity.

But, in our times, the quest, or rather, the pursuit is towards developing a "Grand unification theory of the Universe!"

It is said that the four fundamental forces of Nature are: gravity, electromagnetism, the strong force, and the weak force, and that they need to be unified into a mega-theory that would account for all known physical phenomena.

Has there been any progress in creating a new comprehensive equation that would usher the next totalizing theory of universal integration? Given developments in post-Relativity physics, is the next theory to necessarily unify all fundamental forces, or are there "intermediate steps" still left to be discovered?

An a priori sine qua non necessary ingredient for the discovery of a universal unification theory is the formulated integration of Newton's Laws of Motion with Einstein's theories of Relativity, the sum of which, simultaneously accounting for the substantial Field properties of Electromagnetism that factor in generating a Force, as well as the paramount Motion-properties of Gravity that factor in generating a Force.

An object that is at rest or that is moving at a constant velocity will not undergo any change without additional force acting upon it. A body already in motion experiencing acceleration engenders Force as in the equation: Force equals mass x acceleration, or $f = ma$. When one body or object acts upon another body, a commensurate reaction is generated from that other body, in equal magnitude, but opposite in direction.

From this summary of the Laws of Motion discovered by Dr. Isaac Newton (AD 1643-1727), we can deduce that the property of Force means: To cause to move a body that is at rest, or to change the motion of an already moving body.

Newton's Law of Gravitation states that centers-of-mass "attract" centers-of-mass or that Gravity is an "attractive force" between centers-of-mass, as expressed in the equation $F_g = G\, m_1 m_2/r^2$ depicting the force of Gravity between two centers-of-mass as being directly proportional to the product of the two masses and inversely proportional to the squared distance between them, as "normalized" via the gravitational Constant.

Newton did not conceptualize Gravity as "a tensing theater" of interactive "mass-effects" akin to "magnetic field-force operations" that engender an apparent anomaly as "the perihelion shift" of planet Mercury. Rather, the force of Gravity yielded centri-vectored forms of rotary Motion that tended to model spherical or ellipsoid "paths of travel," e.g., an object leaving earth surface to again fall towards the center of the earth in a curvi-linear manner, after having reached "a peak distance-in-the-sky" away from its surface; a planet "traveling around the Sun" in a trajectory crafted along its ellipsoid plane of Revolution or around the Solar System's center-of-mass.

Undoubtedly, Newton did not define "the force of Gravity" in its intrinsic qualities, but rather demonstrated its applied characteristics in experimental observations or in actual natural events and universal phenomena.

Newton did not "operationalize" the force of Gravity in terms of its "space-curving properties," nor in terms of its "space-bending dynamics." It took Albert Einstein to do that. Rather, Newton depicted Gravity as an "attractive force" between bodies with mass: Centers-of-mass "attract" centers-of-mass.

Then, what is Gravity, really? And how is the force of Gravity different from the force generated by an electromagnetic field?

Dr. Albert Einstein (AD 1879-1955) proposed the Special Theory of Relativity in AD 1905. His General Theory of Relativity was published in AD 1915. Special Relativity explains "the equivalency principle" operating between Matter and Energy as formulated by the equation $E = mc^2$. General Relativity clarifies that Mass does not merely "attract" mass but also "curves" or "bends" space as it moves within Continuum Space-Time.

Energy is the equivalent of "excited matter" and Matter is the equivalent of "congealed energy," as $E = mc^2$ embodies a specific "Energy event," such that: "E" stands for energy, "m" stands for matter and "c" stands for the Speed of Light in a vacuum, utilized as a "multiplication conversion factor" to the second exponent power.

Thus, as things stand in Physics at this time, Gravity and Electromagnetism remain the two most important quantified properties, the sum of which, impacting the way in which Transformation of Energy cycles in measurable iterative patterns, proportions, and quantities, even as "mass frames" constituted of Matter simultaneously and contiguously move, within the "Gravitational Metrics of Moving Bodies within the Curvature of Continuum Space-Time," e.g., given the specific mass of planet Earth, lunar mass-properties of attraction and motion, and its proximal relative distance from the Sun, the Earth completes one Rotation upon its own 23-degree tilted axis during a period of 24 hours; whereas its travel around the Sun along the Solar System's ellipsoid plane of Revolution is completed in the span of a year or 365 ¼ days.

As energy transformation cycles in observable iterative patterns of operation within differentiated conditions respective to parameters/variables/determinants characterizing each "frame of reference," — e.g., Frame of Solar-Planetary Motion Dynamics as exemplified by relatively constant iterative periods of earth Rotation and Revolution, respectively, —the sum of which, fulfilling the required cause-and-effect relationships arising from the presence of "catalysts of Motion" ensuing from the ubiquitous operations of the "Universal Input-Process-Output Mechanism" that typify processes of Energy Transformation," at the same time, numerous "Frames of Mass-in-Motion,"— e.g., Frame constituted of the "gravi-magnetic mechanics" of the Earth-Moon Complex (Newton;) or Frame constituted of "Sun-Planets gravitational space-curving or space-bending dynamics" (Einstein;) or Frame of "galaxies-in-Motion," each galaxy, respectively constituted of solar systems that continuously revolve/rotate around each other, etc... — are acted upon by "Gravity-force equivalents," "pseudo-Gravity motion-forces," electro-magnetic fields, and "Curvature pressure-force Energetics," or "Space Curvature tensor-metrics," all of which, working together in "simultaneous contiguity" to actively sustain the maintenance of Solar System and universal "gravitational metrics of dynamic equilibrium," e.g., the Sun's gravitational radiation energy output in proportion to absorption rates of the Electro-magnetic Spectrum by Planet Earth during its Rotational and Revolutionary Cycles, as synchronized with lunar effects, holistically designed for maintaining the viability of Earth Ecology, as well as of all its life-support systems.

Since the great impetus in Astro-Physics now, is for a general theory that unifies all universal forces, what is the scientific paradigm "inspiring" such efforts?

Regrettably, it is "the theory of evolution," otherwise referred to as "Evolutionism" or "Evolution Theory!"

But, what are "Fundamentals of Physics" as opposed to "fundamentals of Evolutionism?" Are they necessarily "in synch" or do they coincide in integrated consistency for real scientific discovery?

Are astro-physicists active in real scientific inquiry pursuits, or do they engage in rituals that evoke "simply going through the motions" while acting as "the secular priesthood of evolutionist believers?" Has evolutionism become "the new religious ideology" undergirding their pretense of following the rigorous procedural tenets of "the Scientific Method?"

In establishing the Scientific Foundation that is necessary for materializing such an endeavor as discovering a so-called "grand-unification theory" for explaining the whole Universe in its totality, we must develop a "Body of Principles in Physics Fundamentals," which must include all constituent components of the Universe made manifest as "Force," e.g., Gravity-force, Field-force, the sum of which, to factor into its comprehensive formulation: theoretical, mathematical, applied, and technological.

Fundamental refers to "foundation." Foundation is "the base on which something rests; the principle upon which something is built."

How does the Theory of Relativity "fit into" this "grand-unification theory Formulation?" Because of the way in which the Theory of Relativity is applied to scientific problems that touch human cognitive characteristics and psycho-motor attributes, it is commonly held that, even the human Mind appears to factor into observed experimental results.

Consequently, research hypotheses should be aligned with what is observed in Nature rather than with preconceived evolutionist doctrines scavenging "for evidence that fits" a preconceived, already-decided, already-found result, e.g., "cosmic background radiation."

In sum, given the dearth of factual scientific evidence to substantiate them, the things that astro-evolutionists are looking for might not necessarily correspond to what is really occurring in Nature, e.g., scientists have discovered that it is not possible to simultaneously determine via exact measurements the momentum and position of an "accelerated isolated particle."

Thus, at present, there is no validly proven all-encompassing totalizing paradigm that works as a unifying theory of the Universe.

From Newton to Einstein, theoretical development in mathematical physics had been a fruitful adventure that climaxed into leaps of technological innovation. However, this pattern of progressive refinement in "Physical Sciences Theoretics" and in technologically applied Physics has practically come to a dead-end as pre-deterministic evolutionist doctrines have infiltrated the scientific basic-research and research-and-development establishment.

Evolutionist propositions and assertions cannot be scientifically confirmed through laboratory or field experiments, given that all research frames are predicated upon proving the preconceived "big bang theory" of universal beginnings.

The current overarching paradigm that governs scientific inquiry is the theory of evolution as well as all pre-deterministic doctrines appended to its ideological framework of unproven conjectures.

Astro-evolutionists desire that "the past universe" as supposedly represented by "relic background radiation," would factor into our formalized understanding of current universal phenomena, natural events, and cosmic processes. Yet, no experiments have been designed to research these conjectures as a consistent hypothesis; for example, astro-evolutionists are "looking" for relic cosmic background radiation because they've already decided that it "must exist" as a by-product or as a consequent residue of "the big bang explosion" from which they presuppose the Universe to have had begun.

Thus, evolutionists' modus operandi is settled in this format: Pre-deterministic questions are posed; Pre-deterministic answers are provided; and Pre-deterministic proofs are expected either in Nature or in the Universe! But, to date: None to be found or discovered!

Regrettably then, there is a need to broaden horizons, to enrich thought processes, and to "clear the theoretical cul-de-sac," built on "false hopes" as well of "denial of reality," both of which, afflicting the mentality and activities of astro-evolutionists; and the sum of which, has solidified stagnation, stultified innovation, and "deterred" newness of Mind, through a form of "pseudo-scientific fascism" called "the peer review process!"

These fruitless pursuits and aimless activities of astro-evolutionists can be overcome only by returning to the true-to-Reality Scientific Method, a "movement of the Mind" that is so necessary, in order to overcome not only theoretical sterility, but to also initiate the literal resurgence of newly aroused personal motivation and scientific interest in re-igniting the "re-circulation" of the innovative intellectual transformation heretofore responsible for our progressively refined understanding of universal events, cosmic processes, and stellar phenomena.

For, it is not to the advantage and benefit of Science that all conceptive efforts at theoretical discovery should be "tunneled" into only one doctrinal vision-frame: That which is circumscribed by the theory of evolution.

How do "Fundamental Equations," e.g., $E = mc^2$, and their proven-applied scientific principles "interface" in factoring relative gravi-Energy transactions between universal frames-of-reference that would then expand our understanding of the necessary ingredients that fulfill our need for discovering a "mega Physics theory," the sum of which, to conceptually, experimentally, and technologically "hold" the gravitational Space-Time Continuum together, as proven by its validated, reproducible, and applied, true-to-Reality mathematical formulations?

Notably, because of ramifications ensuing from the adoption of the "Uncertainty Principle" as proposed by Werner Heisenberg (AD 1901-1976), and of "the Indeterminacy Principle" as illustrated by Erwin Schrodinger (AD 1887-1961), in "Schrodinger's Cat/ thought experiment," a post-Relativity approach to theoretical physics would aspire to entertain "a differentiated approach" to Physics in general and to Astro-Physics in particular, through the multiple lenses of "diverse perspectives," as embodied in the enriching paradigms that iteratively and contiguously overlap, even as they respectively encompass and predominate in the numerous "branches of Knowledge" that constitute our wellsprings of universal understanding.

Our credo that all "branches of Knowledge" are "related," though even as they belong to albeit "different Trees," the greater sum of which, have their deep roots respectively anchored within the same rich, fertile, arable, fruitful, "mutual-Soil complex," is very well founded, e.g., "Energy is never created nor destroyed but always transformed."

Examples of many "mutually Sourced" enriched databases of Information with apparent "Iterative Patterns of Self-Similarity," prevail in Physiological Biology, Electrolytic Chemistry, Earth Sciences, Astronomy, Psychology, History, Philosophy, Politics, and Religion, so as to empower researchers and theoreticians with observations of "The Principle of Iterative Patterning," as demonstrated in numerous, diverse, and "Differentiated structural Frames of operations" that appear to be "self-similar" but not "self-identical," e.g.,

(a) Electrons revolve around the atomic nucleus in comparison to Planets revolving around the Sun;

(b) Earth core center-of-Mass is made-up of Molted Iron Magma even as Iron is also the metal-base for Human blood that possesses an intrinsic stream of "Oxygen-carrier Cells," the base of which, also being Iron (hemoglobin/Red Blood Cells);

(c) Ecological processes occurring in geo-atmospheric phenomena such as Condensation/Cloud Formation/Lightning/Rain necessary for providing Sweet-Water to the Earth Soil, consistent with agricultural Human needs, are based upon "Hydro-Electrical Conductivity," in the same manner that the Human Body which is constituted of organs and members made-up of "Hydro-Colloidal Tissues," e.g., close to 80% Body-water composition, also utilizes Hydro-Electrolytic Conductivity in order to maintain dynamic system equilibrium conducive to Organism-life, e.g., Ca^+, Cl^-, are examples of Human Body Electrolytes.

(d) Human Body temperature at a relatively constant 98.6 degrees Fahrenheit, above which there might be an incidence of fever, and Earth atmospheric pressure or barometric pressure at a relatively stable constancy of 29.92 inHg (inches of Mercury") below which there might be a forecast of probable precipitation.

(e) Earth has a tilted axis of Rotation relatively measuring 23-degrees of obliquity, just as the Human DNA displays a 3.4 Angstrom angle of structural organization.

Therefore, due to the ubiquitous impartation of "The Organizing Principle" to both cosmic phenomena and Earth ecological processes via the Thermodynamic Mechanism: Input-Process-Output, the sum of which, "operationalizing/grounding/structuring" the Law of Conservation of Energy ("Energy is never created nor destroyed but always transformed.", scientists and researchers have at their disposal numerous universal wellsprings of Knowledge-based information, adoption of which, geared for developing the greatest latitude in exploring important "variegated patterns of enriched understanding," as characterized by apparent "iterative patterns of thermo-dynamic structure similarity," so characteristic of scientific methodology common to all the respectively differentiated branches and fields of Science, the sum of which, to result in beneficently applying the Scientific Method, as designed for impacting the fortuitous "Genesis of new approaches" in Physics and Mathematics, the sum of which, will synergistically innovate the "Science of Theoretics" in the branch of knowledge we call: "Astro-Physics."

****

Mass, Radiation Energy, and Motion: The Electromagnetic Properties of Matter

What is the essential constitutional foundation of the Universe?

Literally speaking, the Universe is made-up of: Electro-Magnetic Matter that can be either as inert or non-radioactive Mass-in-Motion, e.g., Planets; and/or as Radiation Energy Mass-in-Motion, e.g., Stars!

All forms of Matter, e.g., liquid, solid, gas, or plasma and/or combinations thereof, either as inert Space-bodies such as Planets, or as radioactive Space-bodies such as Stars, operate, within the thermodynamic bounds of the Universal Organizing Principle, the basic mechanism of which, being: "Input-Process-Output," even as they "move-and-function" in Gravitational Continuum Space-Time, in "operationalization" of the Laws of Thermodynamics via applied structuring of the Law of Transformation of Energy!

The Universe is, in short, constituted of "Electromagnetic Matter" moving in Continuum Space-Time, the sum of which, having certain "physical properties," variables of which, can be measurably determined in accordance with "the knowledge-metrics" of Human Understanding, e.g., An object "moving" in three-dimensional Continuum Space-Time having depth/height, length, and width as well as surface and volume, contiguous with other parameters consistent with Motion from point A to point B, such as, distance traveled and duration of travel; speed of travel and direction of travel, etc...!

Matter in the Universe is made-up of chemical elements, such as Hydrogen, Oxygen, Iron, Nitrogen, Sulfur, Copper, etc . . . "clumped together" or "aggregated" to constitute Matter-mass bodies, e.g., atoms form molecules; molecules form greater Space-objects or bodies of Matter with mass-in-Motion.

The two primary paramount Universal Forces that cause these elements to "clump together" as Matter-mass bodies in Motion, are referred to as, Electromagnetism and Gravity:

(1) Electromagnetic Field Motion-force; and

(2) Gravity Motion-force.

Every place where Gravity is present, there is also a magnetic field-force inducing centri-vectored forms of cycling rotary Motion, e.g., Rotation and/or Revolution, around a "center-of-Mass-in-Motion."

Basically, chemical elements are constituted of atoms. The atom is the smallest indivisible unit of Matter. In its simplest known form: The atom is made up of particles: the proton having a positive electrical charge; the neutron having no electrical charge; and the electron having a negative electrical charge; the proton and neutron are at the center of the atom, both of which, constituting the nucleus of the atom. Electrons revolve around or orbit the atomic center or nucleus.

Greater agglomerations or "clumping" of Matter are constituted of atoms. These atoms amalgamate to form greater bodies of Matter making up cosmic mass, such as planets, stars, comets, asteroids, etc . . .

The atom possesses the property of Electromagnetism, in that the proton has a positive charge, the neutron has no charge, and the electron has a negative charge. These counter-opposite complex sets of inter-relationships resonating, emanating, and ensuing from particulate atomic electrical charges constitute the basis for the emergence of the property of "Electromagnetism."

Though the neutron has "zero electrical charge," it is not "inert," or "passive," or "inactive." Within the atomic frame-of-reference, the neutron is a "dynamic particle" in absence of which, gravi-metric determinants of molecular structure, such as "the weak force," could not operate with "the strong force" in fulfillment of functional dynamic system equilibrium. Because opposite electrical charges "attract," without the apparently "passive activity" of the neutron as a "dynamic particle," the negatively charged electron would "crash into the atomic nucleus."

For example, Hydrogen is the only element without a neutron in its nucleus, hence making it the first and lightest element possessing specialized properties as the most abundant solar nuclear fuel. Hydrogen's "near atomic weightlessness" contributes to its electron maintaining a "safe distance" from its positively charged proton at its nucleic center.

Hydrogen is also very active on the earth. As an inert gas, it constitutes, however, less than one percent of Earth atmosphere. In bio-organisms, DNA base-pairs are bonded by Hydrogen. It is also present in electrolytic physiological metabolism that is common to colloid-constituted bodies, such as the Human Organism, in that, foodstuffs, such as meat proteins and plant fiber that we eat, are also comprised of nutritious molecules infused with Hydrogen.

Hydrogen also chemically reacts with Oxygen to form sweet or potable water. In the oceans, it exists as a heavier isotopic form: Deuterium.

Deuterium, which is also a nuclear fuel, is produced in the Sun, via a sequence of chain reactions, as Hydrogen is transformed/transmuted/converted by positron mass energy gains, due to "captured neutrons" emitted by "enriched radioactive Mass" during nuclear processes.

Thus, for example, sequential nucleated chain reactions depicting transmutation of Hydrogen into Helium can be illustrated by the following:

H to H^+ is the sequence yielding Hydrogen to Deuterium (Hydrogen gains one neutron, which converts the electron into a positron, hence, the legend: + to denote one positron for Deuterium);

H^+ to H^{++} is the sequence yielding Deuterium to Tritium (It gains another neutron, the sum of which, converting the electron into another positron, hence the legend: ++ , to denote a total gain of two positrons for Tritium); and

H^{++} to H^{+++} is the sequence yielding Tritium to Helium, (It gains another neutron, again converting the electron to a positron, hence, the legend: +++, to denote a total gain of three positrons for Helium.

The interplay of particulate charges and masses works synergistically to potentiate the atom with electromagnetic properties that hold each particle in its proper place and sphere of influence. Each particle is transformed/converted/transmuted into a mass-charge-energy complex that generates "a force dynamic" as it is "reacting/bonding" and "moving/displacing" in Continuum Space-Time.

The atomic property of Electromagnetism "makes allowances" for particulate charge-displacements, hence, in turn, inducing Motion, the sum of which is common to molecular change, e.g., NaCl.

Gravity within the atomic frame "works with" Electromagnetism, to "become" a "holographic resonance" of micro-tensor Force-Dynamics, as engendered by minute particulate centers-of-mass, centers-of-field, and centers-of-Curvature, as they all "intersect-and -interface," to yield "electro-motive magnetic propulsion," e.g., Electrons revolve around the atomic nucleus and can "jump," when "excited," from one energy-level to another, even as they remain at respective proximal distance from the atomic-nucleus center-of-mass.

Due to the "minute load" of particulate masses upon the atomic structure complex, at the atomic or Quantum Mechanics frame-of-reference, Gravity and Electromagnetism are practically "equivalent," as motion-forces that yield electron revolution around the atomic nucleus, the sum of which, facilitating "electron quanta jumps" from one "energy level" to another.

Just as Gravity and Electromagnetism "co-labor together" in order to yield electro-motive displacement Motion-Force, e.g., Electron revolution around the atomic nucleus center-of-mass at the atomic level within the Quantum Mechanics frame, so too do they "co-labor" in order to yield Centri-Vectored Forms of Rotary Motion-Force at the Sun-Planet level, e.g., Rotation and Revolution within the "frame of Newtonian Mechanics" depicting Solar System relationships, in order to yield self-reflexive planetary Rotation, as well as planetary Revolution around the Sun, at respective proximal distance, away from Solar-System Center-of-Mass.

However, at the Solar-Planetary Newtonian Mechanics frame-of-reference, due to "much greater Mass-differentials" between the Sun and the Planets as compared to "micro-Mass-Differentials" between the Electron and the Atomic Nucleus; and due to relatively greater Mass-Differentials between the Planets themselves, e.g., Jupiter has "much greater Mass than Mercury, Gravity Motion-force effects are observably more "divisible" or "separable" or "independent" from Electromagnetic Field Motion-force effects.

For, both "G-Motion-force effects" and E-M Motion-force effects" are accountable for Rotation and Revolution, and hence, their counter-opposite directional Motion-vector, i.e., Rotation is counter-clockwise whereas Revolution is clockwise, from the co-determinant, co-equal, and covalent operations of which, derives, Earth Gravity Quotient, i.e., G-1, affording us, Human Beings, the complex array of flexible movements within an extensible Range-of-Motion.

G-1 Earth-Gravity range-of-Motion-Force represents a flexible wellspring of extensive Forms of Movement summed-up as an ambulatory-displacement Resource, e.g., centripetal and centrifugal Forms of Motion and combinations or variations thereof towards or away from Earth center-of-Mass, e.g., a traveling vehicle; a ballet dancer, a flying jet plane, or a pole-vault jumper.

G-1 relative constancy allows us to enjoy many activities that in totality are extremely "Gravity-dependent." Whether walking, running, or flying a jet plane, G-1, this "natural Resource," as necessary as Oxygen, is embodied within the Range-of-Movement and Range-of-Motion Complex that makes it possible for us, Human Beings, to live and procreate on the Earth, even as we demonstrate all the variegated endowments and diverse gifts we innately possess.

By the richness of the ways in which we express our creative capacities as made manifest in fruitful processes of labor, we diligently engage, from inventiveness and innovation, in industrial and manufacturing production, towards the safe and efficient design, assembly, transport, and operation of technology and machinery, benefits and advantages obtained there-from due to G-1 constancy, thus, allowing us to fulfill all our necessary needs of daily existence.

This "Relative Principle-of-Equivalency Complex" between Electromagnetism and Gravity Motion-Forces at the Quantum Mechanics level, causes centri-vectored electron rotary motion around the atomic nucleus center-of-Mass as well as other less-resolved electron trajectories or paths; making possible also, other forms of "atomic movement" or "particle displacement in Space," consistent with operational-Motion conditions that must define the variegated Forms of Energy that are embodied in particular "states of Matter" or "bonding structures," e.g., an "isolated particle" will change its directional vector and other pertinent co-variables, when bombarded by a stream of radiation Energy, frequencies-and-wavelengths of which, having an "electromagnetic Motion-pressure-force" different/away from G-1, able "to displace it," away from its normal-natural centri-vectored atomic nucleus center-of-Mass directional vector; and hence, Heisenberg's Uncertainty Principle, or the Researcher's inability to know/measure a particle's position and its momentum at the same time; and also, Schrodinger's Indeterminacy Principle, the apparent ability of a particle to be in two different places at the same time.

Nevertheless, G-1 makes possible the reactive molecular processes between H and O to yield H_2O, as well as those accountable for the bonding structures consistent with the Density of Lead (Pb); G-1 also allows Matter to exist in all its Energy Forms on the Earth: as solid (rocks), gas (air), liquid (water), or combinations thereof ("freezing rain;" Mercury is a metal that appears to "flow" or "move" as a liquid).

Electrostatic molecular bonding in formation of greater Matter-mass bodies in Earth G-1 Gravity Force is resolved, via interposition of temperature gradients synchronized with pressure Differentials, the sum of which, constituting co-variables and co-determinants of reaction-formation, as controlled by experimenters and researchers, within production environments commonly thought of as chemical "reaction chambers." The most efficient combustion of a hydro-carbon-based fuel would result in its by-products being: Carbon Dioxide (CO_2) and Water (H_2O), having, in sum, the same molecular structure and elemental composition as those ensuing from our basic life-process called: Human Respiration.

Particulate charge displacement Energies due primarily to bi-polarity transacting within a micro-Curvature environment, i.e., as within the Quantum Mechanics frame-of-reference, are also responsible for electro-conductivity in certain metals whose molecular structure and bonding arrangements coincide with the matter-mass densities required for electro-magnetic interface between "excited" and "congealed" gravitational frames, e.g., Iron being an electro-conductive metal, in the same manner that, due to "electro-magnetic-conductive affinity," the electro-magnetic core-nucleated dynamo of the Sun, finds "Continuum affinity" with Earth ionosphere, in order to generate lightning, tornadoes, cyclones and other geo-atmospheric meteorological storm systems, and/or to engender Aurora Borealis.

Stars contain "excited Matter," basis of which, being nucleated processes of plasma Energy transmutation, whereas, Planets as non-nucleated Matter, contain "congealed Energy," basis of which, being "ground-state" molecular processes, that, upon ignition, yield combustive and/or explosive "reactions."

However, there prevails "Continuum affinity" between the Sun's "nucleated electro-magnetic radiation Energetics," — as "excited Matter" permeating the Solar System, the sum of which, then impacting the pseudo-atmosphere(s) and geological structures of Planets, respectively, (the Earth, excepted, because of a "complete Oxygen-Nitrogen life-sustaining atmosphere"), — and Gravity-sensitive electro-conductive thermo-dynamic processes, such as geo-atmospheric meteorological phenomena that are indigenous to the Earth, though as a Planet, the Earth exists in "congealed Energy Form."

Consequently, $E=mc^2$ represents a specific form of Energy-Matter "relation of transformation," embodying "the Equivalency Principle," that on the Earth, is made manifest, as nuclear fission and nuclear fusion — brings the Sun "down to Earth."

However, this does not make void "the Continuum affinity" between "excited Matter" and "congealed Energy;" rather, interrupts the flow of their "ground-state inter-related transactions," thus causing them to climax into explosive forms of nucleation, e.g., the Atomic Bomb, the Hydrogen Bomb.

Because the Atom is the fundamental constituent of Matter, this emergent "electromagnetic Continuum affinity" propagates as a property characterizing both "ground/congealed state" molecular reactions and nucleated "excited state" plasma chain-reactions, the sum of which, also transferred to "Matter-as-Energy," for "re-translation" into its many different Forms, e.g., gasoline combustion, nitro-glycerin explosion, atomic nucleation.

Paper will burn when ignited, even as water will flow to extinguish its flame. The Sun is a Star constituted of "excited Matter" as nucleated radiation mass Energy, while the Earth is a Planet, as "congealed Energy" constituted of "Geo-Oceanic and Hydro-Atmospheric Energy," serving as fundamental "structures of atomic organization" allowing "ground-state" or non-nucleated Energy, to "transubstantiate" into its many diverse Forms, e.g., petroleum, coal, methane, and wood, being forms of "fuel Energy."

Albeit in "congealed/ground state Form," Earth non-nucleated environment embodies many different properties of Energy Transformation that climax into diverse, usable converted/ transmuted Energy forms, e.g., petroleum is a "fuel," that can be converted into many other forms of energy: methane, gasoline, gas-oil or diesel fuel; "the blowing-force of Wind Energy can drive a dynamo for production/generation of flowing electrical current.

When atoms agglomerate into greater matter-bodies, they form molecules whereby, because "like-charges" repel each other, negatively charged electrons "bond" within "respectable distances" with each other at their outer orbiting energy-levels. Due to their possession of similar negative charge, this electro-magnetic bonding process reveals a property of Motion related to particulate-movement by "electromagnetic charges displacement."

For at the same time that similarly negatively charged electrons are "repelling each other" while still "bonding" at their outer energy-levels "within respectable distances," they are also "attracted" to positively charged protons because opposite charges "attract each other;" yet, this is occurring even as similarly positively charged protons are engaged in "repelling each other."

And it is this constellation of "Electro-Magnetic Tensor Metrics" relatively configured as "push-and-pull," or "attraction-and-repulsion" that, even in the presence of "electromotive particulate displacement," "holds objects together" as body-masses made-up of Matter, such that, functionally and structurally, they are "existing-and-moving" in Continuum Space-Time.

Particles are moving; atoms are moving: thus, matter is moving. Consequently, within the universal expanse, applied structures of mass-radiation-motion give rise to properties of electromagnetism that get "re-translated" into bonding Curvature force Energetics, binding field energy forces, and gravity-vectored-trajectories that embody other forms of motion specific to each frame. These frame-specific energy transformation operations give rise to cycling mass Differentials as each frame fulfills the universal Input-Process-Output mechanism in its own "processing pattern" for functional equilibrium. As these binding energy forces engendered by electro-motive and field phenomena take the form of Curvature motion, frame interactions climax into equivalents of the gravity force, e.g., earth revolution and rotation; Moon orbital motion; a ripe apple falling towards the center of the earth.

Rotary forms of "traveling Motion," or "rotary trajectories" depicting, either a circular orbital path, an ellipsoid orbital path, or a "nearly-circular path of rotary Motion, are primary within the Universe, e.g., The Earth rotates upon its own 23-degree tilt axis in a relatively circular rotary path of Motion, which at times, due to Gravity-mass effects, is marked by "bulging at the Equator" or "slight flattening at the Poles;" and Planets revolve around the Sun upon an ellipsoid plane of Revolution rather than a circular plane of Revolution, due to "centers-of-Mass differentials:" The Sun "does not sit exactly at-the-center of the Solar System!"

Due to "mass-Differentials" between the nine Planets revolving around the Sun in proportion to the mass of the Sun in-and-of itself and due to "mass-Differentials" between the Sun's "center-of-Mass" and "the Solar System's center-of-Mass," the Sun's location within the Solar System "creates," relative to each Planet's position, a perihelion (at the nearest point close to the Sun) and an aphelion (at the farthest point away from the Sun).

Therefore, a straight-path trajectory exists where:

(1) The force of Gravity is weak or exhausted beyond the boundaries of the body-mass in-Motion projecting the Gravity-force in consonance with either the near-absence of a magnetic field strength, or within the magnetic field's "force-strength limits" as also being projected by the body-mass in Motion, the strength of which, being weaker than that necessary to sustain a centri-vectored rotary form of Motion, e.g.,

(a) In such a case, the Moon would simply "fly out of Earth orbit" and would not remain in a relatively circular geo-synchronous orbit around the Earth;

(b) A rocket launched from the surface of the Earth to travel at a straight path until it traverses "Earth-orbit capture range," due to power of engine thrust capable of overcoming Earth center-of-mass gravity-force akin to that causing the Moon to remain in geo-synchronous orbit around the Earth; or

(2) An additional Force is present to overcome the force of Gravity and/or of its corresponding magnetic field strength, the sum of which, would be necessary for causing a centri-vectored form of rotary Motion, e.g., a jet plane having thrust, the power of which, greater than or able to overcome the force of Gravity necessary to cause the aircraft to fall towards Earth center-of-mass, hence, the plane's trajectory being a straight path relatively parallel to the surface of the Earth over which it is flying.

When matter-bodies are in Motion for bonding purposes, they undergo acceleration, which attaches a property of Mass to their weightiness. Accelerated matter-with-Mass increases in momentum or "exponentially-weighted velocity." Thus, electromagnetic matter in Motion has Mass. Mass energy gains are "re-translated" into forms that conform to the electromagnetic properties of each frame, e.g., "congealed" or "excited."

Motion in time requires Space. Atomic particles move in a "compressed Time frame" as "quantum events" measurable only as "waves of probability." Planetary motion in Space occurs in a "dilated Time frame" as "quantum moments" cycling as rotary motion systems. Accordingly, Space and Time are in a universal Continuum. Since matter-bodies that exist with Mass have to be "contained within something," they travel and move in Continuum Space-Time. Universal reality as we know it includes the observed motion systems of electromagnetically "re-translated" mass engaged in cycling energy transformation patterns of functional specialized operations as they pursue thermodynamic equilibrium.

Matter can be transformed into Energy because it is composed of particles that possess electro-motive charges accompanied by field-mass properties that can undergo dynamic chain reactions. Matter is "congealed Energy," and Energy is "excited Matter." There are many forms of Energy due to its ability to "change state" or to be "converted" into many different Forms. This property of Energy conversion is called The Law of Transformation of Energy. Energy is a property of Matter that allows it "to do something," such as burning, working, moving, powering, "tensing," "stressing," "stretching," exploding, etc . . .

Matter has displayed a natural tendency to "clump together" or to be "attracted to each other." Dr. Isaac Newton (AD 1643-1727) described this "attraction force" as Gravity. The force keeping matter-bodies in "attraction relationships" with other matter-bodies, works to give "holding power" to the universe. It is believed that the form of force that keeps atoms together, molecules together and the solar system together etc . . . is a transformation of the gravitational Form of force. In principle, all forms of Force that are activated by mass-in-motion relationships, from "congealed" or "excited" processes, partake of the universal electromagnetic template that permeates matter in all its structured frames.

The vast expanse of Continuum Space-Time is filled with "centers-of-matter-mass in centri-vectored Forms of rotary Motion," thus, precipitating Curvature pressure-force "exertion-points" within "bounded range(s) of Field projection," that are applicable to all regions of Space. For that reason, it is believed that no Space is "truly void;" in that, no Space exists without the presence of "a certain property of Force-Strength," either represented as electromagnetic Force-field and/or as Gravity motion-Force "equivalent," e.g., even Comet Haley follows a centri-vectored Form of rotary Motion due to galactic and stellar Curvature pressure-force(s) acting upon its traveling Mass, the sum of which, being "predictably iterative in pattern-of-trajectory," and hence, Comet Haley's "return" to our Earthly spatial environs every 75 years; predictable, in the same manner that Planet Earth fulfills a complete Revolution around the Sun within a period of one year!

Because Continuum Space-Time "carries within its depths" the projection of exponentially powered gravitational momentum accelerated Curvature pressure-force, e.g., c^2, the super-strength of the Sun's electro-magnetic Field, and perturbations due to inter-planetary Mass-Differentials, then, solar-interplanetary Dynamics also impacts-and-affects Gravity-determinants of a Planet's "orbital rotary-metrics," hence, yielding, for example, the perihelion shift of planet Mercury, or illustrating a kind of "disruption" of Mercury's otherwise, "would-be" relatively constant quasi-circular orbital trajectory.

Solar radiant energy produces a "Curvature-force" that causes matter-bodies with moving mass to develop "rotary motion" around its center-of-Field, center-of-Mass and center-of-Gravity, in relative proportion(s) to the projected strength of Solar-System center-of-Mass. As cosmic bodies-with-Mass in Motion "traveling in Continuum Space-Time," the Sun and all the Planets, together constitute, "an ensemble" or "system of Entities" possessing developed and/or emergent Properties, the greater-Sum of which, transcending properties that merely pertain either to a single Planets or that could be ascribed to the Sun itself, i.e., "a Solar-System ensemble of Properties" that is "greater than the sum of its parts."

Planets revolve around the sun and rotate on their own axes. Galaxies revolve around each other. Thus, the Universe and all its constituents are "always moving." However, objects on earth surface do not normally appear to us to revolve or rotate; they can be "at rest," or "made to move," upon the exertion of additional external forces, e.g., a "resting rock" nested unto the side of a hill at "inertial Gravity," can be pushed down-hill and caused to roll until its momentum is exhausted to a stop.

In the same vein, Gravitational Curvature Pressure Force is also expressed as curvilinear Earth-surface motion and rectilinear Earth-surface motion, that can be along "Earth line-of-radius," or in the atmosphere above Earth-surface at a line parallel to its diameter, and yet, with "suspended attraction" towards earth center-of-mass and Earth-center-of-gravity, e.g., a traveling automobile; or a flying aircraft with motorized propulsion.

Because the Earth is a spherical-bounded Planet operating as a "closed System," and yet, not as "an isolated System," these surface motion Forms complement the "external" or "void-Space bound" motions performed by the Earth, such as its revolutionary travel around the Sun and its rotary motion upon its own axis. Consequently, Gravity on earth surface is articulated in various Forms that differ from "external" patterns of rotary motion, though those apparently "differentiated Forms" of Gravity-expression on the Earth are also centri-vectored rotary Forms of Motion-force, either centripetal or centrifugal and/or combinations thereof, the sum of which, having for its center, Earth Molten Iron Magma Core. Thus, all Earth-bound motion-force Forms, "gravitate" towards Earth-center of Mass and/or Earth-center of Gravity.

Curvature pressure force due to gravitational "push-and-pull" and to magnetic field induced motion Forms, is transformed/transmuted into its many "variant states" that are "offshoots" of the same gravi-metric Form of motion initially developed from "common spheres of interaction." For example, the earth "externally" revolves around the sun and rotates upon its own axis, while the Moon follows an orbital path around its circle or circumference (rotary); within the earth, a plane will follow a rotary path around earth "circum-sphere" to "spiral down" towards its center-of-mass; a ripe apple will fall towards the center of the earth (rectilinear); a powered vehicle on earth surface moves in patterns formed by "apparent obstacles" in its path (curvilinear).

Thus, within the earth, objects can appear to travel in rotary, rectilinear, and curvilinear forms of motion, or combinations thereof, the sum of which can also espouse composite motion-force patterns, but while simultaneously "directed" towards Earth center-of-gravity and center-of-mass, e.g., without a propulsion engine, a flying plane would fall to Earth surface, or rather, towards Earth center-of-mass and center-of-gravity — Earth surface serves as "a barrier" preventing the plane from "reaching" Earth Molten Iron Magma Core.

Within the whole solar system, radiated solar Curvature pressure force energy impacts planets, acting upon them in a way that corresponds to each planet's constellation of variables respective to mass, rotational velocity, revolutionary momentum, gravity quotient, geo-atmosphere structures, core matter density, magnetic field strength of electro-conductivity, etc . .

The physical constitution of the Universe can be envisioned or framed as a complex Entity: comprised of Mass, Radiation, and Motion, all of which being intrinsic properties of Matter "traveling" within gravitational Continuum Space-Time; from the totality of relative interactions of which, arise: electromagnetism, field-force, gravity-force equivalents, and energy cycling patterns. These energy cycling patterns operate within "structural frames-of-reference" that process the Laws of Thermodynamics via the universal Organizing Principle "Input-Process-Output."

"Input-Process-Output" driving the Organizing Principle of the Universe is a "mechanism of functional operation" prompted by "Curvature Force Energetics," to climax into frame-specific Forms of magnetic field induced gravi-metric Motion. Energy cycling patterns are frame-specific as pre-determined by fulfillment of the Input-Process-Output mechanism for functional system equilibrium, e.g., Earth receives solar inputs that induce cause-and-effect relationships between its electro-conductive Atmosphere, its salted Oceans, and its bacteria-housing Soil, in order to yield "ecological properties," the sum of which, sustaining life-support systems facilitating the growth of vegetation and sweet-water precipitation that are nutritional Sources conducive to the metabolic health of Biological Life on the Planet.

The Sun engenders centers-of-mass, centers-of-gravity, centers-of-field, and centers-of-motion while accelerated momentum radiation mass is "re-translated" into Continuum Space-Time Curvature Pressure Force Dynamics.

Mass-in-motion Frames embody many different Forms/Patterns of energy transformation that "cycle" in temporal time, recurring in iterative functional Forms that represent the "re-translation" of the electromagnetic properties of Matter.

Redundancy in the presence of iterative substantial Forms and of functional structural Patterns causes a kind of "quasi-symmetry" in "frame cycling patterns" commonly identified in observed phenomena, due to their common initial structural Source, as having a common quantum mechanical origin: the Atom, e.g., Usually observed phenomena having a common pattern: Electron revolution around atomic nucleus and planetary revolution around the Sun; fruits/flowers from Spring-budding back to seed or degradation and Human Life having a span or duration from womb to tomb; Sun's magnetic field causes Earth self-reflexive axial Rotation and Revolution around the Sun and magnetic field induced by copper-coil in electric fan yielding blade-assembly to revolve in so many revolutions-per-minute in order to provide "cooling of the air" for Human comfort.

All forms of Motion mimic/replicate/duplicate/imitate/re-translate centri-vectored patterns of rotary Motion, but within numerous variegated Frames-of-Reference wherein differentiated energy-cycling properties fulfill the universal "Input-Process-Output mechanism," the sum of which, imparting to Matter/Mass/Energy-in-Motion within the infinite vastness of Gravitational Continuum Space-Time, the "Organizing Principle" that is accountable for the aggregation of Atoms for electron-bonding, molecular change, combustive processes, and nucleated transmutations.

Theoretically identifying, experimentally operationalizing, and mathematically "formalizing" these cycling Energy-frame Differentials as "a normalized-unified Whole," are crucial, to integrating the Universe. From the paradigm of an all-encompassing "General Theory of Motion-Force," all Matter-Energy Cycling Frames structurally and functionally operating within universal Gravitational Continuum Space-Time can be unified, but with integration of all Curvature-pressure-tensing-force Motion-Forms, from Classical Mechanics to Relativity Dynamics.

****

The Continuum Principle

Every Matter-Energy Frame that consists of Mass-in-Motion embodies "Continuum phenomena-&-events Dynamics," the sum of which, embodying cycling thermodynamic processes, "repeatedly iterated" within the Universe, in quasi-similar or apparently "self-similar familiar Structural Patterns," as these Matter-Energy Frames functionally operate within the Field-Curvature of Gravitational Space-Time, e.g., electrons revolve around the atomic nucleus (Quantum Mechanics); planets revolve around the Sun "situated" at the gravitational-center of the Solar System (Gravitational Relativity Dynamics); the Moon, Earth natural satellite, geo-synchronously orbit the Earth as the Earth itself rotates upon its own 23-degree tilt axis (Newtonian Celestial Mechanics).

As universal phenomena, Earth-ecological processes, and cosmic events "cycle in definite recurring periods," matter-mass frames partake in "energy-motion-force inputs" that they process in order to utilize/replenish their resources for achievement of dynamic System equilibrium. Cognitively, things may appear to be "fragmented," "compartmentalized," "separated," "sequestered," or "divided," in the sense that each Frame-of-reference appears to have its own unique identity. But this is due to the complex natural constitution of Energy in its variegated functional transmuted Forms, manifested as both "quanta" and "wave;" as both "unit/part" and "whole/Entity." For example, it is a "stream of quantized photons" that constitutes visible Light.

All natural phenomena, cosmic events, and universal processes "share" or "belong to" the same Gravitational Space-Time Continuum, as characterized by overlapping mass frames that thermodynamically transform the same universal energy from its "essential Form" – as these Frames "flow together" and "operate inter-dependently," in contiguous, simultaneous "Oneness Integrity," the sum of which, "to hold the Universe together."

Each frame's unique identity characteristics engender the emergence of complex Differentials, e.g., in Mass, Velocity, Momentum, Orbital/Rotational/Revolutionary Duration, Temperature, Pressure, Gravi-Tensor Metrics in factoring "Applied General Relativity Theory," all of which, finding "integral unification" with each other, within the same solar system ellipsoidal tensing-stressing plane magnetic field dynamics e.g., such "unique characteristics" as: planet Mercury's perihelion shift; earth "wobbling," and "bi-polar oblation."

Mercury's perihelion shift climaxes as a time-specific "gravitational quantum-event" in Continuum Space-Time Curvature, as the planet is impacted by field and gravity force Differentials which it must accommodate in order to sustain dynamic System equilibrium. This pattern of dynamic response to Curvature force Differentials is self-iterative, in the sense that comparable adjustments by other planets demonstrate parallel re-calibrations to relative gravitational inputs that hold the Sun's ellipsoidal revolutionary plane together.

****

SPIRITUAL CONTINUUM: THE CREATIVE BASIS

OF PHYSICAL CREATED THINGS

Deists believe that God created the Universe but "left it alone to its own designs," to no longer participate in its transactions. However, this credo contradicts the Continuum Principle because natural phenomena cycle in "Temporal Time" or "Thermodynamic Time" — mechanically/clock-measured, non-Eternity Time, within the span/duration of which, Matter-Mass Frames thermodynamically cycle — via processing of iterative structural patterns marked by "rates of change" in its co-determinant parameters and variables, to which is attached, a "sensitive dependence upon initial conditions," such as those dictating a beginning and an end as well as an operational duration, much necessitated by "streaming waves of quantized Energy Forms"; and hence, observed redundancy of Forms, climaxing into the Law of Energy Transformation (Conservation and Entropy).

Continuum observed in "created things" of the Universe and Nature is also prevalent within the processes of Thinking common to the Mind of Humankind who are the most intelligent and creative Beings in the Creation. The Universe is designed to operate in this manner, because after God created "the heavens and the earth," He also created Humankind to whom He gave "dominion" or "legitimate authority" over the "created things" of the Earth, e.g., its natural resources.

In addition, by His birth, life, death, burial, resurrection, and ascension unto the heavenly realm, Christ Jesus Messiah, also re-united and re-integrated "all things," both in Heaven and on Earth (Ephesians 1:7-10, NASB, Holy Bible).

Thus, Continuum predominates in "things' both abstract and concrete; both spiritual and carnal; both ideational-imaginative and objective-material, e.g., the correspondence existing between a mathematical equation and its technological application, such as $f = ma$, such as $E = mc^2$, is not "accidental," but rather rooted in the initial Intelligent Design imparted to the Universe as a form of necessary Logic, based upon and guided by original Physical Laws embedded therein since the time of its very Genesis, for the functional operation of all its Matter-Energy Frames in accordance with "transformational dynamics of Mass-in-Motion."

Continuum prevails as Spirit precedes Matter; as Idea precedes speech; as Thought precedes action. Speech precedes communication. Desire precedes fulfillment. "Things gravitate towards each other" in ways that engender "overlapping operational Differentials," due to special uniqueness of frame properties for pertinent functionality.

It is Spirit that facilitates Continuum between Eternity and temporal Time: between ever-present Eternal Time and Thermodynamic Time (clock-measured Time) where "things created" undergo "cyclical rates of change," as necessarily processed for sustaining functional operation in dynamic System equilibrium.

Continuum implies a spectrum of graduated processes that embody the diversity of pertinent variables whose Relativity proportions thrive to climax into wholeness. Do not seconds periodically recur in order to form minutes that accumulate into hours? Period cycles into Continuum, as Continuum is "embedded" within self-iterative cycles, e.g., a line is a succession of points. As a single point constitutes "a period," numerous successive points "stream as a line," and hence, "the moving pictures effect," as in motion pictures.

Continuum permeates all things spiritual and physical. All things must "interface" via Continuum! As a "complex-Logic programming/encoding Form" driving the Universe, Continuum "instructs-informs the Universe," the sum of which, constituting Matter-mass-energy Frames, together functioning in "temporal Time," as engineered by rotary forms of centri-vectored directional Motion embedded within Gravitational Space Curvature.

We, Human Beings, live temporally! Within the devolution of Thermodynamic Time, as framed by co-determinants of healthy prolongation of Biological Life, we enjoy "a range of lifespan period," as fixed by our Species-specific, DNA-dependent, "initial thermodynamic conditions of Genesis."

In a unified Universe where all things interdependently function for operational integration via relatively overlapping Continuum interactions, e.g., solar radiation "connects" with the Earth so as to engender lightning within G-1 Gravity parameters while producing rain for our crops, our Biological Organism is breathing, as its respiration processes extract Oxygen from the Planet's atmosphere.

It is through Continuum, by Continuum, and for Continuum that all realms interconnect: At the "intersection" of Space and Time, co-labor, rest and motion, inertia and kinetics, matter and energy, mass and electromagnetism, static energy and dynamic energy, "excited" frames and "congealed" frames.

In the same vein, we understand Continuum existing between thought and action, idea and object, abstract essence and physical materiality, inner-desire and external fulfillment.

Therefore, a constructive understanding of the Continuum Principle implies its beneficent application to all aspects of living, including the recognition of its operation within specialized cycling mass frames: from Quantum Mechanics to Newtonian Celestial Mechanics, from Relativity Dynamics to universal "Unification Force Energetics."

Dr. Albert Einstein (AD 1879-1955) discovered the Gravitational Space-Time Continuum. It is Matter that gives "embodiment" to Space-Time, otherwise, even though "still real," they are both void. And Space-Time has Continuum because of the successive accumulation of particulates that aggregate into molecules for greater mass bodies that form functional Frames of thermodynamic operation. Matter-with-Mass-in-Motion within Continuum Space-Time "bends" or "curves" Space.

As a line is a succession of points, Matter gives successive accumulative embodiment to Space-Time by agglomerating particles as Space-bodies such as: atoms, planets, stars, asteroids, comets etc... , while establishing "Differentiated Energy Forms" that operate as "a unity of opposites" within the same "band or spectrum" of cause-and-effect relationships, as apparently "separated or grounded by Space-Time Distance(s)," in identifying these Forms as function-specific Frames-of-reference, e.g., the "Atomic frame" v. the "Solar System frame;" the combustive petroleum Energy frame v. the nucleated Star Energy frame; Earth planetary "Atmosphere-Land-&-Sea frame" v. Sun's nuclear plasma electro-magnetic Radiation Energy Frame. As a form of "congealed Energy," the Earth is a Planet that "stands grounded" or "apparently separated," but not "isolated," at a "respectable distance" of 93,000,000 miles from the Sun which is a Nucleated Star-Energy in the Form of "excited Matter."

Single particulate points of nearly mass-less Energy-in-Motion, constitute the Electro-Magnetic Radiation Spectrum. But they had to have had originally belonged to an ever-present emitting "excited Energy Frame," such as the nucleated solar inferno.

If all points successively constituting a line are regressively removed from that specific line, after the last point is removed, the line is no longer, it disappears; and Space- in-Time results: as void and empty.

Both Time and Space are therefore infinite or have the "property of Infinity" attached to their essential nature and quality. Both Time and Space have a characteristic of "limitless unbounded infinite intangibility," until contoured or encompassed by Matter in its respective Form(s), the sum of which, designed to "embody" both Time and Space. For example, an empty square box made-up of cardboard is said to have "Space Volume" within its four walls that can be measured in cubic centimeters; but for the four walls "confining" or "sequestering" a definite "region of Space," Space could not be "measured or quantized as Volume."

However, at the same time, Space qua Space itself, all of which, both external to the box and empty within the box, is "the container" of said box – it has nowhere else to go! Consequently, under such "embodiment v. container conditions," Space-Time is said to become "the Continuum container" of Matter that in turns gives "Mass-in-Motion" to Continuum Space-Time. Matter-Energy Frames in functional operation constitute Mass-in-Motion that thus "temporalizes" or "makes to exist" in "thermodynamic State:" Continuum Space-Time.

It is the fixed value of the Speed of Light in a Vacuum that gives "bounded finiteness" to Matter-Energy Frames as Mass-in-Motion. For, if it were possible to travel beyond the Speed of Light, then nothingness would have existed. Only "mass-less Matter" could "travel beyond the Speed of Light," which then would mean that such Form(s) of Matter cannot exist; for, the lightest or "nearly most mass-less" particle/wave being the Photon belonging to Visible Light, already "travels at the Speed of Light." Particulates or rays or waves of radiation Energy such as X-rays, Gamma rays, and cosmic rays that "travel at the Speed of Light," are also in "excited Matter-Form," which makes them "nearly mass-less."

Gravity is also bounded by relatively overlapping measurable Matter-mass motion-force quantities. And, at the same time, as the electromagnetic properties of Matter must cycle for energy transformation (and by that, conserve resources for Continuum processes), they impart a "constant limit" to the Speed of Light: at 186,000 miles per second.

Gravity is an emergent property of "moving Mass" as induced by Electro-magnetic Field-Force parameters and determinants of "bounded Matter-Energy," hence, our ability to measure both the range and intensity of its Strength, e.g., Earth Rotation takes 24 hours to complete one cycle. And within spherical boundaries of the Earth: 9.88 meters per second per second is the value that fixes the acceleration of a falling object from Earth atmosphere towards Earth center-of-Mass.

Therefore, it is not "by accident" that the Speed of Light in a vacuum is: 3.0×10^8 meters per second, or 186,000 miles per second.

Mass-in-motion and Light-radiation Energy in-motion, e.g., X-rays or Gamma rays, Visible Light or Ultra Violet, depend upon each other's "Time-measured cycling patterns," in order to operate as units-of-Energy with wavelength and frequency quantities, the sum of which, climaxing in relative equilibrium and "Continuum-streaming" that together factor in integrating the Electromagnetic Spectrum of Radiation Energy as the primary "gravi-metric Input" that then "triggers" Earth "Eco-Environmental processes," accountable for synthesizing cause-and-effect Output-reactions that sustain Earth life-support systems, i.e., Maintaining atmospheric Oxygen to Nitrogen ratio and "inert gases" proportions; allowing sweet-water evaporation from the Oceans that then condenses as clouds in the Atmosphere, the sum of which, "containing" the constituents of precipitation.

In Earth ecological processes inheres "a curriculum vitae" or "essential history" ensuing from fundamental natural phenomena that "overlap-and-resonate-and-traverse" throughout these three interdependent Mass-Frames, i.e., Atmosphere, Land, and Oceans, operations of which, embodying "re-translated forms" of Electromagnetic Mass-in-Motion — (the electromagnetic properties of matter —atomic-particulate charges having "motion-force displacement properties" necessary for forming molecular constituents of compounds, substances, materials, and elements needed for Earth viability as a Life-Planet).

Light radiation energy-cycling and mass-in-Motion cycling patterns embodying electromagnetic properties of Matter are synchronized in "Thermodynamic Time" for gravitational Motion-force cycling as well, in order that stochastic quantized processes undergoing "rates of change," climax into relative Continuum.

Thus, all "quantifiable categories" that inhere within the universal "Matter-Space-Time Frame," cycle in "Temporal Time" or in "Thermodynamic Time," e.g., Mass, Motion, Force, and Energy.

Time qua Time is therefore eternal and Space qua Space is therefore infinite—what was (past), what is (present), and what is to come (future) — neither of which, having a beginning nor an end, nor a boundary, nor a limit.

A beginning, an end, a boundary, or a limit: These measurable characteristics or quantifiable qualities belong only to Matter-Energy structures or frames, operations of which, as Mass-in-Motion with "rates-of-change," are "cycling in Thermodynamic Time."

Time is a reality as "distance traveled between two points," — such as the 24-hour rotational "travel" of the Earth upon its own 23-degree tilt axis.

Time is also "the measured duration" of a physical event, phenomena, or process — such as the duration of an earthly Season as a trimester.

Time, in the absence of "Mass-in-Motion serving as "reference points," is, as well, an "inertial ever-present concept" in the conscious Mind that can operate without "external coordinates" or "physical properties of entanglement" — such as, a seated person reading a book in a lighted room having no windows and devoid of a clock-device.

Thus, it is the presence of Matter-Energy Forms "cycling in Continuum Gravitational Space-Time" as Mass-Frames in Motion that give "the property of measurable duration," "the property of lapse-span," and "the property of quantized qualification" to Time. Otherwise, — that is, without the presence of Matter-Energy undergoing "thermodynamic rates of change" — Time would neither "stream" nor "stop." Time would simply remain "an ever-present inertial in-the-Mind Reality."

How about Space? Space is a filled-reality — as occupied by a chair or a Planet, — as well as an empty-vacuum — such as Space devoid of Matter-Energy with-Mass-in-Motion.

From gravitational or Star Radiation Energy emerges a "magnetic Field Force Strength," the "projected Range-of-Operation" of which, inducing both Revolution and Rotation, "rates of travel" of which, amenable to the "pertinent range-boundaries" of the "Electromagnetic properties of Mass-in-Motion" constituting the "center-of-Gravity" within the Solar System, e.g., That our Sun can only "hold" a specific number of 9 planets in "gravitating range," is "no accident."

Therefore, Gravity, as an emergent property induced by Electromagnetic Mass-in-Motion, gives rise to the different vectoring characteristics displayed by moving Objects, e.g., such as a jet-plane alternating between rectilinear and/or curvilinear Forms-of-Motion, within a 360-degree spherical earthly geographical environment; the sum of which, in the absence of an additional external Force, e.g., counter-centripetal and/or counter-centrifugal Force, will cause all objects to "gravitate" towards Earth-center-of-Mass and/or Earth-center-of-Gravity, wherein is located/situated the "electro-conductive" and "magneto-responsive" Hot Iron Molten Magma Core, e.g., as the last "repository-focus" of atmospheric Lightning, as proven by the well-known "lightning rod," and hence, the generating-Source of Earth bi-polar Magnetic Field.

Within a three-dimensional 360-degree Earth-spherical plane of Motion, we have vectoring directions such as North, South, East and West, as well as, up, down, left or right, or combinations thereof, and in-between positions/locations, e.g., South-East, the sum of which, being representative of numerous multiple Motion-forms that can "take an object," either towards, or away from, Earth center-of-mass and/or center-of-gravity, along its line-of-radius.

In deep-void vacuum Space, these various directional vectors are not experienced in the same manner as on Earth, due to near "zero-gravity" conditions. For example, in a spaceship orbiting around the Earth, Human Beings residing therein, in order to ascertain whether or not they are "upside down" or "upright," would have to utilize "physical coordinates" or "material reference points" that are "internal" to the Spaceship, or that are "inside the spaceship," the sum of which, serving as "a containment vessel."

On Earth, the sky, which is "always there," serves as the focal point for the "up-position;" and the ground, which is also "always there," serves as the focal point for the "down-position."

On the Earth and in the Universe as a whole, given that, in accordance with self-evident scientific Physical Laws, "all things are synchronized" for "perfect fine-tuned" interdependent Organized Order, then, "internally," within ourselves," — given also that we are, the Human Family of God created "unto His image and His likeness," (or secularly, as the Human species), — our DNA and/or "genetic make-up" responds to the Force of Gravity and/or its "equivalents," in such ways that "we self-consciously know" when we are in the "up-position" or "right-side-up position, or when we are in the "upside-down position."

Thus, our "knowing" how to "navigate" or "negotiate" the Force of Gravity and/or its "equivalents" with "the right internal responses," evokes much more than a mere "stimulus-response mechanism," — Otherwise, so-called "adaptations" by this supposed "stimulus-response mechanism" would have "self-adjusted" as pontificated by evolutionists, resulting in "right human perception" when in deep void empty Space where "zero Gravity" or "near-zero Gravity" prevails. But, as facts continually prove, there is no "adaptation" or "adjustment" imposed upon the Human Body for "negotiating" new "internal perception responses" to "zero-Gravity" or "near-zero Gravity."

Rather, this DNA-rooted "genetic understanding" or "innate spiritual knowledge," presents to our self-conscious Minds, our apprehension of transcendent, "synchronized dimensions of Intelligent Design," co-determinant parameters of which, are integral to our "connectedness" with the Organizing Principle prevailing within "the Universal Frame-of-Reference," the sum of which, being, as bounded by the Laws of Thermodynamics governing Energy Transformation: "Input-Process-Output."

Thermodynamic Time, as a concept on the Earth for marking "temporization" of physical events, phenomena, and processes, is "embodied," "concretized" or "operationalized," within our knowledge-and-understanding of "Sunrise and Sunset," as predicated upon the Earth's Revolution around the Sun completed in 365 ¼ days, and Rotation upon its own axis completed in 24 hours or in one day.

Space is vacuum-void-empty until occupied by Matter-Energy in its various self-evident Forms, the sum of which, constituted of atoms and particles aggregating into molecules, to climax, in summation, into "congealed Space-bodies" that "concretize" Gravity (planets), and/or into "excited Space-bodies" that "concretize" gravitational radiation Energy (stars).

As the planet revolves and rotates, in counter-opposite rotary directions, respectively, due to the Electromagnetic Field Force Strength causing Revolution and Rotation, Gravity emerges, "embodying" a force-strength corresponding to the quantity-amount-sum of Matter-Mass "traveling" in "momentum-torque Motion Energetics" within its "Frame of Relationships" vis-à-vis the Solar System "center-of-Mass," as "engrossed-encompassed" within the planet's body.

For example, Earth has a standard Gravity value of 1 or G-1, such that, an object dropped from a certain atmospheric height, for example, will experience "an acceleration rate" as the Motion-force of Gravity on earth surface corresponding to its "Mass-in-Motion co-determinant parameters," measured at 9.8 meters/second/second; whereas the Moon being a quarter of earth size in Mass, has its surface Gravity-force strength measured at only 1.63 meters per second per second. Gravity-force or its "equivalent(s)," is therefore "Mass-in-Motion dependent."

As Space has Continuum with Time due to Matter-Mass aggregation embodiments that fill its three-dimensional Volume, Temporal Time or Thermodynamic Time, also has Continuum with "the flow of Eternal Time," or with the "ever-present-ness or real-time-ness of Eternity."

"Time flows," due to successive embodiment of events, phenomena, or processes, the sum of which, "traveling" or "covering a distance," from one point-in-Time to another point-in-Time, within the "Frame of the Volume of Space." For example, the present, or "the now," as an indication of "status in Time," has Continuum with the past and onto the future, not merely within our knowledge, awareness, consciousness, comprehension, discernment, or memory, but also in "the external constellation" of events and processes, configuration(s) of which, serving as successive accumulation of overlapping "frames-of-reference points," or as physical materials for collective Continuum Space-Time "embodiment" in our Minds, as it "thermodynamically flows," in terms of spans, durations, or cycles that measure or indicate "rates of change."

These "internal markers and external reference points" serve as "mental content" or "mental furniture" that give objectivity to "the interface" between our inner-lives and Earth physical environment, encapsulated within our subconscious or intuitive comprehension as landmarks or coordinates, the sum of which, we apprehend, with consciously retained qualifications of "Continuum cumulative successiveness," in order to form integrated-coordinated and contiguous-connected Sources and wellsprings of consistent knowledge that proves valid, true, and factual in helping us navigate through the Universe with orientation, direction, and understanding, while engaged in "transacting" with physical Nature and the greater Universal Frame.

Eternity is "timelessness" in the "thermodynamic sense" — or, the absence of materiality as "traveling landmarks" that give "embodied correspondence" to our "interface" with, usually, mass-less Space. Empty-void, mass-less Space, is "timeless and infinite" — nothing is "placed/situated/located/positioned or moving in it to give successive and cumulative "anchored understanding," to our conscious Mind, from one point-in-Time to another point-in-Time, and/or from "Distance Differentials" traveled between one physical landmark and another physical landmark.

"Temporality," is therefore, "thermodynamically-driven materiality" as "traveling landmarks" having 'operational duration" in Continuum Gravitational Space-Time.

Consequently, all "quantifiable categories" encountered by our Mind within the Universe will have "physical-material objects-things" serving as "external anchors" to be "temporized" within the "re-translated processes" encompassed by the electromagnetic properties of Cycling Mass-Frames in Curvature motion. All "quantifiable/measurable categories" having "numerically-derived" variables, parameters, and co-determinants, must need-have physical dimensionality for "Mathematical Formalization," — and only Matter-Energy-Mass in Curvature Motion possesses "physical dimensionality."

We have an, albeit, mortal physical biological body. And as such Human Persons, we are still "spirit-beings" created "unto the image and likeness of God: Thus, we must have a name. So must our streets. So must every thing or object we know. Physicality begets physicality. Spirituality begets spirituality.

Humans would be "practically lost" without specific physical-material Time-and-Space coordinates that "interface" with the Mind of our inner-being and our spiritual inner-Person for geographical orientation and Earth-Space navigation. But, in deep-void-empty Space, there are no physical landmarks or reference-points like on Earth to "visualize" or against which "to bounce" Radar or Sonar waves to utilize in the Form of "echo-location." We utilize the Ionosphere and Earth surface as well as artificial satellites, all of which, being absent in deep-void-empty Space.

It is Motion that gives "traveling embodiment" to Space-Time, and it is Matter-Mass entities that embody Motion in Space-Time to give it Continuum for our understanding in "Thermodynamic Time." From sunrise to sunrise, a complete rotational period on the Earth has a 24-hour duration. Rotational Motion gives "Continuum" to days, weeks, months, and years, the sum of which, also "connected" to the elapsed span of Eternal Time.

Otherwise, without Motion, Space-in-Time is totally "static," has no periodic-seasonal or recurring cycles with iterative regularity; and thus, is completely empty-void.

The Thermodynamic Principle or "cycling code" that inheres in the Law of Energy Transformation embodied in universal reality—utilizes the Input-Process-Output mechanism that "operationalizes" the Organizing Principle — as expressed in recurring periods of resource replenishment, from Conservation to Entropy, as utilization and consumption of inherent resources in "Temporal" or "Thermodynamic-Gravitational Time" prompt "the Continuum Algorithm" to overcome the tendency to decay in relation to Eternal Continuum Space-Time.

Eternity or Eternal Time has ever-present immanence whereas we perceive the flow of thermodynamic Time or "Temporality" in quantized periods that contiguously lapse or span from past-to-present-to-future, thereby, establishing our apprehension of "the flow of Time" as a "Continuum."

Eternity and temporality, Continuum and exhaustion, engage in a "tug of war" that causes Mass frames to "periodically cycle" in iterative regularity, in order to achieve dynamic System equilibrium.

The thermodynamic principle embodies the universal Input-Process-Output mechanism, as each whole Energy System fulfills it, within its own specific-structure's cycling pattern. The thermodynamic cycling pattern associated with each specific fulfillment of the Input-Process-Output mechanism for each frame of reference, as "calibrated" by Conservation and Entropy, determines duration of iterative energy transformation periods, e.g., on earth, every day cycles in 24-hour rotational periods.

Thus, we end up with iterative cycles of resource replenishment, e.g. seasonal cycles, evaporation-condensation-rain cycles etc…, that sustain Continuum, as impelled by changes in frame parameters within thermodynamic boundaries that control levels of production, accumulation, abundance, utilization and consumption, while "holding off" possibilities of decay and exhaustion. For example, the yearly harvest represents a mode of "resource replenishment," as manifested in Temporal Time having for a foundation the patterns of Motion activated by solar system ellipsoidal plane Dynamics engineered via gravitational radiation for Earth rotary motion Forms, as well as the Moon's "satellite attachment" to the Earth as a Space Mass-body.

In these Forms of cycling motion are sustained the Laws of Thermodynamics — Conservation and Entropy — processed in Temporal Time in conjunction and connection with Continuum Space-Time as "embedded" in Eternity.

"Temporal Continuum" or "Thermodynamic Continuum" as engendered by reiterative periods of Mass-cycling motion Forms "attuned" to Curvature Energetics, is "embedded" within Eternal Time and Infinite Space, and hence, making the two, "inseparable" in Gravitational Continuum Space-Time.

How are Eternity and Temporality, Continuum and stochastic cycling, to be "normalized," as theoretical constituents of a new paradigm? They must be conceptualized as "enabling constituents" or as "catalytic components" of universal Motion-force Unification that brings both Electromagnetism and Gravity into "Holistic Oneness Integrity."

THE INTELLIGENT PRINCIPLES OF CREATION DESIGN

"In the beginning God created the heavens and the earth . . ." (Genesis 1:1, KJV, Holy Bible).

Stars, planets, solar systems, and galaxies must have had had a "beginning force" at the time of their initial Genesis, the sum of which, "encoding" them with an Algorithm "programmed" for operationalization of the Organizing Principle towards "Continuum Momentum Motion." However, due to the Laws of Thermodynamics, "Continuum" does not mean "in perpetuo."

Given that it is Matter-Mass that has Energy to move due to its essentially electromagnetic particulate constituents — and Matter-Mass did have a "beginning," — then, there cannot be "perpetual Motion", "eternal Energy," or "timeless or non-quantized cycling."

Time is eternal; Space is infinite; — but not Matter-Mass Energy. For, the Speed of Light has a "fixed value" within the "temporal time-frame" or within the "thermodynamic cycling of Energy" contained in particular volumes of empty-void-vacuum Space, which does not "exist all by itself."

Matter has definite units of "weightiness" or "heaviness," as "Space-filling emplacement" that gains momentum-Mass when accelerated by gravitational propulsion or by Human-made engines.

Accelerated mass gains momentum when pushed by additional force. Energy has definite spectral radiation quantities defined as wavelength in centimeters, frequency in cycles per second, and photon energy in electron volt units.

Mass and Energy are quantifiable categories with measurable properties having intrinsic physical dimensions. But Time is "self-sufficient: There are no "Time particles." Neither are there "Space particles;" Nor are there "Gravity particles."

The Atom comprises physical particles imbued with electro-magnetic field-charges imparting "displacement Motion" to its negatively charged electrons. Only physically-anchored particle-based Energy gets transformed into its corresponding cycling frames, because it is an emergent property of Electromagnetic Matter-Mass.

Gravity, as an emergent Force, is absent without the electro-magnetic charge displacement Motion properties of Matter-Energy-Mass. Hence, Gravity, when present, is known by "its effects" upon objects, things, and people, while the Force properties of a magnetic Field are amenable to the oppositely-polarized electric field generating it, e.g., the Sun's magnetic field engenders gravitational rotary Forms of centri-vectored Motion-Force such as Rotation and Revolution, the composite-combined characteristics of which, generating Gravity as an emergent Motion-Force in the Earth, directing-vectoring all objects-things towards Earth center-of-Mass.

Matter-mass-energy thrives in Continuum Space-Time as activated by Curvature pressure tensor-force Motion emerging from Electromagnetism and Gravity that causes it to cycle in energy transformation patterns as configured by frame-specific fulfillment of the Input-Process-Output mechanism.

"Relative Equivalency" between Matter and Energy-Mass in-motion within Continuum Space-Time Curvature, is consistent with "all Forms of Force" — as "compartmentalized" by Astro-Physicists for purposes of study and understanding, — that impinge upon cycling frames, the sum of which, climaxing into dynamic System equilibrium: Magnetic field force, gravity force and its "equivalents," the strong force, the weak force, and electro-magnetic charge displacement motion-force.

Then, how do Time, Gravity and Space "interface" to "re-translate" Curvature motion into "energy cycling force" that holds the solar system ellipsoid revolutionary tensing frame together?

The unfolding of technologically-derived scientific discoveries, have been triggered by the "Quantization of Time."

The Clock is our first mechanical invention, operations of which, to mimic Continuum. It allowed for the first externalization experience of an internal process of Human Understanding: The concept of Time could thus be "concretized" by specific periods of iterative duration, called seconds, minutes, hours, etc...., cumulatively "prized" and "priced," with objective applications to productive life on the Earth. Human Labor, "this sacred ethic" marking our sense of Responsibility: For Self, Family, and Society, could now be said to have "endured" from, for example, 6:00am to 4:00pm, rather than be expressed into such vague imprecise terms as "began in the morning and quit in the afternoon."

But even the Clock, this very "Time-counting mechanism" mimicking Continuum, could not escape from the Laws of Thermodynamics: It had "gears" having a "high entanglement quotient" with Entropy that needed to be periodically wound-up!

Time, previously appearing to be "ever-present" or "immanent-static," was now "dynamically quantized" for both abstract creativity taking place within the Human Mind as well as for its concrete application in the external world as: Technology. Time, now measured with "iterative regularity" became a surveyor that was "rewarding to the diligent" but "discrediting to the procrastinator."

It is "the quantization of Time" that awakened the impetus in Human Beings to pursue knowledge qua knowledge as a worthwhile satisfying interest: It marked beginning of the systematization of mathematical-technological Knowledge as "Science."

Until the invention of the "clock mechanism," the line between "Eternal Time" and "Thermodynamic Time" was not "so well defined; in that, Human Beings measured Time as the conceptual embodiment of abstract apprehensions of "Sunrise" and "sunset," or "day-light and night-darkness," the sum of which, operationalized as "day Time" and "night Time," while indirectly utilizing the rotary Motions of the Earth as "Space traveling landmarks," or as "period indicators."

Time, has no "physical dimensions," such as matter-dimensions of height, length, width or depth. Space has no "particles" such as matter constituents like protons, neutrons and electrons. But Surface is a quantifiable property of Matter. Volume can be measured only in the presence of Matter-Mass occupying and bounding the Space therein contained, configurations of which, allowing the utilization of physical dimensions as variables or co-determinants 'mathematicized" for obtaining computable quantities.

While Space can be bounded by physical dimensions of Matter, Time has no physical limits or boundaries. Time flows continuously from past to present to present, with no "graspable material" to which "physical values" can be "attached," except indirectly, by quantifying the "behavioral operations" of things "as Time is elapsing" as measured by some kind of "Clock Mechanism.

Thus, only structured units of Mass-Energy-Motion cycle temporarily as "frame-of-reference units" to produce apparent "interruptions or cessation in the stream of Time."

Thus, Time, Space and Properties of Force possess physical values only in relation to "working Matter" operating as Mass, Gravitational Radiation Energy, and rectilinear/curvilinear Motion.

In the lapse of Time, scientific progress incrementally refined the correspondence between clock-measured Thermodynamic Time and natural phenomena, durations of which, scientists had sought to measure, e.g., How long does it take the Earth to rotate upon its own 23-degree tilt axis? The answer being: 24 hours or 1 day!

The apparent "spans," "period limits" or "durations" associated with Time, expressed as seconds, minutes, hours and days, are "numerical coordinates" that "embody units of Time," as understood through operations of "clock devices" having "hands" and "numbers."

Or, if the "clock device" is electronic, it displays merely "elapsing numbers" as would a "numerical counter." These "numerical coordinates" denote "periods" assigned specific "durations," which human beings understand as "Time passing." However, the electronic clock device must also have its "measured periods" precisely correspond to Earth rotational Time-period or the Time it takes the Earth to complete one "Rotational Cycle."

Given that the Planet's circumference imitates a circular pattern structure of 360-degree, then each clock-device hour represents 15-degree-arc of "rotational travel" by the Earth upon its own axis: 24 x 15 = 360.

Were it not for this learned conventional understanding of Thermodynamic Time measuring natural phenomena for "precise durations," respectively, the "clock device" would be useless—just a little round machine with one short hand and one long hand rotating over numbers inscribed upon its circumference.

Though as created by God we are spirit-beings in mortal biological vessels, we need physically-associated or objectively-based coordinates for situation, orientation, location, navigation, and direction on the Earth and in the Universe. We are physiologically bound to utilize physical and objective parameters, variables, factors, indicators and coordinates for purposes of identifying, classifying, categorizing and comprehending the world in which we move and live.

Every child must learn "how to tell time," as aided by a properly synchronized clock device. Every child's learning must also include the Roman alphabet-characters or letters, as well as the Arabic Numeral System: 0, 1-9, which are numbers utilized in arithmetic operations, all of which, being the "inner-coordinates of thought-language" that "anchor abstract concepts" in our Minds by associating them with externally comprehensive physical objects, things, and/or natural phenomena, the sum of which, making for constructive and efficient knowledge of positions, sequencing, simultaneity, contiguity, navigation, cause-and-effect interconnections, etc…, as touching "Relationships of Living," e.g., historical conditions of Society, and "Situations of Existence," e.g., "the gravitational-Metrics of Planet Earth in the Universe.

How is Space quantified when measured? Space has no physical limits or boundaries. Physical limits or boundaries, dimensions or quantities, values and coordinates attached to Force, Energy, and Motion in Space are "mathematical re-translations" of Matter-Mass characteristics and qualities possessing properties as expressed in numerically quantifiable dimensions, such as surface, volume, distance, length, height, width or depth having to do with "Space-filling" physical-objective-tangible Parameters identifying energy cycling structural Frames functionally operating in recurring, apparently "self-iterative" periods-of-Time; hence, the CGS system of measurement (centimeter, gram, second); and the MKS system of measurement (meter, kilogram, second) utilized in abstract mathematical operations designed for integrating "Continuum quantities" that "stream" within "structured Frames of Electromagnetic Matter."

These "Frames of Structured Thermodynamics," are embodied in Mass-Energy-Units fulfilling the Input-Process-Output Principle of Thermodynamic Organization, from which ensue "Gravitational Relativity Effects," e.g., more specifically: the Perihelion Shift of Planet Mercury; and more generally: The Solar System Frame of Gravitational Radiation Energy as represented by the Electromagnetic Spectrum in which are "embodied" "Radiation Quanta" that travel as they form "Units of Energy" subsequently "streamed" as rays, waves, and visible light (i.e., Cosmic rays/X-rays, Gamma rays, Visible Light, Radio waves, etc…

What is the foundational constitution of Gravity? Gravity is an emergent force-property ensuing from operations of Electromagnetic Field Force that triggers Rotary Motion-Forms as "attached" to a "Center-of-Mass, i.e., Rotation and Revolution so "framed" by Solar Gravitational Magnetic Field Motion-Force, as to "attach" a "natural Satellite relationship" to the "Earth-Moon gravity complex," the sum of which, having for its "directional vector" the "center-of-Earth-Mass" for the Moon, on the one hand; and on the other, "center of Solar-System Mass" for the Earth; thus "compelling" centripetal-centrifugal Forms of Rotary Motion Force preventing the Earth from "escaping" the Solar System ellipsoid plane of Revolution, just as the Moon is prevented from "escaping" Earth geo-synchronous plane of orbital Revolution.

Gravity causes specific Patterns of Structured Motion —centripetal, centrifugal, or combinations thereof, as "suited" to the particular quantity of Mass, in which is "embodied" a "Framed Unit of Structured Energy," — e.g., Planet Earth — upon which Gravity is "applying its exertion" as a Force, e.g., Due these cause-and-effect relationships between "Differentiated Frames" of Mass, Motion, Energy, and Magnetic-Field Strength, Earth Gravity-force is represented as "G-1."

These "Differentiated Frames" are themselves the gravi-metric products of many primary "Mass-in-Motion Frames" that encapsulate Complex Gravitational Field-Relationships, such as: Sun-to-Earth relationships; Sun-to-All-planets relationships; Earth-to-Moon relationships; and, as "framed" by all inter-planetary "relationships complex," the "Earth-Moon-Frame" as a "structured Mass-Energy Unit" in relation to the complex Frame "connecting" the Sun, to all-the-Planets it "holds" in revolution around its own center-of-Mass, e.g., hence, inducing "Earth wobbling effects" and "Earth bi-polarity oblation."

How is Gravity measured, quantified, and "temporized" as a Force? The accelerating Force-of-Gravity upon an object "falling" towards the surface of the earth is expressed as 9.8 meters per second per second.

In the "category" of Radiation Energy: What happens when a quantity is powered to the second exponent (squared)? The exponential unit causes a "transubstantiation" to occur, the dynamics of which, engendering a complete change-in-Form or "transmutation," as "applied" to the "structured Unit" that is "framed" by Mass-in-Motion, due to the complex combination(s), or due the high levels of "entanglement," of quantifiable proportions/ratios involved, e.g., c^2 "embodies" a different quantifiable "category of Reality" that cannot be the same as "mere" or "plain c."

"Plain c" represents "a unit of Velocity," or of "travel," or of something that is "in Motion:" the Speed of Light or of "visible Light" constituted of "inert Photons," in a "vacuum." But c^2 is not "a unit of Velocity." Thus, by exponentially squaring the Speed, "photons" are "transmuted" into "X-rays?"

Indubitably then, c^2 is not c — though both had initially "originated" from a common iterative Source, or from a common "Energy-Form complex." What is it that "factors" in this exponentially powered mathematical operation, to "transubstantiate," but to also "transmute-transform-convert" c into c^2?

How does c^2 as in "a Star," "connect with," "interrelate with," "interface with," or, " is entangled with" the particles of "an Atom," at the Quantum Mechanics Frame-of-Reference?

What "transmutes" an atom or atoms into "nucleated Plasma?" Is "nucleated Plasma" in "molecular Form," or does it "flow" in a "Differentiated Form?" In other words, how does "Radiation Energy cause-and-effect-trigger" or induce the formation of "cancer" in Biological Organisms such as Human Beings?

MOMENTUM-MOTION ENERGETICS

Momentum acceleration of Mass-Energy-in-motion, gains, in response to "Gravity force metrics," an "expanded Force-weightiness," manifested as "Mass-energy-momentum-motion pressure-force, exponentially increasing as Duration increases in "length of Time," the sum of which, expressed as "distance covered" in "meters/second/second or "meters/(second^2"):

g = 9.88 meters/second/second embodies or "in-loads" momentum, mass, speed, pressure, force, and "tensor-and-stress Metrics," the sum of which, expressed as "Momentum Mass Energetics."

Because of the exponential power of Time factoring in the equation, g-acceleration in the Earth is "transmuted" into a "force-category" that is other than "a simple increase in momentum-speed." This "force-category," is also consistent with the inertial Mass of a jumbo jet airplane standing on the tarmac of an airport, exerted as Momentum-Mass-Pressure-Force "Load-Energetics."

"Load Energetics," is a complex exponential Momentum-Motion-Pressure-Force — because, it is the surface of the ground preventing the jumbo jet airplane from continuing its "gravi-centered" travel trajectory along the radius-line of the Planet towards Earth Iron-Molten-Magma Core.

Electromagnetic Matter-Energy, in Motion, within Gravitational Continuum Space-Time: encapsulates, in a nutshell, the nature of the "Universe Frame." Thus, Matter, viewed from the standpoint of "Mass-Energy in-Motion" brings into account unknown "categories of Force," parameters of which, respectively, expressible only in a "mathematical equivalency process' that considers adding "pressure-tensor-stress metrics" to "the Force-dimension" unleashed during functional operations of momentum-mass dependent "Motion-Energetics," so that "parametric Differentials" can be "normalized" for "Continuum Wholeness Integrity."

Matter has particles: not Space, quantification of which, being associated with the Matter that "fills" it. Thus, Space has "emergent properties" that are "quantifiable-measurable" consistent with the properties of the type or category of Matter "occupying" it.

Every function performed by Matter is "quantifiable-measurable," but not Time. Because, only "Thermodynamic Time," as pertains to "Matter functioning" in Gravitational Continuum Space-Time, is measurable-quantifiable.

Because particle-structured Matter can be transformed into Energy, there is the Law of Transformation of Energy, but not the law of Transformation of Space, (not of a particulate nature) not the Law of Transformation of Time (not of a particulate nature).

Therefore, Time, Space, and Gravity are "measurable-quantifiable" only in terms of parameters engaged in qualifying the operational functions of Matter-Mass Energy within the Greater Frame of Universal Continuum Reality.

Energy transformation is possible because the constituents of Matter as protons, neutrons and electrons, are particles possessing not only physical dimensions but also electro-magnetic charges that engender the "displacement Motion properties" of electromagnetic Mass, e.g., revolving electrons, giving to a "functional Mass-Frame," its "operational energy-cycling duration," from thermodynamically assigned "Temporality coordinates," e.g., Duration of Earth rotational Motion upon its own 23-degree tilt-axis is: 24 hours or 1 day.

Thus, Mass-energy-structures operating as "Units of Motion-Force," e.g., Earth rotation has a 24-hour duration period, the sum of which, operating with its Mass-dependent magnetic field in relation to G-1 Gravity, in order to "keep" the Moon in its orbit as a natural satellite.

These Mass-Energy structures are synonymous with thermodynamically cycling frames engaged in "re-translating," re-organizing, and replicating the electromagnetic properties of Matter into its many wondrously complex, variegated, diversified Forms — 'Energy is never created nor destroyed but always transformed," while fulfilling the requirements of the universal mechanism of the Organizing Principle that keeps the Universe in working operational Order, the sum of which, dependent upon the "Input-Process-Output Mechanism."

These Mass-Energy structures absorb and respond to the "tensor-pressure Motion-Force Metrics" they receive from all sources derived, e.g., Perihelion Shift of Mercury is due to multi-causal factors associated with the Planet's mass; its center-of-gravity; its positional order within the ellipsoid solar plane of revolution; its proximal location in relation to the Sun's center-of-Gravity, center-of-Mass, and Magnetic Field Strength, as governed by gravitational radiation energy inputs from the Sun as well as from the synergism of all nearby Planets, such as Jupiter, impacting its geo-atmospheric structure and composition, etc…

As cold-vacuum Space and the Sun's radiation are held constant, and controlling for Earth Time-coordinates of Duration in relation to its gravimetric Mass properties that determine ecological processes, atmospheric/meteorological pressure, and livable temperature ranges, the relative stability and consistency of its integrated planetary events and natural phenomena, — e.g., Earth weather systems typical, respectively, to each Season within the 365 ¼ days duration, allowing for the planting season, the growing season, and propitious harvest conditions, — emerge from its rotary Forms of centri-vectored Motion properties as manifested in iterative patterns of Rotation and Revolution affording "captive/attractive" Earth-Moon satellite relationships, the sum of which, being replete with inherent resource-generating processes that embody qualities, characteristics, attributes, parameters, as produced by "gravitationally induced Field effects" arising from its numerous "entangled variables" that make its primary natural resources a sine-qua-non for sustaining Earth life support systems, — e.g., An Oxygen-Nitrogen Atmosphere; A proportionally immense volume of ocean waters from which emerge/derive sweet-water evaporation, cloud-condensation, and precipitation as rain, snow, etc…; Polar ice-caps providing not only "cooling effects" during Seasons other than Winter, but which are also a necessary reservoir for "holding" excess precipitation as well as "excess frost-condensation by-products" emerging from "the heat-and-cold complex" resolved as "a unity of opposites" — as characterized by a cold extra-orbital vacuum Space environment "having an urge to merge" with a "heat-absorbing" and "water-retaining" planetary Mass-Energy structure designed to maintain its life-support properties as an Organized Universal Planetary Unit in dynamic System equilibrium, i.e., The surrounding extra-orbital Earth environment is "cold Space" as opposed to its high degree of "heat retention" due to proficient absorption and processing of Solar Gravitational Radiation Energy amenable to its 23-degree axial tilt deflecting radiation at the Poles and allowing "heat capture" at the Equator and at the "Temperate Zones," counter-operations of which, accountable for Earth bi-polar ice-caps as well as the complex meteorological activities leading to region-specific "range-of-Temperatures" for yielding recurring patterns in iterative/cyclical and seasonal phenomena, e.g., Amazon forests' diluvial precipitation cycles; Recurring Monsoon rains cycles of Asia; Cyclical High-altitude mountainous snow accumulation; and iterative area-specific cyclone activity or tornado events causing invention of terms such as "tornado alley."

Revolution and rotation, — in confluence with gravitationally induced ecological processes and geo-atmospheric phenomena as effected by Earth tilted axis. — "co-labor together," due to solar magnetic field dynamics and radiation emission patterns, the sum of which, "crystallizing" atmospheric operations into an electro-conductive Frame-of-reference.

This Earth lightning-generating Atmospheric Frame-of-reference emerges out of the natural "conductive affinity" prevailing between the Sun's Emitted Radiation Spectrum and the inherent capacity of Earth geo-atmospheric compositional structures for "receiving-absorbing-and-processing" its quanta-units of electromagnetic Energy.

The Earth is a "Water-Planet" possessing a Hot Iron Molten Magma Core. Iron and Water together constitute, respectively, highly electro-conductive "catalysts" for triggering or precipitating numerous organic chemo-physical reactions that respond to solar gravimetric Inputs, the sum of which, enabling the Planet to generate its own Magnetic Field as well as a complex System of "entangled" meteorological-geological symbiosis of "Geo-Hydro-Electro-Atmospheric" phenomena, processes, and events that sustain the totality of its "Planetary Ecological Health."

"Gravimetric radiation-affinity relationships" between inner-Star energy properties and inner-Earth atmospheric compositional structures give rise to Lightning. Electro-conductive properties of Earth iron molten magma core yield an earth magnetic field that "duplicates solar Curvature Motion-forces:" as the earth revolves around the sun due to Curvature forces engendered by solar magnetic field-tensing-Energetics, the Moon synchronously revolves around the Earth due to its own magnetic field properties.

Apparent redundancy of Forms, observed in iterative patterns of frame cycling, is intrinsically an embedded universal characteristic, inherently conducive to gravitational Space-Time Continuum Energy Transformation, even as Conservation overcomes Entropy tendencies to degrade, decay, exhaust, and waste.

The Laws of Thermodynamics that animate the Law of Transformation of Energy dictate an ultimate "end" to "all things that physically exist." Like the Earth, we also partake of its elemental constituents: We breathe the atmosphere for our lungs to extract Oxygen there-from; Our body is nearly 80% water-composition; Our blood is also "iron-based;" and chemical physiological processes are activated by pressure-and-temperature ranges that sustain our organic biological lives as "Beings with colloidal tissue," as supported by electrolytes activity.

So, we age analogous to the manner in which iron rusts and aluminum corrodes. However, the extent to which "iterative Energy-Transformation Cycling" engenders decay to negatively impact the effectiveness and efficiency of universal phenomena, cosmic events, and natural processes, is unknown. But, in Human terms, we call it "wear-and-tear" leading to "aging and death."

In addition, "Gravity effects" depend upon earth axial angular momentum energy properties as engendered by its 23-degree tilt that "holds relatively constant," as it journeys around the perimeter of the solar system ellipsoid plane.

Earth distance from the Sun is a function of "The Dynamics of Revolution Energetics," as inter-planetary centers-of-mass, centers-of-gravity, and centers-of-field are in "entangled intersection/interface" for "assigning" each Planet's periodic rotational rate, respectively.

Every rotational angle of "earth spherical condition and status," e.g., In 1 hour of Rotational Duration, a "swath" is "traveled" having a 15-degree Arc distance, is a function of "gravi-metric axial geo-ecological parameters" that determine seasons, durations, temperature gradients, pressure variability, fauna and flora.

Revolution, rotation and solar radiation "never stop moving," but are synchronized into fine-tuned thermodynamic mass-cycling "mutual dependencies," that effect iterative patterns of functional fulfillment, in the "transformative operations" performed by "the thermodynamic mechanism: Input-Process-Output."

The whole spectrum of solar radiation is emitted with cycling electric-and-magnetic fields having complex properties that travel-move, while perpendicularly situated *vis-a-vis* each other. Perpendicularly projected solar emissions as propelled from the Sun's inner-core center of nucleated mass, do not make contact with the Earth at 90-degree angles, because the Earth is axially tilted 23-degrees, which allows Earth bi-polar opposites to accumulate and hold great ice quantities, referred to as "polar ice caps." Thus, all along its 23-degree "axial incline," up to the Poles, safe for the Equatorial Region, solar radiation "only grazes" the Volume-and-Surface of Earth contents.

Though solar radiation is constant generally, the sun itself undergoes periodic changes that impact inner-core fusion processes and convection dynamics, the total dynamics of which, affecting rates of change in "Curvature magnitude," gravity force strength "equivalents," and field-projection rotary-Motion co-determinants, parameters, and variables. For example, every eleven years, the Sun experiences "magnetic-polarity displacement" due to inner-core nucleated transmutation activities, during which expelled solar wind, flares, and plasma rains — coronal mass ejections — make contact with Earth magnetosphere to pierce-through its ionosphere in order to "form Aurora Borealis" or "Northern Lights."

And every twenty-two years, the Sun undergoes a magnetic field polarity change during which numerous instances of coronal mass ejections can effect electromagnetic Curvature, Gravity, and Field disruptions. Magnetic storms can cause ecological turbulence affecting telecommunications transmission and connectivity, and short-circuiting electrical grids and damaging transformers.

Constant initially derived functional properties of Matter in Gravitational Continuum Space-Time Curvature, such as coldness of void-Space and heat from solar radiation, are not "operationally static." They engender Solar System Dynamics that impact the Earth for changes in temperature, pressure, ecology, and meteorological activities.

Polar icecaps and the oceanic hydrosphere, work in conjunction with the 23-degree axial tilt to form "control mechanisms" that climax into livable temperature ranges. Weather systems and life-support systems, lightning and storm systems, earthquakes and tornadoes, are emergent activities that characterize Earth "negotiation/calibration" of absorbed/processed/utilized radiant solar energy, the sum of which, conducive to "geo-cycling patterns," that maintain the Planet in relative dynamic System equilibrium.

"Energy is never created, nor destroyed but always transformed." Apparent redundancy of iterative patterns of Energy transformation predominates in mass-Energy frames that are cycling in Continuum Gravitational Space-Time. Nature and the Universe display many "converted states" of the same Energy form, as each frame cycles under function-specific conditions of Curvature-Motion.

Earth emergent geo-ecological cycling conditions are predicated upon the Oceans, the Atmosphere, the Wind, and Landmass complex, the totality of which, acting in synergistic relations with Revolution and Rotation, gravitational effects of the Moon, magnetic field strength, the 23-degree axial tilt, and phenomena engaged in Earth-core-centric Motions: There is "no other place" for all these thermodynamic activities "to go." Thus, the Earth is a "captive Energy Transformation Environment;" a "closed system," and yet, not "an isolated system." For, Solar Inputs are vital to the Earth that must thrive-and-prosper as a Life-protecting Water Planet.

Together, these "conditions" constitute the variables that factor into causing livable temperature ranges for human and animal life, for fauna and flora, as an emergent property of solar system Curvature Energetics. Namely, the Arctic, Temperate and Equatorial zones of temperature, in coordination with the different topological surface variants of flora and fauna characterize each season, respectively.

For example, a circular pattern of orbital trajectory for sensor-equipped satellites around Earth orbit could be programmed to record ecological parameters within a specific geographical surface-volume unit, e.g., 1 Hour of Rotational Period within 15-degree Arc of "covered distance;" as well as strength of solar gravi-metric electro-motive forces, field radiation densities, rates of ionization, rates of radiation Motion-Force pressure-strength "blasting" the Planet, cloud condensation characteristics, regional temperature and pressure gradients, all of which, could be "mapped" or "matrixed," in order to improve forecasting of draught and diluvium, flood-causing rain periods, as well as for knowing "the best time" to begin planting Season.

A line is a succession of points. Radiation is a succession of quantized Energy units "streamed as a wave." A circle is a straight line cut-in-two and joined at both ends via a 360-angle Continuum.

Continuum implies successive accumulation of single points, "quantized units," or progressive aggregation of single events, phenomena, and processes within Gravitational Space-Time Curvature, the total sum of which, to function and operate as a "Unified Whole Integrity Oneness," according to validly proven scientific Laws of Physics.

Energy cycling and Mass-in-motion evoke "distance travelled from one point to another," as indicating a change in a certain quantity or in a specific "state-of-Matter," e.g., in revolutionary distance, in thermodynamics form(s), in combustive rates of fuel utilization, rates of radiation absorption to rates of radiation processing, etc . . .

Continuum is constituted of stochastic or periodic cycling. Even the Gravity force or its "equivalents" at Earth surface is measured in terms of "accelerated exponentially powered time-distance Mass displacement," or at 9.8 meters per second per second. This implies Mass-in-Motion exponentially accelerated for "distance traveling" as "momentum pressure-force dynamics," gradually increase "to a maximum," in the absence of an additional external Force.

Natural phenomena can be "compartmentalized" as singular quantum events within a movement belonging to a "greater Mass-Motion Frame," or as "a wave of collective occurrences" constituted of "streamed quanta," hence, reference to atomic particles as "quanta" or "packets" of Energy, or as "waves of energy."

"Mental Compartmentalization," or artificial separation of natural phenomena as pertains to the convenience of technologically-derived mathematical operations, does not imply discontinuity or "dis-union" within the universal dynamics engineered by its Gravitational Curvature Frame Energetics.

"Integrated Continuum Oneness" still prevails, because, simultaneity and contiguity are retained via overlapping Frame intersections/interface(s), the sum of which, to operate even through infinite vacuum Space, or even as "touching" or "bordering," Eternal ever-present Time, given our not-knowing the exact "Thermodynamic Time" when the ultimate terminal-final universal Entropy event will occur: The end of all things that exist, or more colloquially, "the death of the Universe."

Overlapping frame cycling contiguity establishes simultaneity, thus, engendering Continuum. Is not the Sun situated at 93 million miles from the Earth, yet its emissions ensure Earth dynamic eco-System equilibrium for prosperous sustenance of Human life?

Duality of form as particle and/or wave qualifying a description of light radiation Energy testifies to the predominance of Continuum, the apparently stochastic or intermittent process-steps of which, have to be mathematically "normalized" via observable Time-bounded, Curvature motion-pressure-force cycling patterns, as delineated by the projection strength and boundaries of magnetic Field forces, Gravity, and gravity-force "equivalents" that engender centri-vectored forms of motion: The Earth has a gaseous Atmosphere that "filters" or "buffers" the destructive impact of ionized radiation Energy, and hence, why the Speed of Light is relatively "faster" in a vacuum, than in water or in any other material having in structural Form, the quality of "a decelerating substance."

Particulate behavior classified as "waves of probability" within the scope of quantum mechanical Energetics is only indicative of composite nucleus-centric forms of motion that must be "reconciled" via alpha-numeric Constants that "connect" Relativity Gravity-variables and "Field conversion factors" into mathematical "proportional equivalency."

However, only particles "isolated" from their intrinsic Nucleus would present such a "probability scenario" to Human Understanding due to the nature of technology utilized in the pursuit of sub-atomic examination. "Isolated particles" do not 'behave" in the same way as "nucleus-revolving" particles. In short, "Light is chasing Light!" And given that when Speed is exponentially squared, a 'transmutation" occurs within the unit of radiation Energy to which a change-in-variables is applies, then, such "interferometric effects" are accountable for the appearance of "waves of probability."

When "Interferometrics Effects" are triggered, co-determinants of dynamic System equilibrium suffer disruptions that then impact how the Laws of Physics operate within the frame-of-reference under examination, i.e., the Atom's so-called "waves of probability."

But, atomic particles of respective elements, do regularly and normally undergo chemo-electrical reactions geared for formation of compounds and substances, under fine-tuned "chamber-reaction conditions" that account for their "mutual affinity" to "react with each other" for molecular aggregation, e.g., Oxygen and Hydrogen "chemo-electrically react" to form Water molecules (H_2O).

The electromagnetic spectrum — radiation bands possessing wavelength, frequency and electro-motive-magnetic units-of-Energy measured in electron volt units — embodies analog Forms of Field force gradients, and Forms of Gravity force "equivalents" that "companion-travel" at 90-degree angles with each other, — "cycles of which," (in hertz-per-second units" — high or low frequency) determine the degree to which they are harmful to Human Beings.

Low frequency-and-long-wave radiation Energy Forms are less harmful than high-frequency-and-short-wave radiation Energy Forms. These micro-mass Frames or "nano-Frames in-Motion" encapsulate Curvature dynamics, field and gravity energetics, the sum of which, operating for differentiating "congealed states" from "nucleated-radioactive states," e.g., Visible Light constituted of photons is the most harmless of all radiation Energy Forms, in terms of its fundamental wavelength, frequency, and energy-Form measured in electron volt units.

Matter-mass frames are constituted of energy cycling structures patterned after this organizational matrix: Particles aggregating to form atoms; atoms grouping to form molecules; and molecules bonding to form greater bodies.

Periods of Energy cycling within Thermodynamic-Time durations, respectively, are different from Frame of Mass-Energy-in-Motion to Frame of Mass-Energy-in-Motion, hence, "pressure Differentials," "temperature Differentials," and "Thermodynamic-Cycling Duration Differentials."

The particle, or the smallest speck of Matter embodied as Electromagnetic Mass-in-Motion, undergoes gravi-metric Field-Curvature pressure-force vectors that cause scientists to describe its motion forms as "waves of probability," due to its apparent electro-fluid dynamics, force-activated trajectories, and vertiginous velocities—hence, the extreme difficulty in tracking its position, momentum, mass and charge all at the same time. Not so for the particle that remains within its specific atomic Frame designed for synchronized molecular reactions; only "isolated particles" bombarded with radiation Energy in "particle accelerator laboratories" display such "waves-of-probability." However, these "probability waves," a terminology "coined" by experimenters, rather reflect deficiencies in their "radiation-rigged" technologies, as well as limitations in the Human capability to ferret-out the complex physical Laws that are triggered under such experimental conditions.

"Waves of probability" implies that micro-Curvature Energetics engender non-centri-nucleic Forms-of-motion, away from the routine display of rotary, rectilinear and curvilinear "patterns of displacement." However, in the presence of radiation-based technologies, "omni-vectored and multi-directional forces" will appear to impel "new trajectories" for "circumventing" revolution around the atomic nucleus. For, many counter-vectored Forces would be found to be impinging upon atomic particulate trajectories, due to the complex interplay of the strong force, the weak force, Gravity force, and magnetic-Field force, and to "external Energies" introduced via spectrum-based electronic instrumentation that modify and/or disrupt dynamic System equilibrium pre-established by regular particulate parameters and co-determinants.

Particulate trajectories articulated as specific Motion patterns, are still in Continuum with planetary rotary forms of Motion and with radiation Motion-force Energies projected by the Sun.

Particular Frame cycling conditions are emergent properties derived from a shared, yet "mutually differentiated Electromagnetic affinity," the sum of which, accounting for Differentials in force, Curvature, gravity force "equivalents," and field pressure vectors that engineer Motion.

Particle or wave — This "duality of form," due to "entanglements" in mass-charge-momentum, motion, force, and Frame-specific reaction patterns, — presents problems for computational analysis. The atomic Frame embodies both "programming" and "computation" characteristics akin to a Frame-specific "operational algorithm." Quantum Mechanics is neither "anarchic," nor "random," nor "chaotic." Nor does the atomic Frame display random "waves-of-probability."

Rather, it's merely "quantum complexity," as governed by specialized micro-metric physical forces, among which, are nano-Gravity force effects and nano-Electromagnetic field-force effects.

Limitations imposed upon apprehension of "particulate behavior" by electronic technology in which Human consciousness appears to factor, compound the problem, as the human Mind itself must participate in the process of "quantum understanding."

An essentially synchronized "Motion-force algorithm" at the Quantum Mechanics level, does have characteristics that participate as well, in experimenters' interpretation of observed natural phenomena.

Curvature pressure force dynamics incorporate many different forms of curvilinear, rectilinear and rotary Forms of Motion displaying convoluted trajectories and paths, the sum of which, appearing to display "probability waves," because, under "compulsive nucleus-centric forces," particles cycle in "compressed time" due to entanglements of field, gravity, mass-motion Energetics, and charge displacement forces.

Even "as torn by conflicting multi-vectored forces," particles are compelled to revolve around the nucleus, hence their convoluted trajectories embodying composite force-vectors and magnitudes interpreted as "waves of probability."

Whole Earth gravitational-field "Curvature-Mass-in-Motion Complex" embodies not only the concept of "travel" in Continuum Space-Time but also the reality of "gravimetric Motion-Force."

As earth Mass moves, "everything moves with it; it leaves nothing behind." The Earth also "displaces" or "pushes-through" Space volume, with its landmass, atmosphere, and oceans "still attached," thus engendering torque, momentum power, Mass-energy and Motion-force. Force, Mass, Motion and Momentum are active in Revolution and Rotation as "gravitational tensing" and "field stressing" that give Continuum Space-Time "Curvature elasticity," or "Curvature tensility," which allows "an orbital band-width" within which, the space shuttle can circle the Earth at 117,000 miles per hour.

"Converted states" in all "Curvature-metric Forms of measurable expression" or "Curvimetrics," are in Continuum with "other converted states" within frame-specific conditions of functional cycling that engender Differentials, e.g., a gasoline combustion engine is a "curvi-metric analog state" to the controlled gas-explosion forms of rocket propulsion; a burning piece of dry wood radiating infrared heat displays "curvi-metrics" in an "analog state" to the enriched rods of Plutonium that radiate cosmic rays, waves and particles, the sum of which, utilized for producing steam that serves as means of propulsion for turbines geared for generating Electricity in nuclear power plants.

Integrating co-dependent, co-determinant, overlapping "Frames of Structured Functioning" into "a unified theory" that accounts for all the unique thermodynamic Forms of the same, but transformed Energy, requires the superposition of alpha-numeric Constants to "normalize" or "reconcile" frame-Differentials for "relative equivalency," between analogous converted states.

Earth Revolution is planetary "travel" around the Sun, and Rotation is planetary "travel" upon its own axis, and hence, the concepts of "sunset" and "sunrise," the operational sum of which, turning "cycling functions" into a Reality: Seasonal cycles, atmospheric pressure, and range of livable Earth temperature(s) within the Framework of G-1 Gravity Force, allowing us to "move freely" upon the Planet in fulfillment of our daily necessities, i.e., creative productivity.

The apparent "stochastic" recurring patterns that structure "self-iterative cycling motion Energetics" that qualify dynamic ecological System equilibrium, embody processes in which participate, all Earth constituent components, operational interactions of which, climaxing into frame-Differentials belonging to Continuum gravitational phenomena in 'relative entanglements" that reign in "holding the Universe together" as an Integrated Unified Whole.

****

The Current Impasse in Post-Relativity Theoretical Physics

What is a theory? Webster's Dictionary describes it as: *"A formulation of apparent relationships or underlying principles of certain observed phenomena which has been verified to some degree,"* or *"that branch of an art or science consisting in a knowledge of its principles and methods rather than in its practice; pure, as opposed to applied, science, etc."*

The Theory of Relativity formulated a mathematical description of Newtonian gravitational "attraction-repulsion force" in terms of Curvature or "bending" of Continuum Space-Time by cosmic bodies.

Where gravity is sensed, felt, experienced, or known, rotary forms of motion exist, e.g., Revolution and Rotation, as precipitated by field Force Dynamics and Gravity Force Energetics.

For example, as the earth possesses a magnetic field, it also engenders gravity force equivalents within its spherical mass, and externally, for satellite orbiting motion. An aircraft flying in earth atmosphere experiences force vectors having magnitudes and directions that temporarily resist core-centric motion forces, e.g., downward weight, upward lift, forward thrust and rearward drag.

Gravity, Field, and Curvature Dynamics, are in "entangled relationships" to climax into rotary patterns of motion, e.g., revolution and rotation. Stars and planets exert gravitational energy upon each other within their respective fields of force, range of which, possessing strength limitations, e.g., It is not "accidental" that the Sun's Magnetic Field has a "range-of-strength" accountable for "keeping captive" only 9 planets — rather than 12 or 15.

Inter-planetary gravimetric forces and field-induced forces also give rise to "particular quantum events," as each planet "re-calibrates" or "re-negotiates" its centri-vectored parameters that are accountable for dynamic System equilibrium, during which there are "disruptions in dynamic System equilibrium, such that, The Law of Entropy must also fulfill "its requirements" that are consequential to Energy Transformation as carried-out by the "Input-Process-Output mechanism," the sum of which, all "framed Mass-Energy structures" must undergo, as necessitated by their respective cycling patterns of Energy Conservation, e.g., perihelion shift of planet Mercury; Earth "wobbling;" and Earth bi-polar oblation, are such examples. And these "Relativity effects" are "counted as" disruptive, Solar-System-generated perturbations that "de-stabilize" the Planet's functional parameters responsible for dynamic System equilibrium.

Mass does not merely "attract" mass, but Mass "bends or curves" the Space within which Mass acts and is acted upon. Attraction-repulsion dynamics are not exerted in "linear reciprocity," but rather as "gravitational tensing-and-stressing" of the Space within which interactive Mass-Energy Frames move.

Curvature Energetics factor into the "re-translation" of the electromagnetic properties of Matter, as Mass-in-Motion undergoes operations "processing" energy cycling Differentials that impinge upon its regular patterns of structural Motion as it "negotiates tensing-force(s)," the sum of which, having diverse numerous Sources of Origination.

Protons and neutrons "cling to each other" as electrons revolve around the atomic nucleus. The Sun and all Planets "coalesce" into an ellipsoid Curvature plane with the Sun at its "Gravimetric-and-Field-tensing center."

Curvature pressure Energetics is centri-vectored in Force direction — things must "gravitate" towards a cycling center of Mass-Energy having rotary forms of motion. The space shuttle displays "spiraling motion patterns" after lift-off, due to convoluted pressure force dynamics inherent in fuels ignited for explosive combustion, even as it is supposed to have "a straight trajectory" into outer Space. A dry leaf with a quasi-circular form falling from its tree stem in autumn will display a "spiral" or "spinning" or "rotating" form of Motion towards earth surface. Even a Person whose parachute failed to open, will, while falling to Earth surface, also display a "spiraling-concentric-spinning-rotating Form of Motion.

Every thing, wave or particle, object or body, falls under this gravitational Continuum of Curvature motion-energy exchange between cycling mass-Energy frames. Gravitational Mechanics, Relativity Dynamics, and Curvature Energetics complement each other in generating "gravimetric Differentials" between "centers of Mass," "centers of Field," "centers of Gravity," and centers of Motion, cycling intersections of which, embodying emergent "overlapping" and co-dependent, co-equal, but opposite patterns of "centri-vectored force-turbulence" that must then be mathematically "normalized" or "reconciled" for "proper fitness" via operational application of certain proven-valid Constants, into a self-evident formulation of their integration in the universal "gravi-equilibrium dynamics" of Continuum Space-Time.

Curvature pressure force is ubiquitously distributed among all cycling mass-Energy frames, "physical dimensionalities" of which, are "engrossed" in displaying motion forms pertinently and correspondingly calibrated for "navigating through" or "negotiating" interconnected regions of "overlapping effects/influence," the sum of which, climaxing into differentiated tensing-and-stressing Energetics that yield cumulative Relative Continuum Interconnectedness.

Thus, Continuum Space-Time has "no disconnects." All phenomena, events, and processes are "held together" or "coalesce together" in conformation with the Motion-Force Principles that animate and drive Gravity Mechanics, Field Dynamics, and Curvature Energetics.

There are Fundamentals of Science and Fundamentals of Theory. And they intersect in applications of the Scientific Method as mobilized for technological discovery and creation.

However, Fundamentals of Science are "non-negotiable." Therefore, "adjustments" or "adaptations" can only be made within the Framework of Theory Derivation, — A Theory must be "derived" from factual self-evident Truths scientifically observed within reproducible experimental conditions that respectively and pertinently climax into the same results when performed or undertaken by other independent disinterested research Scientists— until certainty in the validity of its facts, self-evident Truths, and reproducible experimental results, is attained, so as to "align" its Theoretical Framework with the Scientific Fundamentals that confirm its applicable proven validity, in respect to the Field of Knowledge to which it pertains.

EVOLUTIONISM v. SCIENCE

What are the "fundamentals of evolutionism" as opposed to the Fundamentals of Physics? How do evolutionist doctrines impede progress in "Mathematical Theoretics?"

Evolutionism describes "mutational processes" leading to the "descent of Humankind" from the apes. But random mutations would disrupt Energy cycling equilibrium and "disconnect" Space-Time Continuum.

In addition, evolutionists also coined an expression termed: "punctuated equilibrium," which is an unproven concept, the sum of which, positing the random appearance of "new species." "Punctuated equilibrium" is again a process, the sum of which, "disrupting" the flow of Continuum speciation from the same "common-to-all-Species genetic DNA pool," in that, it introduces "random breaks" within genome-specific DNA-speciation that has been "flowing-forward" in "Thermodynamic Time," since the very "first days of Creation," or since the very Time of Bio-Genesis.

Given that the initial "common genetic DNA pool" had initially identified, once and for all, all definitely resolved living Species, even those still unknown to "Human discovery," then, such "random speciation" is presupposed to have arrived from "random mutations" that can be accountable only to a "different genetic source" — not originating from the same common genetic DNA pool — giving rise to so-called "biological missing links," which to this day, have not been "revealed" by the so-called "fossil record."

A short-version of "their take" on Astro-Physics is summed up in an "imaginative tale" having evolutionists saying: That the Universe began with a "big bang explosion" which caused all cosmic bodies to be moving and travelling at great velocities, due to which the Universe is expanding; and after which remained "cosmic background radiation" with the capacity to represent/embody "the past Universe," when Human Beings peer into faraway Galaxies, such that, they are "seeing the Universe," not in the Present, but rather, as it existed "billions of years ago."

Thus, Astro-evolutionists suggest that radiation or light that we observe at distances light-years away from our Milky Way Galaxy, is always displaying the Universe as it was long ago, not as it is in the Present.

Evolutionist astrophysicists predicated the "big bang theory" upon the theory of universal expansion, which is supposed to explain "the beginning of all things in totality," for all extant Reality, including the beginning of Life on Earth, generally, and more specifically, the "godless random genesis" of Human Life in the Universe.

Needless to say that this evolutionary theory of universal beginnings constitutes a paradigm, doctrines and tenets of which, contradict and oppose Scientific Creationism — the Creation of the Universe by Almighty God as described in the Holy Bible.

The first act, i.e. concocting an evolutionary beginning that denies the existence of God, of necessity will have been extended to rejecting the fact of Biblical Creation. Following the line of reasoning upon which is predicated "the theory of big bang origin," the implication is that a non-existent God could not create something that exists. It was a simple step from rejecting God's existence to rejecting His creation of the universe, to then inventing a universe that "created itself" or "began all by itself."

"Banking" on the "big bang" evolutionary conjecture as a given premise, — evolutionists do "earn their daily bread" through their official professional pontification of those scientifically unproven doctrines — the God-denying scientists, now trained in secular philosophy, could not explain how the "initial explosive matter" got there in the first place. What could have given rise to Matter, out of nothing, to cause it to explode all by itself, to then engender the emergence of Life? A chaotic Matter-dispersing explosion, giving rise to the most complex Reality ever known!

Astro-evolutionists could not ignore already proven valid laws governing physical reality, such as gravitation, electromagnetism, Relativity, the law of transformation of energy etc . . . Thus, whatever they would concoct, had to be within the schematic theme of already known scientific theories. However, the only place where exists a "Theory of Origin" is the Holy Bible, in its first Book, entitled, "Genesis" — even before Darwinian Evolutionism, published in AD 1859.

The Holy Bible states that God created the Universe. Astro-evolutionists, therefore, concocted a competing "theory of origin" rooted in "religious atheism." But their "universal starter" could not easily fit into "its running engine" — how could the great, randomly dispersed chaotic force of the so-called "big bang explosion" lead to the awesome organized complexity displayed by natural phenomena and universal processes? For, if the universe had "evolved" into its present state, then its "initial conditions" could not be operating to this day. Yet, they are: There is no scientifically proven-valid way to "differentiate" between "old radiation" and "new radiation."

Radiation Energy is always coherently sourced from the same physical phenomenon from ever-present materially operating electromagnetic-based processes, e.g., from a Star operating within the uninterrupted flow of Thermodynamic Time as "an ever-present Continuum" from "past-to-present-to-future," up into the Present Time in which we live.

Thus, taking "the logic of Evolutionism" into account, we deduce that: "The Past Universe" could not still exist in our current Time frame, in the Present, in order to be detected by our electronic machines as "relic cosmic background radiation," for its initial cycling frames "would have mutated" into present forms, as its "intermediate relics" would have disappeared, the sum of which, as already stated previously, e.g., a random "big bang explosion from nothing," would constitute "disruptions" or "interruptions" in Gravitational Continuum Space-Time, which would overturn universal dynamic System equilibrium into disorganization and disorder.

However, the Laws of Physics that do keep the Universe in integrated wholeness-working-Order will not permit this farfetched unproven "scenario" to be "transformed," by propaganda and fictional narratives, into factual scientific self-evident Truth.

Evolutionism precludes Continuum. Interrupted processes or "arrested events" which necessitate so-called "intermediate species" or "missing links," and co-called "punctuated equilibrium," violate fundamental principles of universal constitution, which explains why there has never been found any proof for their occurrence.

The theory of evolution and all its "doctrines of random mutation" contradict the interactive flow of co-determinant variables that would sustain Continuum between Forms for energy transformation, including Biological Forms of Speciation.

In addition, to our proven scientific knowledge, all pre-natal "random mutations" represent a precursor to the discovery of a biological deformity, or heralds the declaration of an incontrovertible congenital disease, e.g., spine bifida; Down's Syndrome; infant cancer; fetal alcohol syndrome; club feet, etc . . .

Charles Darwin (AD 1809-1882) intended Evolutionism for Biology and not for Astrophysics; nor for socio-economics. Evolutionism was extended to Physics, due to Astro-evolutionists having "a socio-spiritual interest" in dethroning the authority of the Holy Bible in the framework of Human affairs.

For example, that the Universe is "expanding" is an extrapolated offshoot of the Theory of Relativity. As there was no explanation for the theory of universal expansion, astro-evolutionists utilized "mutational doctrines" to posit the "big bang explosion" and by that invade Physics as a "branch of Science."

The "big bang explosion" hypothesis was developed by Edwin Powell Hubble, (AD 1889-1953), an Astrophysicist who surmised that only a "big initial explosion" could have unleashed a force powerful enough for supposed continued "universal expansion" — which could then account for Motion in galactic Space and solar system environments; that is, for both Matter and the Space within which Matter is moving, traveling or "expanding."

Astrophysicists do not agree, however, whether Space is moving also—the void-entity upon which or within which Matter moves; or whether or not Space itself is finite or infinite. Nor do they provide any explanation for the underlying background structure which would make it possible for both Matter and Space to be in perpetually sustained Motion due to a hypothetical "explosion" that occurred supposedly "billions of years ago."

How would current gravitational Motion, be scientifically differentiated, from the so-called "expansion Motion," in the absence of a comparable "initial base-line galactic Motion pattern?" Who has "the blueprint" for past universal Motion patterns with which to compare its "expansion Motion" in our Time? All cosmic bodies are/have been traveling at relatively constant rates of Motion, subject to "slight variations," the sum of which, sustaining their respective cycling distances from each other. The Earth is/has been situated at a distance of 93,000,000 miles from the Sun.

Given that all planetary bodies and stars are moving around and away from each other in cycles of Space-Time Motion; and given that, as star systems, cosmic bodies are "traveling" within galaxies, and around each other, Astrophysicists assumed the Universe as a whole is "running away" from "something." And what could that "something" be? Where the explosion of the "big bang" is supposed to have occurred, which Astro-evolutionists call "the universal epi-center."

All matter-bodies are presumed to have dispersed with great momentum due to the massive explosion force thereof, to then "cool" into planets. Then, why would the Sun have remained as a hot inferno of nucleated plasma condensate?

The Theory of Relativity posited the assumption of universal expansion — Astro-evolutionists hypothesized there must have been an "explosion," and "big bang" was born. Since no Astrophysicist has ever witnessed the "big bang," nor proven that it really took place, it is an "myth-making assumption" based upon the theory of universal expansion. The circular reasoning is channeled into evolutionist ideas that always escape to the "big bang explosion" as an explanatory scheme.

We know that the Earth is moving through Space with the Matter it holds — the landmass, the oceans, and the atmosphere; and if Space itself is moving, what is it moving through? And what in it is moving?

The most favorite version is that the Universe is like "an expanding bubble" on the whole surface of which stars and planets are "moving" — as the surface of a balloon will expand the more air is injected into it. Other views surmise that stars and planets are moving within as well as on the surface of the still-expanding "bubble."

In summary, Astro-evolutionists started with the theory of evolution, passed through the Theory of Relativity, encountered the theory of universal expansion, supposed a "big bang explosion," to finally arrive at an "epicenter" or point of origin at which the Universe must have begun, from which all stellar and planetary bodies and galaxies are "running away."

Continuing further "down the road" of adding more presuppositions, Astro-evolutionists "reasoned" that an immense explosion like the "big bang" must have generated a lot of heat; that, consequently, there must be "relics" of that heat-radiation-Energy in the universal expanse; and that this "remnant heat" of the "big bang" radiating from the universal epicenter of beginnings called "cosmic background radiation," ought to be quantifiably observable and measurable, albeit through electromagnetic-spectrum-based electronic equipment and machinery.

Again, seizing on this presupposition, Astro-evolutionists plunged "head first" in trying to detect this "relic cosmic background radiation" with electronic instrumentation, particle telescopes, spectrographic and spectrometric machinery.

To this day, there has not been any verifiably reproducible proof of any evolutionary assumptions regarding the Universe, from its supposed "beginnings" to hypothetical "mutational processes" from which "descended" the extant reality we observe.

How could a "missing radiation relic" be found when by evolutionist doctrine, the processes that gave rise to its emissive conditions would have had "mutated" and disappeared "as missing links" to engender "the present Universe"?

Evolutionist doctrines violate the Continuum Principle; hence, there cannot be unification between the past and the present, and between the present and the future, which disproves the proposed notion of "disappeared/missing mutational Forms."

ASTRO-EVOLUTIONISTS AND THE THEORY OF RELATIVITY

A magnetic field will generate an electric field; an electric field will generate a magnetic field: "Energy is never created nor destroyed but always transformed." Framed units of radiation Energy, Mass and Motion constitute the fundamental infrastructure of universal Curvature, from which emerges "tensing-stressing pressure-force," as those Frames "re-translate" the electromagnetic properties of Matter into so many analogs of thermodynamic cycling.

Motion, therefore, originates from gravitational "nucleated" star radiation that projects Curvature Energy Force via electromagnetic Field properties that engender great concentric "displacement-movements" upon planetary bodies, e.g., Revolution and Rotation.

From such great "Field-induced Motion-force interactions," emerges a co-linear, co-equal, but oppositely vectored Form of "Curvature Motion pressure-force," such as, Gravity, e.g., Earth Revolution around the Sun is "clockwise;" Earth Rotation upon its own axis is "counter-clockwise."

From these two opposite magnetic Field-induced, but concentric rotary Forms of Motion, emerge the complex properties of Gravity-Motion-Force, that not only "allows" the Earth to "retain" its Moon as its own natural satellite, but also "afford" us, Human Beings, "the greatest latitude" in "range of Motion:" upon the Earth, in Earth Atmosphere, and in Earth Oceans.

Thus, Gravity is an emergent property of electromagnetically generated field-induced Curvature pressure force Energetics giving rise to centri-vectored Forms of rotary motion. As the earth revolves and rotates, its center-of-mass and center-of-gravity cause all objects to be "attracted" towards its core along its line-of-radius, from their inertial or kinetic position, this core-centri-vectored force being identified as Gravity force.

For example, a plane flying in Earth atmosphere thrives under vectored forces that form a "gravi-metric 360-degree envelope" around its moving mass: downward weight or Gravity, upward lift, forward thrust, and rearward drag, the totality of which, giving to the pilot "ample latitude" to navigate through the Air with much "flight-control" and "ease-of-movement."

It is the power of the engine and the various means of navigational control, such as ailerons and flaps that keep the aircraft in the air, in counter-vectored resistance to the "attraction Force" or "Gravity force" persistently "pulling" the aircraft towards the core-center of Earth Mass and Earth center-of-Gravity. Barring the presence of these "flight control tools," the aircraft would "fall" or "gravitate" towards earth center-of-Mass and Center-of-gravity.

The Earth possesses a magnetic field force acting in conjunction with its great moving momentum-mass-force to engender geo-synchronous Moon orbital motion. As measured from the rotary motion frame-of-reference, e.g., Revolution and Rotation, Earth surface Gravity is an emergent property of "Continuum Curvature mass-Dynamics," as engineered by the Sun's electromagnetic Field, for solar ellipsoid plane tensor-stress Energetics.

Gravity Mechanics, Field Dynamics, and Curvature Energetics, along with Mass-parameters activated in the planetary-lunar frame-of-reference by Solar System gravitational radiation, operate by both in-tandem sequencing and parallel simultaneity, to induce the cause-and-effect relationships and interconnections that generate Earth rotary concentric Forms of Motion, — namely, Revolution and Rotation.

Revolution and Rotation embody "composite applications" of in-tandem, and simultaneous, attraction-repulsion-Force, engendering directional vectors that cause Earth surface objects to move towards its core — to which all things and objects would directly gravitate, unimpeded and unstoppable, were the solid Earth-ground surface to not have existed.

Centripetal force vectors overcome centrifugal force vectors to keep objects like trees anchored into the ground. Star momentum Curvature pressure-force dynamics projected as solar magnetic field, holds planets in their respective place-ordering or positional ordinal sequence, e.g., Earth is the third Planet from the Sun.

Likewise, solar magnetic field-Curvature tensor-stress Energetics "hold" planets, including the Earth, within designated distances from each other. For example, all-along the ellipsoid plane of Revolution, as Earth cycling mass-in-Motion Energy-frame engages in "negotiating" emergent Gravity-force "equivalents," its accommodations, adjustments, and adaptations, — e.g., wobbling; bi-polar oblation; occasional Earthquakes, tornados, hurricanes, tsunamis, and cyclones; Ocean-tides caused by the Moon, etc…, — "re-calibrate" parameters, variables, and co-determinants of dynamic System equilibrium, the sum of which, accounting for motion patterns of objects or things "traveling" over its solid surface, in its oceans, and within its atmosphere.

The extremely potent magnetic field projected by the Sun is accompanied by radiation emissions, equivalents of the gravity force, and mass-in-motion Relativity variables of Curvature tensor-pressure.

Thus, radiation, magnetic Field, and Curvature pressure force are all analogs or "normalized equivalents" that interactively "reconcile" the intrinsic "gravitating property" that inheres in Continuum Electromagnetic Energy Transformation, thus, inducing all Space-bodies to "coalesce" or "aggregate" with each other, so as to form, for example, a Solar System.

 Planet Mercury displays a gravitational anomaly called a "perihelion shift," due to planetary "negotiations" and "re-calibrations" of rates-of-change in its own application of the Input-Process-Output mechanism pertaining to its specific Mass-Energy frame-in-Motion. These "rates-of-change" ensue from interplanetary momentum variables, Gravity-force "equivalents," and positional Curvature-determined Forms of "concentric tensor-pressure Energetics" acting upon it as it "strives" to maintain dynamic System equilibrium, operations of which, as configured by the relative constellation of co-determinants superimposed as "centers-of-Mass," "centers-of-Field," and "centers-of-Gravity."

By necessity, all interplanetary and solar-planetary "relations-of-Force" must "resolve" as concentric, rotary Forms-of-Motion, that holistically and "satisfactorily fulfills," respectively, each Planet's own unique structurally-framed operational functions that cause it to be specifically identified as an integrated "whole System" in Dynamic Equilibrium, e.g., Earth maintains dynamic System equilibrium throughout all its intra-component interactions and/or inter-constituent operations, processes, phenomena, and events: — after an earthquake, the landmass will "reset itself" within "the newly derived" parameters "making for" dynamic System equilibrium, with Oceans and Atmosphere, "responding-in-kind."

Interactions with other Planets where motion-frames, Gravity-force "equivalents," and magnetic field-force strength, intersect along the Sun-induced ellipsoid plane of Revolution, create "overlapping areas of gravitational turbulence" that must get "re-translated" as cycling-mass Differentials, hence, the perihelion shift of Mercury.

Intersecting-overlapping areas of gravitational turbulence in "Relativity interactions" modify solar-induced interplanetary Curvature pressure force Energetics impinging upon "Mercury's system-structural complex" consisting of: Mass-in-Motion as "Momentum-Energy Force."

****

THE THEORY OF RELATIVITY AND GRAVITY

Given that the Theory of Relativity proposes that Space-Time is "curved" or "bent" around massive bodies that reciprocally exchange "Gravitational Motion-Force Energy," then, a deduced application is that, a beam of light will "bend" or "curve" when passing in the vicinity of a Star or a Planet that is structured as a "unit of Mass-in-Motion" and generates a commensurate or proportionate, corresponding, gravitational field.

The acceleration rate of an object "falling" towards Earth center-of-Gravity is calculated to be 9.88 meters per second per second; the Speed of Light in a vacuum is measured as 3.0×10^8 meters per second, or 186,000 miles per second.

Light radiation energy plays a crucial role in the mathematical formulation of Special Relativity Theory. $E = mc^2$ describes how "congealed energy" is transformed into "excited matter." Matter is "congealed energy," as gravitational Space-Time Curvature pressure force dynamics act upon the thermodynamic process to engender Forms taken by the matter-energy inter-exchange complex. The above equation of Energy Transformation denotes a "nucleated Energy process."

Because there is a "limit" to the Speed of Light in the known Universe, then, there are also "limits" imposed upon gravitational Space-Time Curvature pressure force and converted energy Forms. Because as c^2 is utilized as a "conversion factor" between mass and energy, the speed of Light, 3.0×10^8 meters per second, is held as a Constant: Energy is "conserved" as matter-mass is "conserved," because "Energy can never be created nor destroyed but always transformed."

Mass is inversely proportional to c^2, as c^2 is inversely proportional to Mass. The ratio of energy to mass, has direct equivalence with c^2 whereas the ratio of mass to energy is has direct equivalence with c^{-2} or $1/c^2$.

Therefore, the ratio of energy to mass determines whether a cycling mass-motion-energy frame is in a "congealed state" (a planet) or an "excited state" (a star), in relation to the fixed value of the Speed of Light as 'the limiting standard' for all converted Forms in accordance with the Law of Energy Transformation. This property determines how energy forms, either as combustive fuel or nuclear fuel, are "consumed," and "how much work" they can be potentiated to produce.

Hydrogen has no neutrons in its nucleus. Hydrogen (H) cycles from its H^+ positron as Deuterium, then to Tritium (H^{++}), then to Helium (H^{+++}).

Why does Hydrogen cycling stop at Helium? Hydrogen is the lightest element in the Periodic Table of Chemical Elements — Entropy compounds the fact that Hydrogen must gain neutrons for transmuting its electron into a positron; because its nucleus consists of 1 proton and because it has no neutron, Entropy confines its cycles of nucleated chain reactions to "limit" the extent to which its nucleated reactions can continue.

For example, Strontium-90 is a heavier radio-active elemental isotope than Uranium-235, and would embody differentiated explosive phenomena and quantized processes as encapsulated in chain reaction rates, material energy density, and radiation energy rates of propagation.

What would that mean for earth atmospheric composition, hydrosphere structure, landmass and thus, for Earth biosphere, ionosphere, and ecology?

Things are "relative" to each other in accordance with a standard—in the cosmographical and geo-sphere frames of reference, this standard is Light.

Why do particles "shed" their mass energy gains by emitting cosmic radiation when accelerated or are "increasingly excited" even to a fraction of the Speed of Light?

They are "reacting" to the electromagnetic bombardments impinged upon them by experimental equipment and machinery. Matter-mass, radiation-Energy, Gravity-force, and

Curvature-Motion are quantifiable "Frames in Continuum Space-Time," all of which, being in "reciprocal tensor-stress relations," where interchangeable-interactive, and inter-exchanged Energy-forces are exerted upon each other, in order to trans-form "congealed Energy" into "excited Matter."

What exactly is "accomplished with Matter-mass" by the constant value of the Speed of Light as exponentially multiplied to the second power, for conversion to mc^2?

Gravitational energy comprises the whole spectrum of wave-particle radiation in seen or detectable forms, such as light and heat, and unseen but machine-registered forms such as X-rays, such as Alpha and Beta particles and Gamma rays.

Curvature pressure force, gravity and field analogs of mass tensing, are expressed in motion patterns that embody frame-specific energy cycling dynamics, for "congealed" or "excited" processes.

Gravity could not be made manifest as having "a particle Form;" if Gravity occurred due to specific particles such as so-called "gravitons," then our electromagnetic spectrum-based research instrumentation technologies would have already discovered such "gravitons." Why? Because all the converted Forms which Matter and Energy can take in accordance with the Law of Transformation of Energy, are extant in the current Universe for Human observation and sentiency, as well as for electro-mechanical detection.

Only "excited Matter" emits radiation energy. Gravity is an emergent property of Force, and hence, we say "The Force of Gravity." Gravity is not an object or a thing composed of protons, neutrons, and electrons – Gravity is not a form of Matter and does not belong to the "Matter category" of material-concrete objective things that exist.

Therefore, given that "Gravity" is a force, and not a "converted Matter-Energy form" as "excited Matter," then, "gravitons" do not exist.

The "Equivalency Principle" as delineated in the equation E = mc^2 clearly establishes that Energy must have an active original ever-present "Continuum Source" in the Form of some type or category of Matter, e.g., dry wood when burning; Hydrogen as nucleated fuel; or petroleum as combusted gasoline.

Atomic particles: the proton, neutron and electron, are not "emitted" by the "congealed atom," though they participate in chemical reactions under specific pressure and temperature controls.

Solar spectral emissions display the positively charged Alpha particle which is the "excited analog" of the positively charged proton; display the negatively charged Beta particle which is the "excited analog" of the negatively charged electron; and also display the neutrino which is the "excited analog" for the charge-less, neutron.

No proton decay or proton emission has ever been observed or detected. Hydrogen cycling within the solar nuclear inferno stopping at H^{+++} supports energy Conservation, as atomic nuclei constitute the mainstay-component of fusion materials due to the strong force holding proton-neutron relationships.

"Continuous neutron capture" conserves atomic nuclei as "compressed fusion units" that serve as radiation fuel for the Sun. As the extremely potent magnetic field engenders centripetal Curvature Energetics that climax into sustained Continuum nucleated plasma condensate chain reactions, the strong force always overcomes emissive convective processes that re-direct fissionable materials towards the coronal region.

Thus, due to the strong force, atomic nuclei always "replenish their supplies of neutrons," via "neutron capture," this dynamic engendering energy Conservation as centripetal inner-core vectored forces overcome centrifugal corona-vectored forces, due to magnetic field projection-strength Energetics.

Excess radiation Energy is released into the "heliosphere" encompassing the Solar System, as the Sun achieves dynamic frame cycling equilibrium. Contraction and expansion, compression and dilation, fusion and fission, are in "Transmutation Relativity entanglements" as Conservation and Entropy counteract each other via cycles expressed in counter-vectored force dynamics, as engendered by the Sun's center-of-mass, center-of-gravity, and center-of-field.

As the Sun "negotiates" these gravi-Curvature-field cycling Differentials for dynamic System equilibrium, resultant tensor-and-stress metrics cause it to rotate upon its own axis every 27 earth days.

The Theory of Relativity implies that "force effects" can be accounted for by formulating acting variables in terms of either of the elements involved in "relations" of exertion. It marshaled the "Equivalency Principle" between Matter-mass and Energy, as the conversion factor, the Speed of Light to the second power is "held" with constant value, e.g., $c = 186,000$ miles/second.

For example, if there is an elevator moving sideways, e.g. to the left instead of downwards, with a velocity powerful enough to exert a force which would peg a man's feet in the direction opposite to that velocity, e.g. to the right, the man would experience the exerted force as "a gravity equivalent," that is, however, not towards the center of the Earth, but rather in a direction opposite the vector of that velocity-force; that is, toward the right side of the elevator. Both the velocity-vector Force and the Gravity-vector Force would then be co-linear, co-equal, but opposite in direction: Velocity is experienced towards the left of the elevator; and Gravity toward the right of the elevator.

It is posited then, that there would be no distinction between "this rightward gravity equivalent" and real Gravity towards the center of the Earth, because these forces would have become "relative," in the sense that the "right-ward gravity equivalent" is an opposite, but co-linear and co-equal "action," countering the "the-leftward vectored velocity Motion-Force."

In the elevator, the "right-ward Gravity force vector," is opposite to the "left-ward velocity Force vector;" while "in co-linear relationship," they must necessarily be co-equal; however, it does not imply that they are "self-identical."

For, the problem is that when a man is "upside down," he is upside down. If he is "sideways," he also knows and feels that he is "sideways."

The Theory of Relativity proposes that the velocity-force pegging the man's feet to the rightside of the elevator (now the 'equivalent" of the elevator's real bottom floor,) would become the relative equivalent of true "Gravity." However, in the elevator, if the leftward velocity Motion-Force suddenly stops, the man will fall to the real-bottom floor of the elevator, to which, true-and-real Gravity-Force will direct him as "a unit of Mass-in-Motion), which means that "the rightward elevator gravity-force equivalent" is the self-same-identical to the real Force of Gravity towards the center of the Earth. For, on the Earth, the real-and-true Force of Gravity is always towards Earth-core center-of-Mass.

The property of "upright sensation" is innate to human biological-genetic psycho-motor constitution. For, even in the absence of Earth G:1 Gravity-force, when in void-empty Space, astronauts still know that they are "upside down" when their feet are opposite the side of the craft considered to be its bottom-floor.

Consequently, though in the elevator there would be "relative equivalency" in "force-strength value" between the substitute or "equivalent G-force" and real Earth-core center-of-Mass-vectored Gravity-force (as the elevator's real bottom floor), the man would be consciously aware of his spatial situation, location, and position, in that, real-Gravity Force is concentric and centri-vectored, whereas in empty-void Space, there is no center-of-Gravity being generated by a center-of-Mass-in-Motion for "attracting" the astronaut's body thereto, into a rotary Form of Motion.

Because things are "in relation" with each other physically, vis-à-vis a standard value held as a Constant, a change in one variable causes or necessitates a change in another—which explains why an astronaut outside of Earth's center-of-gravity and center-of-mass, experiences blood starvation in his bottom-half (legs and feet), with the greater blood volume accumulating in his upper body part and head.

In addition, his bones decrease in density due to the loss of minerals and the lack of real-and-true G:1 Gravity-force induced maintenance-growth, that would have been conducive to optimum immuno-metabolic health and hormonal-organic functionality.

In short, we have to "carry the Earth on our backs" when we leave its ecological environment as engineered by gravi-metric, field, and Curvature pressure-force Energetics. Without G-1 gravity force vectors (in magnitude and direction) acting upon our organism, Energy processing is "reversed" in that Entropy comes to gain predominance over Conservation.

Earth oceanic navigation, e.g., "navigating by the stars," according to spatial constellations in the heavens presents relatively fixed "celestial anchors" for geo-nautical orientation. However, deep empty-void Space offers no such fixed "mileposts." In vacuum Space, the long-utilized geographical quadrant of Earth coordinates, or for "in-Earth-traveling vectors," i.e., North, South, East and West, do not apply, due to the vast expanse of Space, near zero-Gravity conditions, and the absence of landmarks such as Earth magnetic Poles, as necessary for positional fixation or navigational grid-of-coordination.

Because even the Human Body has its own center-point of Gravity to be respectively accounted for when in-Motion, then, normal spatial perception and normal homeostatic equilibrium depend upon gravi-metric co-determinants of regular natural healthy Human constitution, such as: properly oxygenated blood flow, and sensed-and-felt apprehension of "proximal center-of-Mass indicators," as "registered" in the Human brain, the sum of which, being predicated upon the self-experienced and self-acknowledged sensation of being in the "up position."

In deep empty-vacuum Space, due to "conditional Relativity interactions" that impinge upon dynamic homeostatic equilibrium, there would have to be artificially created gravimetric operational parameters that mimic Earth center-of-Mass, Earth center-of-Gravity, Earth center-of-Field, and G-force "equivalents" emerging from Curvature tensor-pressure variables, the sum of which, activating or triggering psycho-motor-sensory "feeling-in-my-skin" or inner-intuitively apprehended kinds of "proximal cues" in the Space-environment, which, would then be conducive to Earth-like normal-natural "feelings of good health."

THE THEORY OF RELATIVITY, GRAVITY, AND THE ELECTROMAGNETIC SPECTRUM

It is in nuclear physics that the Theory of Relativity has demonstrated its strongest applicable validity, as earth-bound experimental fusion and fission processes have confirmed the Law of Transformation of Energy.

Scientists discovered that the electromagnetic spectrum, to which "visible Light" belongs, "carries information," e.g., Human voice is "carried" through insulated wires or fiber-optics cables, or via "wireless transmissions" common to the framework of telephonic technologies, from one end of the "telecommunications matrix" to another.

That micro-electromagnetic Mass-in-Motion, such as photons and electrons, "carries information" in the same way that hemoglobin blood molecules "carry Oxygen" is a fact of bio-Physics, and of Physics-proper. Devices, appliances, and instruments predicated upon this Principle took the Forms of the telegraph, the phonograph, the radio, the telephone, radar, sonar, the moving picture, and then television, to be followed by the fax machine, the cell phone and the digital video disk recorder/player, micro-wave towers, and telecommunications satellites.

In short, our whole "telecommunications matrix" and "electricity-generation grid" are "anchored" to this Principle, i.e., radiation Energy "carries information," the sum of which, has been discovered to be "attached" to all electromagnetic spectrum-based technologies. Research-and-Development enterprises, medical-hospital establishments, power plants, military institutions, business corporations, transportation industries, etc…, all, rely upon this Principle in order to structure "command-and-control" methods of organizing production-and-distribution logistics "to flow along with" the efficient delivery of goods and services, respectively.

Minute electrical charges or impulses, electrons and photons are involved in all these processes, as well microwaves, in all of which, both electricity and magnetism are actively united. Other developments count nuclear energy, microwave ovens, copiers, microwave transmitters, fiber optics technology and computers, meteorological electronics, particle-accelerator structures, etc… All these products/conveyances/establishments avail themselves of the same elemental-constitutional expressions of the "wave-particle," in the form of emissions that are "converted Forms" of the same Energy — electromagnetic spectrum radiation in all its representative embodiments or manifestations: from visible Light to infrared, from X-rays to radio waves.

In addition, Energy flows in "packets or quanta" which "carry information" as an apparently "discontinuous" stream, that is still, "a Continuum wave," the sum of which, can be albeit "dissected" into its constituent-components, within the units of which, particles can be isolated for singular identification, "loading," and measurement.

That Energy comes as particles or waves; and that it "carries information," have contributed to our ability to design machines and create devices for a multiplicity of purposes and functions, all of which, allowing for a great amount of diverse and enriching "command-and-control processes," the epitome of which, being the electronic computer, based upon a "microchip" entitled the CPU, relatively amenable to "crunch" information/data/numbers, operations of which, constituting and standing-for: "Control-Processing-Unit."

Hence, the electromagnetic energy spectrum as radiation can "carry information," e.g., voice, data, audio-visual via numerous invented devices, machines, instruments e.g., iPad, cell phone, television; and we can "control" how we utilize it (e.g. turn it on or off) by constructing machines and devices that exploit its particle-wave characteristics — radio waves carry voice and music, or sound, only; television waves carry both sound and moving pictures; and when "conjoined" with the World Wide Web, both the Personal Computer and Television technologies, via satellite or fiber optics, for examples, can perform all on-screen-only abstract "output operations," or as "hard copy," "3-D prints," or in a Form to be utilized as necessary for private, business, industrial, or public use.

Electronic machines have great versatility for presenting opportunities for "combined operations:" (1) Because of the way in which all electronic machines can be encoded, i.e., via binary numbers 0 and 1; (2) Because of their "affinity" for certain "responding" to certain Forms in which "lines of instruction" are written, e.g., HTML; (3) Because such "instructions" for performing "output" are "carried by" or "contained" within "quanta" or "streams" of electromagnetic radiation that are "loaded with various Forms of information," many possibilities exist for "combining" many of these machines together electronically by "linking them as a Network," the sum of which, being very efficient at performing diverse numerous

Forms of "output," via the universal Organizing Principle, mechanism of which, being embodied within its operations, as: INPUT-PROCESS-OUTPUT, e.g., for resolving complex mathematical equations; for "bundling variables" used in weather systems forecasting by meteorologists who methodically gather and assemble them for analysis; and (4) Because of the "standardized methods of data/information inputting" via common instruments, equipment, or devices, e.g., via a Personal Computer keyboard, or via laser, microwave, or infrared decoding/reader devices, such as those utilized at cash registers, aviation automatic pilot systems, and/or cell phones or other "wireless electronic machines."

In a way, given that electronic machines and the variegated Forms of "information" their operations are "computing," as "processing units," are based upon the electro-magnetic spectrum: Then, if is self-evident that "Light is chasing Light," or that "Radiation is chasing radiation," the sum of which, giving rise to many problems, especially in Research Experiments.

For example, when scientists attempt to measure the position and momentum of an accelerated particle, they encounter a "stumbling bloc" giving rise to "interferometrics," the cause-and-effect mechanisms of which, are said to be amenable to "experimenter-observer effect," or due to measurement techniques.

But, more certainly, "the Science of Interferometrics" emerges out of the necessity to address these problems at their root-causes: Because all these electronic instrumentation technologies as well as the "information" they are "processing," are utilizing the same electromagnetic Form of Energy, e.g., An electron microscope is being utilized for viewing a particulate operation in an experimental setting that has only electronic machines in the totality of its experimental accoutrement or "hardware/software complex."

The discovery that "Light carries information" or that "Radiation can be loaded with information," or that "Information can be contained within Light-Radiation," so as to travel, for examples: Through the atmosphere, trough a metallic wire, or through specially processed glass-materials called "fiber optics," is an undeniable "scientific wonder" constituting a "Cybernetic Tautology."

Such a factual scientific "tautology," is directly connected or related to reasons why astro-evolutionists have come to presuppose that "relic cosmic background radiation," would also "carry information" from the "past Universe" as it was billions of years ago.

Astro-evolutionists desire to detect "cosmic background radiation," which is supposed to be "remnant energy" from the so-called "big bang." Distance between stars or planets are estimated in "light years." A "light year" is the time-distance period it takes Light-Radiation Energy to travel in one year at 186,000 miles per second or 3.0×10^8 meters per second.

It appears that a "subconscious transference" or "deliberate subterfuge" has taken place from witnessing factual scientific operations, as above-noted, of the "cybernetic binary principle" that radiation "carries/contains information," or can be "loaded with information," the sum of which, climaxing into a presumptuous fallacious formulation that feeds the preconceived and pre-established evolutional expectations of astro-evolutionists, as a fait accompli, that the presumed or presupposed, so-called "relic cosmic background radiation," has also "persisted," throughout billions of years of lapsed Time, so as to "continue" to "carry past universe information" for us to observe to this present day.

This subconscious transference of application has fueled the quest for detecting "relic cosmic background radiation," in order to then assume that discovering or detecting it, will somehow reveal the precise Time at which the Universe might have begun.

The reasoning is that the "relic cosmic background radiation" being also electromagnetic in Form, must also "carry information," but about the past, not the present.

It appears that this "logic" arose, due to the fact that light travels at 186,000 miles per second. Thus, astro-evolutionists assume that by the time it reaches us, the "information" it is presumed to be "carrying," is already the past.

This assumption comes from the reasoning that since, given the speed of Light and given the distance of the Earth from the Sun, then, solar radiation takes eight minutes to reach the Earth, then, accordingly, Light-radiation-energy "traveling" a much longer distance, such as "miles-light-years," from a "place" that it is supposed to have originated, called the "epicenter of the Universe," would take even much longer time-in-traveling, before it would finally arrive to us on Planet Earth. Consequently, astro-evolutionist "logic" concludes: As it is "carrying information" from light-years away while "traveling" towards us for billions of years, then this "information" that "the relic cosmic background radiation" is "carrying," has to be "from the past."

Astronomers go as far as reasoning that were the Sun to suddenly "go out" or "die," we would not know about it until eight minutes later; and that, were we to look up into the heavens during these elapsing eight minutes, even though the Sun had already "gone out," we would still "see it in the sky" because its Light-radiation-rays Energy would still be "traveling" towards us until these eight minutes would have had elapsed.

But, given "the Equivalency Principle," Matter must also be present in order for Energy to be present — Thus, according to $E = mc^2$, energy cannot exist without the presence of Matter in one of its many variegated Forms.

Thus, given that "Energy is never created nor destroyed but always transformed," and given that Energy cannot exist without the presence of Matter, then, Light's apparent instantaneous Motion as Energy, must-needs have an ever-present "Star source" with nucleated plasma reactions-processes as Matter, in order to emit or project its "traveling" rays.

Not only that "relic cosmic background radiation" cannot exist because no astro-evolutionist and no electronic device can ever differentiate between "old radiation energy" and "new radiation energy," but also, given that all observed radiation energy is from ever-present-current real-time Matter-Sources, then it cannot "carry information" from "the past."

In addition, c^2 in Einstein's equation $E = mc^2$ represents not simple accelerated velocity of Light, but momentum traveling gravitational-tensor-stressing of Curvature radiation Mass pressure-force exponentially potentiated to the second power, accounting for nucleated electro-motive magnetic radio-active Mass-in-Motion Energetics that drive planetary Revolution and Rotation.

The Sun is not "from the past," as its thermodynamic cycling existence is independent of the distance, perception mode, and time-frame of the observer.

Given that the Sun constitutes the ever-present "real-time, On-source" for radiation Mass motion-force-energy as "projected" by its magnetic-field strength within our current Time-frame, then, Curvature of Space-Time cannot "reverse" the forward vector of Time.

Evidently, the astro-evolutionist assumptions are flawed for many reasons. For were it true that Light constantly and continuously travels even in the absence of a real-time ever-present Source, we would be able to "beam our present" to some other peoples in distant galaxies, after which we would turn-off the light Source, and which they would then interpret as "their past" due to the extreme distance that Light would have to travel before it reaches them; simultaneously, their apprehension of "our present" as "their past," would become "our future" by the time we would learn about it.

So, just as "relic cosmic background radiation" coming from their "cosmic neighborhood," would be the Universe "as it was then," our "present cosmic region" would have had become "their past."

Additionally, we ought to have been able to preserve our "olden structures" in perpetuity though they would no longer exist in the reality of the present, simply by first preserving the light they had reflected in the past. The "preserved light" would then possess the capacity "to materialize" those "long gone structures" in the very substance that was hitherto destroyed.

When driven to their logical conclusions, evolutionist presuppositions do not make scientific sense. When would the present ever begin if distance of the observer from Light Sources would determine the Time-frame of universal phenomena?

Everybody would "remain in the perpetual past" due to flawed theoretical assumptions based upon evolutionist doctrines affecting the perceptual mode of all observers.

How can we be looking "at the past" when we look into the heavens? And how can we assume that all the Lights we see form imprints of the universe as it existed "billions of years ago?" If all Light-Sources in the heavens are "from the past," and given that "Space-beings" would see us and our Solar System as "their own past," then how do we get our own Light-Sources, such as our Sun, to illuminate us "in the present?"

Proximal distance of Human Beings from cosmic phenomena does not determine their Time-frame. Time-Differentials are factored into observer frames of experimentation in order to account for variables, such as Motion, that might distort results. But we cannot "piggyback on Light" and travel "back in Time" to the past to actually see and experience the Universe as it existed "billions of years ago."

The so-called "relic cosmic background radiation" which astro-evolutionists are seeking and pursuing, cannot be isolated from current stellar radiation to be identified as "a relic of the big bang," and thus, does not exist.

Physical phenomena require cause-and-effect mechanisms/connections/relationships within the same Time-frame for "keeping Continuum" as factually observed; thus, stellar lights or "points of lights in the heavens," observed within Continuum Space-Time, are informing us now regarding present inter-galactic conditions, and not as they were billions of years ago, and not as they were when the Universe initially "self-created" by the so-called "big bang explosion."

Otherwise, all "light Sources" that allow researchers to peer into outer-Space would become "relic cosmic background radiation" since there are no "old sources" that can be differentiated from "new sources."

Scientists cannot simply rely upon their own personal opinions or perspectives that happen to be aligned with their friendship with colleagues of the same set-of-beliefs, the sum of which, being called "Peer Review."

The Scientific Method is a framework of impartial and independent principles founded upon rigorous procedures for conducting experimental research as well as upon forthright methods of post-experimental analysis of honestly collected data, from which, validly proven conclusions and results that are obtained, must be reproducible by other independent, impartial, and disinterested research scientists having no conflicts-of-interest or bias that could prejudice their own results and conclusions, either for favoring or disfavoring previously obtained conclusions and results.

A principle of the Scientific Method is that a theory's predictions must be experimentally produced via pertinent research experiments, the sum of which must be reproducible, in order that the theory's "degree of falsifiability" or "range-of-application" can be ascertained, as well as the physical conditions under which specifically derived results and conclusions validly apply, e.g., Newton's framework of celestial mechanics presents no verifiably proven-valid explanation for the Perihelion Shift of Planet Mercury; whereas Einstein's framework of Relativity does.

Meanwhile, under the dubious hospices of "Peer Review," tremendous sums of money, time-consuming research, material expenditures, and highly valuable Human resources are being devoured by the pursuit of this sole obsessive assumption of the "Scientific Establishment" concentrated at public colleges and universities: That the so-called "relic cosmic background radiation" can be detected to then "prove" when the Universe had begun in the past from the so-called "big bang explosion," the "epicenter" of which being the supposed initial past-Source of the pre-established "relic radiation expectation."

Is the so-called "relic cosmic background radiation" projecting Carbon-14 for "age determination?" This is a "non-sequitur!" Yet, every impulse for astrophysical research has been absorbed into these presuppositions while alternative explanations of universal beginnings are viciously ostracized.

A tenet of objective scientific inquiry is that scientists be open to all explanations of physical Reality that might address contradictions not resolved by the evolutional paradigm.

Astro-evolutionists should not be so closed-minded as to stifle debate, obstruct discussion, and thwart the rise of new theories that are "pregnant with scientific promise." Even those that appear not to fit into their "paradigm of preference" could reveal patterns in physical relationships not predicted by the "evolutional frame."

Evolutionary scientists are promoting theoretical sterility and scientific stagnation while protecting "their turf" and defending their vested moneyed interests by zealously gearing all research grants awards, towards "proving the big bang explosion theory of universal beginnings." Hence, reasons why, Science, that must be based upon validly proven physicals Laws, has come to "a stand-still."

Why are fundamental astro-evolutionary assumptions flawed? They lack internal logical consistency and analytical content coherence, the sum of which, being accountable for the complete dearth of their applicable validity.

Evolution theory pronouncements rely upon mere conceptual constructs for which no scientific proof has ever been provided, and yet they are professed as if they were proven facts. These fundamental astro-evolutionary assumptions are unscientific for the following reasons:

1) Non-intelligently directed Light radiation energy cannot "carry" or "contain" past "information" that can then be "brought to us" on the Earth. Television networks and radio stations can re-broadcast an even recorded in the past. But these recorded past events would not be recurring or happening again at the same time of their re-broadcasting within the Time-frame of the Present. Time is uni-linear or uni-directional. Time can only go forward. Mass-Energy "spheres of bending-influence" cannot "curve Time to the past" in the same manner that they cause rotary forms of motion — Curvature of Space does not imply "Curvature of Time" because Time is "embedded" within Continuum Gravitational Space-Time Curvature. We cannot go back in Time. Forget about science fiction movies or science fiction fantasy novels! We are talking about real Physics here. The possibility that Time itself might be "curved" does not imply that its forward-vector can be reversed — for, Time has computable values, but only within the iterative processes of thermodynamic cycling by physical Mass-frame that are in Motion. Time as a universal category intrinsically "embedded" within Continuum Space, and hence, Continuum Space-Time, is dependent upon Mass-in-Motion "dimensionalities" and "modalities" of iterative cycling in accordance with the Laws of Thermodynamic Energy Transformation, which has fundamental inherent ranges-of-influence, parameters, boundaries, variables, and limits. And therefore, Time cannot be mathematically "curved" or "bent" in a way that would reverse its forward vector for so-called "time travel." The Speed of Light having its own numerical limits at 186,000 miles per second, is the limiting boundary upon the vectoring of Time-dimensional modalities: Time only goes forward from the past into the present into the future. Only Mass-in-Motion frame-properties of physical dimensionality can be "curved," or "bent," such as within the "bounded regions of field influence" impacting the Space-vicinity or proximal-area within which Mass is moving as "attuned" to "Relativity interactions" with the "Thermodynamic Fundamentals" of energy cycling operations. However, not to "go backwards in Time." Analogs of Curvature pressure force, such as a Magnetic Field and a Gravity "equivalent," engender force-motion properties that cycle in Continuum Space-Time. Continuum Time in simultaneity with Space-Curvature is dependent upon initial conditions of energy cycling that "move forward in Time." As a "physical distance-unit of process-measurement," consistent with "Thermodynamic Fundamentals," Space-Time is quantifiable only within a "same Time-frame-Continuum context" inextricably "chained" to Mass-bounded processes

that display iterative cycling periods having a beginning and an end, but only as they "go forward" or "continue to advance" into the Future. From the experimental frame of the observer, "the past" consists of "contents of memory" and not of an "actual return to the same Space-Time data and conditions." Thus, we remember "the Past" but do not physically return to the past to re-fill the "past-Space-time-frame" wherein the event must have had had occurred. Though inextricably connected to real validly-contexted events that prove they did certainly occur, "The Past," as we remember and commemorate it, becomes then "a mental-heart-felt operation" of the Human spirit whose only anchor must be the physical vestiges, residual landmarks, or "material footprints" within the geographical boundaries and dimensions of the Planet that firmly corroborate those pertinent real events, as left behind for us to witness. The physical vestiges from the Past are real; but we are making such observations in the Present. In that same manner, the "points-of-Light" we observe in the heavens come from real-time ever-present Star-Sources shining forth their radiation Energy as produced by the nucleated processes taking place therein "in-the-Now."

2) All forms of electromagnetic Energy, e.g. Light as physically detectable wave, particle, ray, or beam, must have a real-time physical "on-operating Source" existing within the same Time-frame for both the Source and the observer. Frames-of-motion Differentials must be accounted for in ways that do not contradict the Continuum Principle and the Law of Energy Conservation.

3) Light does not continue to propagate without an active real-time "on-source" to sustain its emitted wave-particle ray, or beam substance, as radiation which we can observe, see, feel, or detect via mechanical or electronic means. "Disappeared mutational forms" contradict Continuum. Continuum propagation necessitates ever-present radiating emissions that substantiate particle, wave or streams of energy that must counter Entropy, absorption, reflection, diffusion and refraction. Possibility of "Time-curving" only implies "fast-cycling Time-compressed processes" with "convoluted patterns of motion-force," and not a change in the forward-vector of Time.

4) Since light is physical, to be manifested in Space-Time, it must have a physical, real-time, ever-present, now-existing, "on-source" to maintain and sustain its beam as a stream of energy quanta or "packets." There is no light without a physical real-time "on-source," and therefore, light can never represent the past, be from the past, or embody the past as it was "billions of years ago." Due to Curvature pressure force dynamics acting upon cycling mass, light is absorbed or refracted, reflected or diffused, and it takes a continuous stream to sustain its substance in Continuum Space-Time from a real-time physical source, in the absence of which, it does not exist. Absorption, exhaustion and termination would have resulted long ago as there would not be any so-called "relic cosmic background radiation" to detect, measure and quantify.

These facts prompt us to ask certain crucial questions regarding evolutionary presuppositions about the beginning of the Universe:

1. Does "cosmic background radiation" really exist? (Did the universe begin with a so-called "big bang explosion," the supposed relic remnant radiation of which can be identified still, and differentiated still, from extant cosmic radiation?)

2. Just suppose, for the sake of discussion, that a "big bang explosion" did occur, then is there really a "relic radiation" that "made it unto the present" as it "carries information" from the past—the past Universe as "big bang theorists" claim? Or, more precisely:

a. Why would the so-called "big bang explosion's radiation-in-Motion dispersal rate" continue to "expand ad infinitum" into "the present," or "into every Human-generation's present?" When does the apparently "infinite Present" ever become "the past" in order that "thermodynamic Time" continues to "travel into the Future?"

b. What type of "information" does light "carry"? Of what substance is it made? What are its constituent elements? When — under what physical cause-and-effect conditions — can Light-radiation-energy be said to be "carrying information," or to be "loaded with information," or "to contain information?"

c. Why do electrical devices and electronic instruments have an "on and off" switch, and why must they be connected to "an electrical source" in order to work?

3. Suppose again that "cosmic background radiation" exists — Light-radiation-energy of the electromagnetic spectrum — and suppose that it "carries information," then, can it carry information about something that is no longer there? No! Because the Universe as it began "billions of years ago" would no longer exist: Remember, evolutional theory claims that it "evolved" into the present. If "mutated forms" disappeared as "missing links," while emergent forms supplant their operations, why would the so-called "big bang explosion" even display a "relic cosmic background or remnant radiation?"

4. Because of "the Equivalency Principle" established by the Law of Transformation of Energy by the equation $E = mc^2$, no energy-Form can exist without a particular Form of extant Matter. For when Matter = 0 (When Matter is absent or is not present), then $E = 0$ multiplied by c^2 such that $E = 0$. Thus, Light-radiation-energy in Gravitational Continuum Space-Time, needs an ever-present physical "on-source" constituted of Matter, or as a Form of Matter, existing within the same Time-frame as the Light-radiation-energy that it is emitting or producing as a material-Source for the Electromagnetic Spectrum, the sum of which, to be perceived-apprehended by us, Humans, in real-time. Under Gravity-based tensor-stress Curvature pressure-force Motion, no extant Light-radiation-energy, observed in the present, can be from the past:

a) Its Source must be "radiating its emitted energy units" in the Present, as we observe it, "in the Present;"

b) Even if radiation could have "continued to travel" without its material-physical Source-of-Origin, due to the Laws of Thermodynamics, Entropy would have triggered many ways in which operations of the Input-Process-Output mechanism would have nullified, cancelled, "diluted," or neutralized such radiation; thus, it would have had diffused, refracted, diffracted, exhausted, and/or absorbed by extant Space-bodies "populating" the vast expanse of the cosmic Universe;

c) In addition, due to its "missing or disappeared Source-form" that would have had been "transmuted" into present Forms to then radiate in our current Time-frame according to evolutional doctrines of "punctuated equilibrium," then, the "transmuted

Forms" that would have represented "interruptions" or "disruptions" of the Continuum Light-radiation-energy streams of rays-waves-particles constituting its emissions, would not have "remained embedded" within "the Gravitational Continuum of Time-Space" for "survival into the Present."

5. What "leap of faith" does it require to believe it possible "to travel back in time to billions of years into the past" simply by "piggybacking" on presupposed "cosmic background radiation," from the assumption that one "sees" the universe in its "past stages of evolutionary beginnings," by simply looking into the heavens? Why would the Past exist in the same Forms, if those forms are supposed to have "mutated/transmuted" into the Present? In the first place, if "past stages of evolutionary beginnings" would have still existed for observation in our Present Time-Frame-of-Reference, this would contradict the presumption that "trans-mutational processes" of the "missing Past" were to have had been accountable for presently observed "current Forms of cosmic phenomena." And, if there had been "trans-mutational processes" engendering "missing links," then why would have those Forms still persisted unto the Present as current sources of radiation? Thus, no "mutation/transmutation" from a so-called "big bang singularity" ever occurred; and Forms being observed in our "Present-Time-Frame" have been in existence, as formed, as framed, and as operationally structured for iterative patterns of thermodynamic Energy cycling, ever since the initial Time of universal Creation by Almighty God. No "evolution" ever took place.

6. How could the Past still exist, even in the Present, countering the "doctrine of mutated states" or of "transmutational Forms" that must necessarily disappear so as to then be re-interpreted by astro-evolutionists as "missing links?" This would imply that "evolution stopped," to then begin again, to then become "on-going;" and it would also infer that "things are still mutating" even within the "Present Time-frame." (Time is uni-linear, uni-directional — Time only goes forward in sustenance of Continuum, Energy Conservation, Operational Cycles of Thermodynamics, and Conservation of Angular Momentum.) Evolutionist assumptions are counter-paradigmatic since evolutionism implies "the disappearance" of previous Forms as "missing links" in order that "mutated types" or "trans-mutational Forms" can continue to thrive into the future, as if something must be missing or be "taken out" of the stream of Time in order for the Past to "conform" to the flow of Continuum Space-Time, to then arrive at the Present. That does not make scientific logical good Sense. For, out of that disjointed "reasoning," astro-evolutionists would have to also believe that the "past Universe," having had "evolved" into the "present Universe," would not still be there for us to observe in this Present-Time-Frame, the sum of which, would have had to have been its "trans-mutationally evolved Future." Given that there is no "Past Form" that could have "attached" to the stream of Time leading to our Present Frame or "Frame-of-the-Present," because such "Past Forms" would have become "missing links" having no actual bearing upon the uninterrupted flow of Thermodynamic Time, then no Continuum could have ever existed.

It is instructive to note that the type of "cosmic background Light-energy-radiation" that astrophysicists are trying to detect is infra-red radiation—plain old heat. In their experiments, they send up into the atmosphere a rocket which propels a sensing mechanism that only travels a few seconds above the Earth. During those few seconds, the helium-cooled thermometer takes several temperature readings which are to reveal a wide band or spectrum of temperature differences. The reason why the thermometer is cooled is to eliminate the possibility of registering temperatures that would fall outside of the "cosmic background radiation" range — in order to "isolate" the excessively high temperatures which would characterize the "big bang emissions."

The last of these experiments were conducted in 1991. Per astro-evolutionist interpretations, preliminary results revealed that the range of temperatures had followed the "predicted curve" of the "cosmic background radiation." But what is the range of the "big bang cosmic background radiation" as opposed to the range of temperatures that can be currently recorded above Earth orbit which are indistinguishable from solar radiation heat?

What this experiment revealed is, that a space rocket equipped with a thermometer registered a spectrum of temperatures that fell in the range or plotted curve of temperatures within which all known measurements of extant temperatures in Space will fall.

In short, there is no way to isolate, separate, sequester, or distinguish so-called "old space radiation" or "old relic big bang radiation" from Present Gravitational Continuum Space-Time radiation. The temperatures recorded by the probe are those of presently-Sourced radiation — current radiation that is emitted by "real-Time-now" celestial bodies in the Present Space-Time Continuum. It's plain solar radiation; and it is not "from the past."

In addressing the second question, we know under what conditions electromagnetic radiation "carries information." Our radio and television sets, our telephones and DVD players work that way. Data, voice and pictures are "carried" by streams of electromagnetic radiation either through wires, fiber optics, or the atmosphere, to reach our electronic devices that "re-translate" them into usable Forms for our senses to understand.

We've seen all the examples given above ranging from the telegraph to telecommunication satellites, whereby electromagnetic radiation and Light are emitted by real-time, ever-present, extant "on-Sources" that have an electrical power-Source with an "on-off switch."

Once a satellite is inoperative, it cannot transmit; once a Light-source is turned "off," it cannot emit photons. Since clearly, electromagnetic emission is a physical event or process, i.e. a dis-continuous stream of wave-particle "packets of Energy," then it must have a physical Source or material wellspring to sustain it for operations within the same Time-frame.

Therefore, when we look into the heavens, into the "deep of void cosmic empty Space," the Light-sources that we witness do exist in our Present Time-frame; and they are not "from the past."

The fallacy of the supposition that, were the Sun to "go out," the light that reaches us would actually represent the Sun itself as it was eight minutes earlier, comes from a misrepresentation of what Light accomplishes in the process of Human vision, and from a misapplication of the Laws of Thermodynamics.

In the same sense that a Line is a succession of points: Light-radiation-energy is a "succession of photons." Light is the medium via which vision takes place. Without the presence of Light, we cannot see. The Sun, which is a "ball of nucleated radiation Energy," is not "physically embodied" within the Light rays it projects; the Sun is the physical Source or wellspring of the Light. When we look at the Light projected by the Sun, we don't see the Sun in that Light-stream of photons.

The Sun possesses cycling mass-Energy, as the Light it projects is being continuously absorbed, within the framework of gravitational motion force dynamics, the sum of which causing it to undergo Entropy. Given that Light-radiation-energy does not propagate without an ever-present real-time physical Source, e.g., a Star, and given that the stream of Light rays would continue to "travel" for eight minutes after the Sun had "died out:"

(a) Because the Speed of Light does have a limit that cannot go beyond 186,000 miles per second; and

(b) Because the Earth is 93 million miles away in distance from the Sun: Then, were the sun "to die out," the fact that Light that had been emitted from it within an eight-minute time period would still be visible as it "travels" towards the Earth within these eight minutes, the Sun's disappearance from the heavens would be instantaneous, not elapsing in eight minutes.

"Traveling" Light-radiation-energy does not "contain," "embody," or "hold" in its stream, the physical object-Source from whence it comes. Thus, the Sun would have had already "disappeared from our view," even though, due to Earth-distance from the Sun and due to the limited Speed of Light, its rays would have still been visible eight minutes after "its death."

In the same vein, had the "big bang explosion" really occurred, the Light-radiation energy presumed to be "traveling from it" when the "big bang Mass-energy" projecting that "relic background radiation" would have had already disintegrated, dispersed, and/or absorbed by numerous Space-bodies populating the cosmic Universe, our viewing of that Light would not cause it to "contain" the presumed "big bang explosion" as it had occurred so-called "billions of years ago."

Since the Light was not being originally emitted by the Universe presumed to have been "created" by the "big bang explosion," but was being emitted-projected by the "explosion itself," then, this "traveling relic cosmic background radiation" could not "embody," for us to view, the Universe "as it was billions of years ago." For, it is not the "created Universe" that emitted this so-called "relic radiation' but the "big bang explosion itself," which no longer exists in its initial physical State, given that it is the disintegrated-dispersed materials from its explosion that then "cooled" in order to "create" the Stars and coalesce into the Planets, etc...

Thus, if the so-called "relic cosmic background radiation" existed, to then give us a view of the Universe as it was "billions of years ago," we would have to only see, perceive and visualize a perpetual display of "a long-ago big bang explosion," and not a well-ordered organized universal complex governed by scientifically proven-valid physical laws! Perpetual explosion in the heavens! How nonsensical is this!

A movie projecting people moving on a screen reflects an artificially projected image-representation of those moving people, but this artificial projection is not the people themselves.

The projected light-beam is constituted of mass-less photons, and not of materiality equivalent to the objects themselves; once the "switch is off," pictures fade away and totally disappear from the screen.

The fallacy that Light from distant galaxies is reflecting, representing or embodying the past Universe is analogous to the fallacy that parallel lines meet at infinity. That is an optical illusion voiced as an oxymoron or a contradiction in terms. For, parallel lines are equidistant at right angles; otherwise, they are not parallel. How could parallel lines ever meet if, by definition, they do not meet and can never meet, for otherwise, they would not be at right angle equidistance? The false deduction— "meeting at infinity"— comes from the realization that our perceptual vision has distance-defined limits, which cause us to utilize the microscope and the telescope.

In addition, even if the Universe is held to be or believed to be "a bounded-spherical expanse" into which parallel lines are projected "at infinity," they would return to their source, following "a rotary arc trajectory," but still as parallel lines that retain their 90-degree equidistance, in the same manner that Comet Halley follows its "curved path" every 75 years as it returns to our solar system.

But Space cannot be "solidly bounded." Nor does it have "spherical boundaries" like a planet that has atmospheric-landmass-seashore limits. Space is immensely infinite in limitless expanse. Space has Continuum Curvature due to cycling Mass-energy frames projecting "magnetic field curvilinear arcs-of-force" that engender rotary forms of motion, e.g., electrons revolve around atomic nuclei; planets revolve around stars and rotate upon their own axes; comets follow iterative "rotary arc trajectories."

"Infinity" does not mean "limit" or "boundary," but only an indeterminate numeral number denoting a massive value-amount that is so big as to be incomprehensible to the Human Mind.

Thus, "infinity" would not violate the 90-degree equidistance that constitutes the property of parallelism. "Curvature geometry alone" does not determine Motion-force trajectory patterns, velocity, and angular momentum.

But "Natural Force dynamics," as engendered by electromagnetic field arcs-of-force and by "equivalents" of the Gravity Force determine "Curvature geometry."

The geometry of Space-Time is "curved" not due to mass dimensions, vectors and coordinates alone, but due to force properties of electromagnetism that climax into rotary forms of motion-energy, from which emerges the force of Gravity and associated "equivalents."

Electromagnetism and Gravity work together in producing a "trans-sonance of commutable forces" co-equally synthesized into "commonly shared Force-equivalents" that reinforce, prolong, and expand each other's "range-of-influence," especially within the Quantum Mechanics frame-of-reference wherein, due to micro-Masses of particulate bodies, Electromagnetism and Gravity forces are so "mathematically symmetrical" as to be equivalent and indistinguishable.

Upon the Earth, this "quasi-equivalent Force-trans-sonance" emerges from the "marriage" of Electromagnetism and Gravity, the sum of which synergistically operating, so as to give rise to as many variations of movement-and-displacement as can be performed by a "moving body-Mass," such as from the vantage point of an aircraft, within a 360-degree "spherical envelope" serving as its pivotal fulcrum, to either "hover" or execute controlled "in-flight maneuvers" that are performed to keep it in the air, such that its engine thrust and avionics are synchronized to collaborate in bolstering the aircraft's resistance against the pull of Gravity towards the Center-of-the Earth.

In the same manner, a Human Being can walk, run, or jump in the air, while having complete control of his or her movements, whether upon Earth surface, in a flying aircraft, or in a sailing ship.

Thus, properties of Motion-force "embedded" within the electromagnetic energies of Matter account for Gravitational Curvature pressure force that then shapes and molds "the geometry of traveling/moving Mass" in Continuum Space-Time.

Electromagnetic fields cause rotary forms of motion that then climax into Curvature geometry of Space as Matter-mass is "molded" into spherical forms having a pre-designed affinity for centri-vectored patterns of motion.

Are not all planets "rounded spheres" revolving in helio-centric motion and rotating in self-referential axial patterns of motion? As force determines geometry, the strong force determines centri-vectored motion-force cycling patterns of Energy conservation.

There is redundancy of Motion Forms in the Universe, not because of geometry, but because of gravitational and magnetic field forces that engender Continuum Curvature for thermodynamic cycling of Energy transformation. Geometry is an emergent property of overlapping electromagnetically-induced Force exertions that climax into Curvature tensing pressure Energetics, due elemental affinity for molecular arrangements, respectively, possessing differentiated organizational structures as designed for specific functional purposes.

For example, designing blades for an electric fan requires a structural Form that is in consonance with the rotary forms of motion that the electromagnetic field engenders. Geometrical shape of the blades is determined by the magnetic field force into which they will have to revolve in order to ventilate the air.

PHYSICS, VISION-BASED HUMAN UNDERSTANDING, AND INTERFEROMETRICS

Natural phenomena operate within boundaries that have ranges of effectiveness, as limited, by the fixed value of the speed of Light (namely: 186,000 miles per second) that causes mass frames to periodically cycle in iterative patterns of processing that sustain Continuum Curvature.

For example, visual observation has limitations and boundaries that cut off its effectiveness, because, like all of our senses and all things in Nature, human vision is limited to a certain range. So is our hearing — bats and dogs and some other animals perceive audible sounds at ranges which Humans cannot. And distance curtails the effectiveness of our visual acuity. That's why, when things are too small to be perceived, we utilize the microscope, and when too far to make out their details, we utilize the telescope.

In addition, now that we have instant telecommunications devices, we could station relays of human observers at "outer naked eye" horizon or "infinity" to tell us that indeed parallel lines do not meet, e.g. railroad tracks, and that they never meet, even though our weak limited vision due to the structural Curvature of Earth "circum-sphere," and "eye ball curvature," might misperceive that the tracks appear to "meet at infinity."

Parallel lines do not ever meet for by definition they are equidistant at right angles. However, because the earth is a bounded sphere, if railroad tracks were to be tied endlessly within the circumference of earth geography, as parallel lines, they would "return to their point of origin," while yet "conserving" their right angle equidistance.

Thus, with the telescope and binoculars, the distance at which parallel lines seem to converge would recede or grow farther. In short, it turns out that "infinity" is nothing more than the outer limits of "naked eye horizon."

As a mathematical construct, infinity accounts for, not quantities without limits, but rather, quantities that are so great, that, within the "Force-bounded Universe," they would yield impracticable applications.

We live in a Universe, physical Laws of which, compel Matter-Energy Forms to operate within certain limits as defined by their "ranges-of-effectiveness" or "ranges-of-influence," respectively. Space is infinite, but Mass-in-motion within Gravitational Continuum Space-Time is finite, while Space-Time "quantifiable categories" are measured as "re-translated dimensions" or "thermodynamic characteristics" of Matter-mass physicality.

The Speed of Light is finite — 186,000 miles per second in a vacuum — because, as a Form of Energy, due to the "Equivalency Principle," it must inevitably depend upon Forms of Matter that respectively yield the Electromagnetic Spectrum, e.g., a Star.

Light is the medium—the wave-particle or spectral photon—that is emitted, and then reflected or absorbed, by objects upon which its beam falls. It is this light that registers in the rods and cones of the eyes, focused by the ocular lens on the back of the retina, and which is transformed, from a reflected upside-down image, into minute electrical impulses. These impulses travel up the optic nerves to the occipital lobe of the brain, and then relayed to the cortex which integrates this whole process of sensory perception into our conscious understanding — by converting the upside-down retinal image of the perceived object into the internal-mental experiencing of the sensed object, as a real thing "out there;" thereby converting the reality of the seen object that is sensed or perceived, felt or touched, into a concrete Form of objective knowledge that can then be grasped by the Human Mind.

The Light reflected as an image of the object, e.g., an apple, is not the object itself. Light that is radiated by cosmic Mass, e.g., a Star, has to embody processes of the cycling Mass-Energy Frame, the sum of which, are "in the Present." But, the objects that we perceive in their true-to-real colors, e.g., a green leaf, still exist, even in the absence of Light. However, the original material Light-Source affording us to perceive or visualize them, must also belong to our current Time-frame, which has "unbroken Continuum," in order that they are objectively sensed and perceived in objective Forms that are amenable to measurable abstraction, to procedures for application, and to quantifiable analysis.

Some stellar bodies emit light, like the stars, and others reflect light received from Stars, like the Moon. Light, however, cannot be absorbed or reflected by things having no physical substance, hence, Light traveling in a vacuum at a constant speed of 186,000 miles per second. There is a slight reduction in the Speed of Light traversing a medium consisted of sea water. During the night, the Sun still exists, in-the-Now and still continuously shines. Its time-measured cycling nucleated Energy processes are independent of the Frame of the observer.

Human Beings depend primarily on Visible Light for perceiving and viewing objects and things. But Visible Light is only one method among many ways of detecting or seeing, or sensing objects. Seeing via means of Visible Light is the Human biological method. It is the human way of seeing or apprehending objects that really exist externally.

Objects are also detected or "seen" by mechanical devices using other wavelengths of the Electromagnetic Spectrum that cause them to have a Form of "artificial sentiency." Radar, sonar, infrared lasers, ultraviolet devices, magnetic resonance imaging or MRI, ultrasound machines, the CATSCAN machine, fiber optics technology, etc…, all utilize wavelengths of the electromagnetic spectrum as emissions that "carry information," such as sound, picture images, or data, as related to things that really exist. These electronic devices then convert or "re-translate" the electromagnetic emissions into a Form that our range-of-vision and our range-of-hearing can process, e.g. print or hard copy, video on-screen images, and/or humanly audible sounds.

However, these audio-visual data have to be "interpreted" against the real objects whose properties they are "re-translating." For, if a difference exists between the things or events we register via our senses and the real material existence of these objects in reality, that difference must be explained in consonance with objective standards of known reality. For example, results of magnetic resonance imaging or MRI for disease identification must be "interpreted" in light of real health conditions of the person and real knowledge regarding human anatomy, physiology, chemistry, and immuno-electrolytic metabolism before a definitive diagnosis can be declared.

When we sense objects via projected representations such as previously recorded by devices for "differentiated perception," or "simulated objectivity," e.g., a moving picture on a screen, then a reality-check for authenticity is required. For example, astro-evolutionists are interpreting cosmic phenomena as "reported" via electronic machines positioned in Space, such as the Hubble telescope; they don't have "direct access" to the things being studied for analytical understanding, and thus, there is no process of independent authentication. Astroevolutionists already have a preconceived "big bang explosion mindset" through which all data are being filtered, hence restricting their capacity for unbiased analysis. The same situation prevails when a particle is being accelerated in a cyclotron as attempts are made to simultaneously determine its position and momentum, due to "Interferometrics" between particulate constitution, instrument functional properties, and Human vision, all of which, utilizing in common, electromagnetic-impulse-based encoding and decoding systems.

A recorded event is different from the actual occurrence of the event itself. Currently emitting cosmic space bodies could not "record the Past" to then re-transmit it to us in the Present. Distant Stars are not "recording devices" but are "excited" mass-Energy frames, thermodynamically cycling in Gravitational Continuum Space-Time as ever-present Sources of emitted radiation.

Objects are sensed by the Human Mind, either directly-naturally via our retina and optic nerve, or "artificially" via electronic or mechanical devices that project representation of them to our retina and optic nerve. When "the connectivity" between the representation of external objects and the objects themselves which are seen, detected or observed, or sensed, is direct, and is not altered or severed, nor modified or adulterated, then, we know that there is Continuum between their reality and our sensation of them; and reality-checks would only confirm their authentic factual existence.

"Continuum connectivity" must exist between the human brain and the objects being observed via the retina and optic nerve even as electronic or mechanical devices are being utilized for perception. And standards for authentication must exist in order to prevent emergence of false interpretations.

Direct natural vision or machine-assisted perception for real understanding, must remain faithful to the form and substance of the objects being represented, during perceived or observed experiencing of their reality within the same Time-frame — there must be preserved, coherent and consistent standards of fidelity for the type-of-Reality, e.g., a molecule of Mercury, being observed in order to compare "interpretations" or "apprehensions" of corresponding data being evaluated, respective to the Form of Reality under observation.

There is no evolutional "standard of authentication," given that "relics" of either "biological missing links" or of "cosmic background radiation" that must be discovered, are presupposed to have had disappeared eons ago, but to yield "mutated forms" that would then supposedly "re-appear in different Forms," into the Present, e.g., From the apes to Human Beings: there is a "missing biological link" between the apes and Human Beings which is yet to be "discovered" in the so-called "fossil record."

Thus, according to the evolutionist's paradigm, there is "no Continuum-connection," nor any "direct reproductive processes" between the Past and the Present as required for maintaining Continuum from operations of the Input-Process-Output mechanism in accordance with fidelity to sustaining dynamic thermodynamic System equilibrium, i.e., such as parentage between a child and his or her two parents who are: one Man and one Woman; and such as between the child's parents and their forebears who had preceded them thousands of years ago.

Fidelity to the Principle of Continuum is a sine qua non component of the Law of Transformation of Energy in that Mass-Energy Frames thermodynamically "cycle" as they "re-cycle" their "previous outputs" into "new inputs" in order that functional operation processes "begin anew," e.g., Each calendar year, Earth seasonal cycles proceed from Spring to Summer to Autumn to Winter, and, to-and-from Spring to Summer to Autumn to Winter, ad infinitum, as each season's respective "outputs" co-determinedly overlap to become "new inputs" for the "on-coming new Season."

Consequently, all evolutional doctrines and derivative tenets are "articles of faith" for which no direct cause-and-effect proof can ever be discovered, analyzed, recognized, and ascertained, in order to substantiate the Theory of Evolution as factually and validly scientific.

It is therefore illogical to assume there could be Continuum between "the big bang explosion" and so-called "relic cosmic background radiation" for us to detect, since their source would no longer exist for supposed on-going connectivity.

According to conflicting evolutionary doctrines that contradict Continuum, "missing links" could not thrive into the Present Time-frame to "power" currently observed radiating stars with nucleated Energy, as they would have had "disappeared" while their non sequitur "unlinked and disconnected mutated Forms" would have had supposedly become the Universal Reality that we now perceive.

Thus, Light from distant galaxies come from stellar radiation sources that exist in our own Time-frame, and not from a conjectured "Past Time-frame" that supposedly "carried information" into the Present as the Universe existed "billions of years ago."

Analogous illusion-effects exist when drug-induced optical conjectures are compared with objective reality. Illusions also occur when false belief proceeds from illogical assumptions that yield false deductions. For example, we know that if the chemical processes undergirding biological vision are altered by drugs or alcohol, then, illusions, hallucinations or fantasy-visions can occur. Likewise, if the electron streams being processed by a television receiver for "lines of definition" are of a "pre-recorded" form or "not live," then, images and sounds can be electronically manipulated and altered, even during re-transmission or re-broadcast.

Given that television is an electronic device that also serves as a "semiotic/symbolic medium," even real-time transmissions or re-broadcasts of events can be unrepresentative of "present Time-frame processes."

Sometimes, recorded background pictures of landscapes or "stages" are prepared in-advance, with props and contraptions that are substituted for an apparent out-door scenery as visualized or observed on the television-screen by viewers.

By virtue of the fact that the representation of the object or event must pass through an "electromagnetic stream of radiation," the sum of which, that "re-translates" data into "pixels" or "picture-cells" that can be modified, replaced, altered, or adulterated with "the touch" of electronic command-and-control mechanisms and instruments, then, a broadcast is not necessarily "live;" and thus, nor is it necessarily within the "Present Time-frame," e.g., In the movie "Forrest Gump," the principal character is shown to have shaken President Kennedy's hand, though, in historical reality, the President had already died since the year of our Lord Nineteen Sixty-three.

In fact, it is at the mere "discretion" of the broadcasting television network or station to customarily display a warning on the television screen to viewers that the event or events being portrayed or pictured thereupon has been "pre-recorded."

Because Energy can be transformed into its many states or Forms, respectively; and because it also "iteratively cycles" in Temporal or Thermodynamic Time" as it "carries information" by having "a beginning" and "an end," then, the "loaded information" or "carried data," can be "turned on or off."

Thus, these "picture-cells or pixels" allow technicians a great amount of flexible control that in turn enables them to possibly tamper with electronic renditions of the broadcast, whether "live" or "pre-recorded."

Television networks have counterfeited pictures on their electronic mass communications media in ways that make if difficult for the viewing audience to determine whether broadcast on-screen images are true-to-real, or factually representative, projected renditions as "Present Time-frame events," or as electronically reproduced, or previously recorded spectacles.

Network television personnel also perform "visual tricks" by interposing already filmed or already video-taped, pre-recorded scenes, to then intersperse them within "live" broadcasts without informing the viewing audience of these altered, modified, and adulterated conditions and circumstances.

Consequently, it is required that broadcasting organizations inform the viewing public when a mass media communication process involves pre-recorded events. Sometimes, reporters appear to be broadcasting "live" and with "continuity-of-engagement" in the program being treated while they are merely placed in front of a screen upon which pre-recorded material has been projected as background. Since the audience watching the electronic medium cannot make distinctions between "live" or "pre-recorded" broadcasts unless informed by the network's personnel, then, the electronic transmission of the image of a real object can be mistaken for real-time observation of the object itself within "the Present Time-frame."

As a television commercial asked, "Is it live or Memorex?" Viewers are thus given the impression and appearance that reporters are broadcasting "on-location" but which is really "an electronically-induced illusion" as the reporters are always in the network's studios.

Therefore, according to examples listed above, errors in analytical judgment can occur due to false belief from illogical and unscientific assumptions, e.g., such as evolutionist doctrines; or can be due to drug-induced mind-altering perceptions, e.g., such as by alcohol; or due to electronic manipulation of represented renditions of observed reality, e.g., such as by pre-recorded pixels.

Consequently, "standards of authentication" for physical and social Reality in which we live, but which we cannot directly observe, e.g., if a Person resides in Florida and is viewing on a television screen events regarding the Official Seat of our federal Government, e.g., a Session of the Senate of the United States, yet, our great nation's Capital is in Washington D.C., these "standards of authentication" must always undergo verified substantiation as required by "Reality-checks" that confirm the viewer's analytical deductions or experiential results.

What kind of "Reality-checks" can be performed when the objects under evaluation for "authentic Real-ness" are distanced at millions of light-years away, such as "Star clusters" laid-out in "constellations" located in far-away Galaxies that are situated within the "deeper remote recesses" of outer-vacuum Space?

Science, in the forms of laboratory or field experiments, is a visual-sensory experience. Thus, it is not "accidental" or due to "random chance," that astro-evolutionists persist in "pursuing" the elusive "relic cosmic background radiation" by evoking astral sightings in the heavens, even as they are relying upon electromagnetic-spectrum-based equipment, machinery, and instrumentation.

But how must we evaluate the authenticity of the apparent "Real-ness of a thing" that is "outer" to our inner-personal apprehensions or external to our bodily "boundaries?" We can clearly "feel," "sense," "identify" and ascertain our bodily limbs or physical members. However, when the object or thing is situated at a distance from us that is measured, not in inches or miles, but in thousands of millions of light-years, then, the visual-sensory experience is more complex and more remote; and hence, our reliance upon humanly invented machinery, equipment, apparati, and "sensing methods" that serve as enhancements to our "electronically assisted" sensory-visual experiences.

When the eyes "register" an object "out there" through the medium of Light — where direct-connection between the object and our visual cortex is authentically sustained for apprehending its accurate or precise representation in our Minds, — then we can ascertain the degree to which objective fidelity of our process of understanding to the object's essential characteristics has been consistently upheld; for, the eyes see a real thing, and the Mind's understanding, climaxes into real objective knowledge. Under such uninterrupted "Continuum conditions," the "information carried" by Light, i.e. images "carried" by streams of photons to the retina; or when images focused by the iris, that are "carried" to the retina via means of electronic machines using the electromagnetic spectrum, are "registered" as authentic by the Human Brain — "its substantial contents," are not only real, but they are also objectively provable as valid and reliable, due to methods of independent Reality-checking that can be judiciously performed by all Human senses that coherently, consistently, firmly and directly "unite" inner-experience with outer-Reality for "Oneness Integrity."

Human vision for understanding is a physical process having electrolytic and chemical dimensions made possible by photon particles of light-reflection. But human understanding also involves "Mind inner-thought movements" or "movements of Mind," proceeding from abstract spiritual discernment to objective analytical logic, while we are engaging in extrapolations from externally observed phenomena. Even blind people can think, speak, read, write, and teach from their creation of abstract conceptual theories that correspond to physical relationships extant within the organizational structures of natural phenomena, e.g., why Earth has seasonal cycles and the functional purposes they fulfill by their periodically recurring iterative patterns of "ecological framing."

Radiation we observe from cosmic Space bodies is reaching our Galaxy and cannot be from the Past, given that the cognitive connectivity between our understanding and their factual emissions is occur in the Present Time-frame, as originating from ever-present physical "on-Sources" of real-Time spectral emissions, — in that our established standards for "Continuum Reality-checks" consistently go-on to confirm that no eventual interruptions have happened to "displace" their thermodynamic cycles "out-of" our Gravitational Continuum Space-Time Frame that has been proceeding from Past-to-Present and Present-to-Future.

Therefore, evolutional "biological missing links" or "relic cosmic background radiation" that persists in contriving interruptions and disruptions, the sum of which, purporting to "insert" breaks in the chain of events characterizing the Frame of Universal Gravitational Space-Time Continuum, will be proven as "categorical fabrications" that will not stand the rigorous tests of scientific scrutiny.

For, strict adherence to the Scientific Method precludes any pre-emptive non-scientific categories of Mass-energy relationships, e.g., in Bio-Thermodynamics, pretending to "randomly stop the flow of Gravitational Space-Time Continuum" to then "randomly restart it:" That is not possible! Because, when the true-to-real Time-flow of Continuum is factually "interrupted" or "stream-displaced," permanent terminal Entropy prevails in overcoming the processes of Conservation, i.e., "It's time to die."

The Laws of Thermodynamics that inhere in the Law of Transformation of Energy affirm that, barring a divine Miracle, in any pre-established System that has been progressing in order to achieve dynamic System Equilibrium, permanent interruptions or irreversible disruptions in Continuum only announce the dawning of entropic end-Death!

These examples illustrate that, in order for external reality to be perceived in its genuine or unaltered Form and objectivity, the direct connection between the real object and the "Present Time-frame" must be in Continuum relationships; and that, processes of Human sensory perception, knowledge, and understanding must not be altered, modified, adulterated, tampered with or severed.

Evolutionist conjectures such as "big bang explosion," "biological missing links," and "relic cosmic background radiation" are "conceptual categories" that "sever Continuum" between the Past and the Present as they also sever Continuum application of the Law of Energy Conservation.

Hypothetical past radiant Forms of Energy cannot be differentiated from current radiant Forms of energy: for the Law of Energy Transformation states that "Energy is never created nor destroyed but always transformed." Inexistent Light sources from "billions of years ago" cannot emit radiation for us to acknowledge in the "Present Time-frame." Light from the Past is only fantastic fiction.

The whole process of how a theory "makes it" into "the scientific establishment" lacks Reality-checks and sense-making methods of validity authentication. Astro-evolutionist "gate-keepers" prevent constructive criticism of the process via which professed theories become accepted as valid: They call it "peer-review."

"Peer-review" makes no room for the Principle of Reproducibility that is inherently and infallibly intrinsic to the Scientific Method.

"Peer review" for "big bang derived theories" of universal beginnings revolves in a circle; it only amounts to collective "amicus curiae" affirmation of prescribed doctrines within which evolutionist thinking can be expressed. Because they all have the same philosophical outlook and evolutionist mindset, they only "rubber-stamp" each other and there is no independent, dispassionate, and impartial oversight of this unscientific process.

Why is there a "veil of peer pressure" compelling all scientists to "toe-the-line" with the evolutionist thought?

****

REALITY-CHECKS AS REQUIRED BY THE SCIENTIFIC METHOD

We have forgiveness of sins in Christ Jesus and the power of renewal for reinvigorated prosperity. The "veil" in the Temple of Stone that housed "the Holy of Holies" was torn in two when Christ died. And Christ removed that "veil" totally by his burial, Resurrection from the dead, and Ascension unto heaven, to then reveal in us "the mystery of the knowledge of God" by endowing us with the miraculous gift of His Holy Spirit.

We, God's children in Christ, now have bold and confident access to "the Holy of Holies," to enter into the presence of our Heavenly Father, without shame, fear, guilt, regret, or condemnation. (Mark 15:38-39, 2 Corinthians 3:17-18, KJV, Holy Bible).

Spiritual discernment in accordance with God's commandments inspires works of righteousness on the Earth that confirm our Creator's good will for our lives. Freedoms of thought and expression are God-endowed inalienable rights of Human Beings created "in the image and unto the likeness" of our loving Creator. Spiritual liberty for moral righteousness is a divine gift that transcends all temporal phenomena that appear to escape human understanding.

Validation of scientific theories must have unbiased Reality-checks, such as those required in fulfilling the Principle of Reproducibility, in order that illusory end-results do not control research activities, e.g., that there exists "relic cosmic big bang background radiation," that current radiation is from "past cosmic bodies" as the universe existed "billions of years ago."

Unproven opinions, such as the so-called "big bang explosion," as well as the past occurrence of supposed "biological missing links" or "intermediate species" has gone unchallenged for too long, while they amount to mere conjecture.

The evolutionist mindset is an impediment to true-to-reality perception and objective understanding of distant universal processes, given erroneous assumptions that undergird research methods and pre-formed opinions regarding results. Astro-evolutionists are looking for "relic background radiation" as a "remnant" of the so-called "big bang explosion" which they assume took place at the very time of universal beginnings. Thus, they are not engaged in the scientific testing of a theory of "big bang explosion," but rather in manufacturing methods of detection that are designed and geared in advance to prove its presumed existence.

By the foregoing analysis then, it is scientifically logical to conclude that there has never been any so-called ex nihilo "big bang explosion" that supposedly began the Universe as we know it. Consequently, there could not have been any "relic cosmic background radiation" from the "past." And it could not have "carried information" even if it did exist (for the sake of discussion) since there would not have been any ever-present "on-Source" to sustain or "back-up" or give substance to the observed radiation, within "Present Time-frame," — the whole assumption about the falsely observed "past radiation" being an illusion. It would also become a "subjective illusion" in that, belief in its occurrence had been generated by false assumptions — such that the assumption that a "big bang explosion" took place was also erroneous.

Subjective illusions occur when mental conceptualizations, symbolic representations, or inner-mind apprehensions of apparent reality have no basic anchor in external facts, but rather originate from train-of-thought analytical deductions that are pegged to an already prescribed inaccurate frame of reference, e.g., the theory of evolution.

Additionally, even if, for the sake of discussion, observed stellar radiation could "carry information," the information observed or detected could not be "from the past," because Thermodynamic Time is a physical unit of distance-measurement for rates-of-change Mass-Energy Frames that are in motion. Thermodynamic Time is "embedded" within the true-to-real Gravitational Space-Time Continuum.

Time only goes forward, into the future, from extant physical processes due to the Law of Energy Conservation and the Law of Entropy, as executed by the Universal Input-Process-Output mechanism, through apparently stochastic, but Time-measured thermodynamic cycling, the sum of which, must sustain Continuum dynamic System equilibrium. Past physical processes that would have had already disappeared could not project extant stellar radiation energy to supposedly embody the past Universe as it is presupposed to have had existed "billions of years ago."

We cannot return to our former "state-of-Being" by "physically reverting" to childhood: That is biologically impossible! Thus, aging, is irreversible! For, DNA-rooted biological-physiological growth can only "progress" and not "regress." "Dwarfism" can be treated with growth-hormones in order to increase "height-growth;" however, "Giantism" cannot ever be "reversed" for a return to a "shorter-height."

Even if it is assumed that Time itself could be "bent" or "curved," its frame of reference would also have to belong to extant gravitational Curvature processes, events and phenomena, and not to conjectured past processes that no longer exist, but only supposedly projected into the Present by so-called "relic cosmic background radiation."

Continuum Time-frame mass cycling in Space-Time Curvature Energetics requires Relativity interactions within contiguous Frames that consistently interact through overlapping but distinctive gravi-metric Differentials, which they all share, belonging to the same Form of energy.

Imaginary or speculative "Time travel" internally experienced through self-induced "altered" or "hypnotic states," such as those common to "transcendental meditation," is not proven travel in true-real Time. No displacement-in-Space is occurring: The Human Person who is meditating, is literally "going nowhere!" These imaginative experiences are merely feats of abstract mental memory or subjective personal conditioning, because the Person imagining so-called "Time travel" goes nowhere but remains at the same place, in the same location, and within the same Time-frame that "the act of meditation" is taking place.

Therefore, the Light we witness in the stellar universal background is being emitted in-the-now, in the present, by Stars or reflected by other bodies, such as planets and other space bodies that are real, and that do exist in "real-Time," in the ever-Present Continuum-now, such that the Light is "live," and not "pre-recorded by cosmic background radiation." The witnessed Light is not "from the past."

In addition, Light does not continue to travel once its presumed Source is extinguished; it is always absorbed or reflected by other cosmic bodies in real-time gravitational Space-Time Continuum.

In other words, Sun light would not continue to reach us up until the eighth minute were the Sun to have "gone out." Due to the Law of Entropy, the light would have been absorbed and-or reflected by all the planets and other space bodies at the velocity of light — 186,000 miles per second. Thus, whatever, a so-called "relic cosmic background radiation" would reveal, even if it did exist, would always be about the Present and not the Past.

To understand the fallacy of the proposition that somehow the so-called "big bang explosion" had occurred at the epicenter of the Universe, and that "cosmic background radiation" is still being emitted from it and would reveal the Universe as it was "billions of years ago," let us assume, for the sake of discussion, that in some distant galaxy there were some "alien beings" imagining the same theory or supposition.

These "Aliens" would also have to believe, that ourselves on the Earth, belong to a place that is at the so-called "epicenter of the Universe;" they would also have to believe that the light being observed from our own galaxy and our solar system represents "the past Universe," being brought to their region of Space, from where we are, millions of light-years away.

So, our galactic "light show" would become "their" long-coveted "cosmic background radiation;" and our own region of Space would have become their "epicenter of the Universe" from which all Space bodies would have had been "running away." We would be situated "as their past Universe," just as astrophysicists in our own region of Space on the Earth, now presuppose "their region of Space," as reflected by oncoming light-radiation, to be "our own past Universe."

What a convoluted unlikely scenario! As we have amply demonstrated, the whole scenario would be illogical and unscientific. Either we exist, are real Beings in-the-Present-Now, or we are not; either all Stars and other planetary bodies we witness now, exist in the Present, now, or they are fiction; either the universal radiation being emitted now comes from current stellar bodies that exist now, in the Present, as ever-continuing "on-Sources," or it is an illusion; either there is real-time in-the-Now-radiation, or there is no radiation at all.

But we do know that we are real, in-the-now, in the Present; and that the radiation that we witness now, exists in-the-present flow of the Thermodynamic Time-Frame in which we live.

As "embedded" within the Framework of Eternity, Thermodynamic Time — by which we measure rates-of-change in the processes pertaining to Mass-Energy Frames in Motion, — is made manifest as "running" from its Past to its Present onto its Future, as wondrously structured by all the Laws of Physics accountable since the very moment of Creation for its operations as the Material-Physical "embodiment" of Universal Gravitational Continuum Space-Time Reality.

This awesome "Embedded Temporal Reality" composed of Matter-Energy Mass Units/Frames/Structures in Motion, means that: Every thing, event, phenomenon, or process, exists within the current Universe as we know and apprehend it, — which has no so-called "epicenter," — the sum of which, being a Three-dimensional substantive material Reality constituted of numerous, but specifically limited physical Frames-of-Reference that "overlap each other" while undergoing thermodynamic changes within Gravitational Continuum Space-Time, the causes-and-effects of which, we experience from Sources, e.g., Stars, planets, asteroids, comets, etc…, that operate in the "Frame of the ever-Present," or in "Real-Time.

Thus, for us Human Beings, it is Thermodynamics that is the fulcrum around which pivot all our scientific activities that are design to yield functionally purposive technological applications for industrial production and for inventive mechanical modalities.

And, given that we are real and do exist; given that the Stars and planetary bodies do exist now; and given that universal radiation is being emitted now from these stellar bodies, then the "big bang explosion theory" of universal beginnings and all "attached paraphernalia" of presuppositions, assumptions, and preconceived expectations that are predicated upon its evolutional frame-of reference, are false.

For, what evolutionary astrophysicists presuppose or imagine about Light-radiation-energy to be "true" to us, here in this Solar System, would also be "true" to other Beings or "alien beings" who might exist "out there," or to whosoever would happen to be "peering into outer Space" from some other specific region of the Universe, such as previously illustrated in our hypothetical example.

But that is not possible. For then, every one's "alien world," as existing millions of light-years away from their particular region of Space, would then become the "epicenter of the Universe" and would be displaying events that had occurred "in the distant-remote Past Universe." No Beings, Human or "alien" would ever experience "living in the Present" while having "Hope for the Future." For, we would, as it were, always be "in the Past."

Thus, as illogical as it sounds, all Beings, human and "hypothetical aliens," would be "living the past" in a "past Time-frame,"within "a disrupted/interrupted Continuum" that can never be "linkable to the Present" nor can ever "flow into the Present." All universal "Beings" would never be able to "break-out into the Present."

"Reality" would be, only "virtual," in the sense that, no "Being" would ever be able to distinguish the difference between "true-factual Reality" and something they had dreamt while sleeping, or imagined "while daydreaming." And the analytical process whereby "factual memory" can be ascertained as "true-to-real-events," would not be possible. For, Time itself would have become a "dead-ended stillness." Movements of thoughts and ideas" within the Human Mind accountable for making new associations, or for comparing experiences within Space-Time in the absence of Continuum, so necessary for the renewing of knowledge of "external reality" would have had come to a standstill. For "Beings" who constantly believe they are "seeing the Past" while pretending to "have a Present:" Where would the Present begin?

Let us review the principles of axiomatic knowledge that we have established thus far,

(1) That the so-called "big bang beginning" from nothing did not take place as imagined by evolutionary astrophysicists;

(2) That an event or process in Time has a uni-linear or uni-directional progression, from its Present that increasingly becomes its Past, to then continue-to-flow towards its future, and to the stages of which, no one can return or "travel-back in Time;"

(3) That, for us, Time (e.g. seconds, minutes, hours, days, weeks, months, year etc…or light-years) is a consciously experienced and recognized, internal and external reality, as expressed in terms of physical units of "temporal distance" designed for measuring "temporal event-displacements" apprehended as "true-to-Reality by the Human Mind. Consequently, "Temporality" as opposed to "Eternity," exists for us, Human Beings, as "continuously flowing Thermodynamic Time," but, the sum of which, containing or presenting the flow-of-events we are "passing through" or "living through," as "divisible/separable quanta of experiences," e.g., we ate breakfast today at 10:00am, after which, we visited the Museum of Natural History, remaining there for two hours and thirty minutes. "Thermodynamic Time" or "Temporality" is "embodied" in a conceptual framework of understanding wherein is "embedded" the "Principle of Transient Pre-Occupation," e.g., we temporarily attend to feeding the cows, then we return to our regular activity of gathering the Harvest, the sum of which, is objectively superimposed over the Continuum Gravitational Space-Time within which the real Universe exists.

Because Time is not a physical-concrete-material category of things that exist, then "constituents of its measurement" are "external" to its essential Frame, e.g., clocks that come in numerous Forms having differentiated platforms for "telling" or "measuring Time — analog clocks; electronic clocks; atomic clocks, etc... Hence, cause-and-effect processes that factually and incontrovertibly establish, that, unlike Matter that is constituted of the Atom which itself has three primary particles: There are no such things as "Time particles;" and there is no such thing as "Transformation of Time" but transformation of Energy, such as in the equation $E = mc^2$. Therefore, not only that so-called "Time travel" to "the Past" of the Universe, as it was billions of years ago by "piggy-backing' on "relics of radiation" is impossible, but also, Space-Time cannot be "bent" or "curved."

Rather, structured or framed Mass-Energy units that are in-Motion "project" certain Forms of Forces, such as Field-Force, and Gravity-Force, that in turn cause other Forms-of-Matter/Energy to espouse "curvilinear paths or trajectories" that are amenable to the range-of-influence being exerted by these Mass-Energy Frames in the vicinity respective to their "moving Mass," e.g., a Light beam passing near a Planet or Star might "bend" or "curve" in its "traveling path" or "orbiting trajectory."

Because Space-and-Time essentially constitute an indivisible Continuum, then, no mathematical operation that cannot be performed upon Time can be performed upon Space. Given that Time cannot be "curved" or "bent," then, neither can Space. A Continuum means no permanent breaks, interruptions, or disruptions — Dynamic System Equilibrium must always be maintained, sustained, and restored, e.g., Perihelion Shift of Planet Mercury.

(4) That there is no "law" for "transformation of Space," that would allow "curved Time" to return to its past cycling initial conditions in order to project already "dead missing structures" or "intermediate forms" that supposedly had already "mutated into the present." There is the Law of Transformation of Energy, because of the Principle of Equivalency between the two physical-material categories/frames/structures as "tangible Units" possessing other physical constituents, can be subjected to measured rates-of-change.

Due to the fact that Energy properties are amenable to Matter/Mass-in-motion displaying thermodynamics cycling patterns of operations from which emerge "physical Curvature dimensions" as Sources or "points-of-Origin" for other Relativity variables that can be objectified, then, Mass, Radiation, and Motion projecting gravitational or field-induced Curvature pressure force dynamics, are Relativity-properties of Matter, e.g., the Sun as a cycling mass Energy frame, possesses the physical parameters and co-determinant variables from which all objective-applicable Space-Time categories obtain mathematical and practical embodiments, e.g, Electromagnetism (electromagnetic properties of Matter such as energy, the strong force, the weak force); Gravity force; field force; Curvature tensor-metrics; Curvature pressure-force Energetics.

(5) That Time is unified, continuous, universally contiguous, and forward-linear;

(6) That Time-and-Space lie in a Curvature-Motion Continuum within which are "embedded" all Forms and all properties akin to thermodynamically cycling Matter-mass;

(7) That "time-travel" back into the past, as a reversal of the path, vectoring trajectory, or navigational direction of Time, is impossible, — given the existence of Thermodynamic Time as a temporal distance-expression of a unit of travel-measurement between two "points-of-reference," e.g., one beginning-point and one end-point, the sum of which, occurring within the real Space-Time Continuum (as non-Thermodynamic or Eternal Time — "der ding an sich") — wherein the Curvature of Gravitational Matter-energy Space-Time Continuum that only "goes forward," is embodied in the "Ever-present Time-frame" of universal processes, events, and phenomena (as Thermodynamic Time);

(8) That the radiation being detected or recorded in the Universe is not so-called "relic cosmic background radiation" but is radiation that is being emitted in-the-present-Now;

(9) (a) That, within the Continuum Space-Time framework, Energy-Light-Radiation constantly emitted by a specific ever-present "on-Source," will be absorbed, reflected, diffracted,or re-radiated by other body-masses extant in the Universe;

(b) That in the absence of such a specific "real-time on-Source," no such Energy will continue to be emitted, in that Continuum necessitates an "actual stream" of successive accumulation, production, utilization, and consumption, — and not "arrested interruption" or "suspended flow" such as in the Forms of "biological missing links" or of "intermediate Species," the sum of which, that flagrantly violates the Continuum Principle, the Thermodynamic Principle, and the Principle of Dynamic System Equilibrium.

(10) That a theory like evolutionism whose fundamental tenets include the doctrine of "missing links" or "past intermediate forms" that "mutated" or changed but "disappeared, but still "produce present forms" violates the Continuum Principle while displaying contradictory application of its concepts. The "intermediate Species" or "missing links" that "mutated to form the Present" no longer exist for detection or discovery in the so-called "fossil record;"

(11) That there is no such thing as "old radiation" as opposed to "new radiation" or as opposed to "present/current radiation;" infrared is infrared from whatever Sources derived; and the Source must be self-evidently proven to be a physical ever-present "on-Sources;" for, no "old radiation" can exist outside of the present gravitational energy-matter Space-Time Continuum. Consequently, so-called "old relic cosmic background radiation," is not distinguishable from present on-going radiation; and therefore, does not exist. No such physical differentiation can ever be made. To presuppose that Light's encoded information about external objects is the same as the physical Reality it represented in the Past though that physical Reality has already disappeared, is to deny the facticity of absorption, diffusion, diffraction, and reflection to which light is constantly subjected due to the laws of thermodynamics, especially the Law of Entropy. And to equate this Light with the real objects our eyes see and our minds perceive through its visible spectrum as a medium via which information is "carried," is tantamount to declaring that a picture beamed on a screen by a movie projector is the object itself. Embodiment of past universal processes cannot be "contained" within present cosmic radiation. Motion on a screen is not motion in Reality — is not material Reality with substantiated coordinates of actual movement. Projected motion on a screen is an electromagnetic pre-recorded representation of real motion. The illusion of motion on a screen is created by the after-effect of light-wave particles reflecting and processed in the rods and cones of the eyes — due to a "differentiated delay" between the Speed of the Movie's Frames and the Speed at which electrical impulses proceed from the back of our retina to the visual cortex for integration as inner-abstract understanding of the event being projected on the screen; and hence, "slow-motion" and "fast-motion" or "time-lapsed photography." The same after-effect exists if a person focuses vision on a glowing light bulb, a burning lamp or bright television screen — the after-effect is an after-image of these objects' "shell" or "frame," that persists for a few seconds after the eyes are removed from perceiving them. The after-image disappears in these instances once the eyes no longer focus on the light rays bouncing off the bulb, lamp or television screen. However when there is a succession of projected frames, such as in a motion picture movie "at the right speed," then, there is apparent Continuum in the stream of after-effects; hence, producing the phenomenon or illusion of screen-projected continuous motion simulating a resemblance to

Real-motion. The moving picture is not the object itself. It is a two-dimensional projection upon a screen.

The above-explained after-effect produces the illusion of motion in a movie as a certain number of picture frames overlap each other before each frame's respective after-effect upon the optic nerve is exhausted and yet is allowed to blend or connect with the next oncoming after-effect of following frames. As the number of "still picture frames," e.g. 30 per second, is adjusted within a fraction of a second before the previously induced cumulative after-effect disappears, successive after-effects overlap each other and the picture "seems" or "appears" to be "moving." But the whole projection process can be summed up as merely a picture that appears to be "moving." The picture has no substance or reality whatsoever and the light beam representing it needs a solid two-dimensional background or frame on which to focus just as light rays entering our eyes need the back of the retina on which to focus the incoming image of an object. The things represented in the "moving picture" are not the real three-dimensional things from which the bounced-off light rays or photons were previously recorded but are projected two-dimensional substitutes for them.

In fact, there is no time correspondence between the two — the projected screen representation and the real three-dimensional objects. A motion picture can be replayed or re-shown 50 years after the real three-dimensional objects represented therein have been destroyed. Not only is there no time correspondence but the three-dimensional Continuum of substantial reality between the pre-recorded projection and the objects has been broken.

Then, is the "information carried" by the projected light beam on the screen physically real? No, it is not; it is still a non-substance, a recorded image on a screen, just as the image on the back of our retina is only an electromagnetically registered datum to be translated into minute electrical impulses towards the cerebral cortex for integrated recognition — but it is not the material Three-dimensional object-of-substance itself.

Additionally, there is no "Time-frame Continuum" correspondence between the projected images, the process of watching, and the Timed-occurrences depicted therein. As a movie can be seen more than 50 years later, both the celluloid frames and the projector must exist; still, it is a two-dimensional rendition and not Three-dimensional real-time occurring reality that had taken place years before.

Likewise, so-called "relic cosmic background radiation" as a supposed "remnant" of the conjectured "big bang explosion" violates "Space-Time-Frame Continuum" in that, not only the "explosive Mass" no longer exists, but the expected "radiation remnant" is impossible to differentiate from current Star energy, either via Human senses, or technological media.

If that were possible, then, the so-called "relic cosmic background radiation" would only "carry information" about the chaotic explosion itself, and not about a well-organized orderly Universe as exists in the Present Time.

The theory of evolution proposes a different Form of energy from the form that now provides the universal substance for all energy transformation phenomena, i.e., a "pure Energy singularity." As previously explained, none such can exist, given that Energy is Matter-dependent.

Therefore, the presumed "point of singularity" that caused the so-called "big bang explosion" could not belong to the present Space-Time-frame radiation Energy spectrum. Evolutionist presuppositions contradict the Continuum Principle by interposing "missing links" between the Past and the Present that can never be "recovered," while yet expecting a regular Continuum in the flow of Thermodynamic Time that would climax into "today's Present." Though such a "conjectured operation" is possible in thought processes, it is impossible "in the real world."

And thus, in the same manner that "biological missing links" contradict the First Law of Bio-Genesis (That Life proceeds from Life of the Same Kind), the conjectured "big bang explosion" and its "companion fabrication," the so-called "relic cosmic background radiation," sever the direct integrated gravitational connectedness that must prevail between cycling Mass-in-Motion Frames that have been functioning in "Continuum Relativity interdependence," the sum of which, for the purpose of iterative well-ordered Transformation of the same Form of Energy wherewith the Universe must have had begun: "Energy is never created nor destroyed but always transformed."

The foregoing treatise demonstrates that evolutionist assumptions regarding Light, human understanding, and universal phenomena are not congruent with Thermo-gravitational Mass-Energy Curvature Cycling Dynamics that have been functioning within Continuum Space-Time such as, that which leads, to our current Present Space-Time-Frame.

For, concerning radiation-emitting thermodynamically cycling Mass-Energy Frames, in order for the physical representation of an object by light rays to be real at the time the eyes are seeing it, the object itself must exist in-the-Now, in three-dimensional Space; the principle of direct connectedness must suffer no permanent interruptions or "missing disruptions" between the eyes and the object, between the Mind and external apprehension, between mental processing and the object's bounced-off light rays such as those being re-translated by electromagnetic-spectrum-based devices engaged in recording them. And "knowledge acquisition" or "apprehension of knowledge," must submit to immersion or "embedded-ness" within "Continuum Present Time-frame," such that real-direct processing prevails between external phenomena and Human understanding; conclusively meaning that, no so-called "relic background cosmic radiation" can exist to then "travel" to us from a presupposed universal epicenter to "embody" or "represent" the Universe as it existed "billions of years ago."

Evolutionists are incorrect in assuming that Light-radiation-energy observed from the heavens is from the Past to then "carry information" from universal processes as they occurred so-called "billions of years ago." Since "seeing a physical object" is an electro-conductive physiological-physical event or process, then, the abstract Mind as anchored to the Human Brain, does possess a colloidal endowment or natural means possessing objective integrity, coherence, and fidelity, in re-translating the represented image of the object, as it directly proceeds from the seen object's true Three-dimensional Reality.

Astro-evolutionists have no standards for authentication of universal Reality; they only desire that their presuppositions, preconception, and presumptions, be accepted as "articles of faith," operations of which, from their own frame-of-reference, mimicking or simulating the way in which religious doctrines based on faith, require only belief, which is ironically, the charge they usually press and wage against believers in Biblical Creationism. But extant physical phenomena logically align with, and do not unnaturally contradict Scientific Creationism.

Evolutionism and Physics: Pseudo-Faith, and Quasi-Plagiarism

The event or process of perception-understanding must be grounded in the material substance of universal Reality — it must be physically connected to directly experienced phenomena in a frame-of-reference that is marked by an ever-present, real-time, "on-Source," in integrated association with the Gravitational Matter-energy Space-Time Continuum, as imbued with a strict reliance upon genuine standards of authentication confirming its truth, reality, validity, reproducibility, and applicability as dictated by the Scientific Method, via proven results obtained from earnestly conducted laboratory experiments and/or field experiments.

Things are measured in accordance with a Standard, e.g., in meters, in inches, etc . . . In accordance with the Theory of Relativity, the Speed of Light in a vacuum, is the Standard that limits all physical processes, phenomena, and events to Mass-Energy cycling patterns of Conservation and Entropy, the sum of which, climaxing into dynamic Motion-force System equilibrium, e.g., Earth iterative daily rotational cycle and yearly revolutionary cycle.

The Science of Physics and of astrophysics more specifically, have to dissociate their theoretical activities from evolutionist conjectures based upon "peer review," in order to remain connected to mathematics that will yield applicable technologies from authentically conducted-obtained experimental results. The logical understanding of physical processes, phenomena, or events is an a priori necessity. The theoretical platform from which mathematical conceptualization is launched must provide the fundamental impetus for maximizing the potential for equation discovery. The evolutionist Frame does not provide such an impetus; to the contrary, its conjectures are counter-mathematical, and therefore unscientific.

A mathematical equation is a shortcut to symbolic understanding; it is knowledge in a nutshell that summarizes a process in a numerical or relational language. Some mathematical propositions might however not yield any conclusive terminal useful equation. The theory of evolution is not a physical phenomenon, process or event; it is a fallacious biological theory of origin. Its tenets have been extended in pseudo-scientific application to physical events, chemical processes, and natural phenomena, but not in a way that substantiates its validity in the true-to-Reality light of known cosmology and witnessed cosmography.

In such a case, Evolutionism is an "article of faith" that has no direct connection to any mathematical equation, chemical formula, or real authentic method of genuine scientific verification: Evolutionary hypotheses are not amenable to either laboratory experiments or field experiments.

Evolutionism is "story-telling" that is fraught with imaginative conjectures and contorted textual fabrications that yield no knowledge-based conclusions for guiding Science towards fulfilling Human yearnings or for designing needed technological applications. Things like "biological missing links" and "relic cosmic background radiation" may sound intriguing, but Scientists who must "toe the line" with Evolutionism, still have "a vast void" in their very souls, just as the Science of Physics is at a mathematical standstill!

The crisis in post-Relativity theoretical physics is amenable to the distortions of logic brought-about by importation of the theory of evolution and infusion of its doctrines into branches of Science commonly called "hard Sciences," which have an affinity for mathematical formulation or equation.

The evolutional "articles of faith" filter out alternative approaches to Scientific Theoretics, valid proofs of which might already exist within the known database of Physics, Biology, Chemistry and Astrophysics. For, without the right paradigm and the correct point-of-view for approaching natural data available within the extant physical universal Gravitational Space-Time Continuum, no substantial mathematically fruitful advance will ever be made.

And, given that in our day, Physics data-interpretation is overarched by the evolutionary paradigm and the perceptual lens its "attached doctrines" or "informing tenets" engender, this pseudo-scientific complex fosters a kind of "quasi-religious fanaticism" in evolutionist adherents who "put all their eggs in one basket," — to prove "big bang explosion theory."

The theoretical physicist who attempted to give an open hearing to alternative theories of the universe is Dr. Roger Penrose ("The Emperor's New Mind, 1989; Shadows of the Mind, 1994). However, while being open to considering Scientific Creationism, the Biblical account of universal beginnings, he maintained his evolutionary paradigm as an *idée fixe,* and consequently, the mathematical formulations he contrived to keep this "article of faith," filtered out belief that "the intelligent design" inherent in the Universe makes sense, because God, our Omniscient Father in Heaven, created it.

In Dr. Penrose's scenario, Entropy has predominance over Conservation, but in a sense that does not deny that there is a greater purpose for human living and for universal Reality in general. To Dr. Penrose, human thought processes, though "non-algorithmic," yield far deeper richness in universal understanding than formalized mathematics could achieve. He thought the computer to be more attuned to algorithmic output than human consciousness, but too deterministic to rival human analytical powers of scientific reasoning.

Still, Dr. Penrose could not dissociate himself totally from the "esprit de corps" engineered by the evolutionist "federal grant-writing establishment." Dr. Penrose still clung to the doctrine that the Universe randomly started all by itself, or "created itself" via a "big bang ex nihilo explosion."

Though coming short of believing that Human Beings are uniquely endowed with certain "mysterious" creative powers, Dr. Penrose concluded that we just have not yet existentially processed the presumed "natural selection agenda" advocated by Charles Darwin and propounded by biological evolutionists, as the self-justifying "genetic algorithm" which is presumed to be imprinted in "human nature."

Though Science appears to be driven by a code of objective open inquiry, astro-evolutionists tend to consider it "heretic," to even open their minds to the consideration of other theories that do not fit into the evolutionary paradigm. Doctrinal appendages that have been grafted unto mathematically-driven "hard sciences" like Physics are held "almost sacred," as if astro-evolutionists were fearful they might be "debunked" as unscientific if they were scrutinized for real validity.

But objective inquiry is supposed to be directed by a research platform that adheres to the Scientific Method, which includes discarding theories that are unproven or un-provable, or not amenable to "falsifiability" and/or "reproducibility — that is, theoretical mathematics not geared towards the applicability of the physical sciences to practical technological innovation, the sum of which not yielding any substantive validity through reproducible research and experimentation, ought to be summarily discarded.

The theory of evolution is such a theory: it yields no validly scientific or technologically practical Forms in theoretical mathematics — the theory of evolution began as socio-biology, to then be unwarrantedly extended to apply to all branches of proven-Knowledge, including Physics and Astrophysics.

The Science of Physics does not have an intrinsic-essential needed requirement to include the theory of evolution in its paradigmatic framework in order to yield mathematical equations that can be substantiated by experimental results and technological application.

A hypothetical process of "disappeared-missing evolutional past," is not needed for mathematical formulation of observed physical phenomena and natural events. Why would Physics need contrived "remnant relic radiation" indicating long-gone processes that supposedly "exploded" billions of years ago in order to proceed to theoretical universal unification?

As a mathematically-driven branch of Science, mathematical theoretics in Physics is sufficiently logical, self-motivating, fruitful, and inspiring for empowering true-Scientists with the data-resources in their pursuit to establish Continuum with already proven and already validated physical laws and theories, such as the law of Gravity, the laws of Motion, and the Theory of Relativity.

However, logical self-sufficiency as attuned to true wellsprings of real universal and natural data is not enough for a branch of Science to progress beyond the applicable limits of its most current paradigmatic framework.

The inspirational impetus that drove the Human soul to scientific discovery but has nearly disappeared since the time atheistic evolutionists began to assault the Judeo-Christian heritage — more specifically, the Law, the Prophets, and the Gospel of Jesus Christ — that had thence motivated Scientists, so that their endeavors and interests would impel their scientific activities into constructive, paradigm-building theoretics.

Astro-evolutionists are not only fighting against God but they are also fighting the Spirit of God "within themselves," within their own heart, within their own mind, within their own soul and within their own spirit. In their own eyes, to prove the so-called "big bang theory" will also disprove Scientific Creationism, as well as the very existence of God. And hence, it is imperative that Human Beings weigh the necessity to have a spiritual foundation of moral-spiritual values that actively sustain loving, lawful, just, and peaceful relations between nations, towards interactions that support creative productivity and promote economic prosperity.

Astro-evolutionists have attempted "to hijack mathematical Physics" for their own enlightened interests, atheistic purposes, and irreligious objectives. But they have not been able "to deliver" any validly proven technologically applicable formulae or equations. For, Physics is a "hard Science" in its own right, in that theories are validated through reliable and reproducible research procedures and experiment results. And until astrophysicists free themselves from the myth of evolutionary beginnings, research methodology that is "fed" by the theory of evolution will remain as unproven "articles of faith" yielding only inflexible discipline-oriented "turf-protecting" and opinionated but fruitless "peer-review" theoretics. As scientists continue to "toe the line" in practicing adherence to evolutionary doctrines, Physics will fall short of theoretical richness and mathematical innovation.

How can the field of knowledge called "Physics" regain its by-gone strength, integrity, and leadership as a "hard" or "mathematically-driven" Science, — when the paramount qualities and characteristics that sustained its scientific connectivity to real mathematics for understanding natural, physical, and universal reality, — to the examples of, $f = ma$, $F_g = G \, m_1m_2/r^2$, and $E = mc^2$, — have been taken away by evolutional conjectures having no proven material substance in true factuality?

It is a well-known truth that not every concept, however elegantly formulated in terms of theoretical mathematics, yields a "sense-making" equation-based solution. Not every formulated equation produces an objective, valid, provable and reliable scientifically applicable result that is reproducible by research and experimentation.

As an abstract alpha-numeric, quantitative language, the symbols and relationships of which can be invented out of pure speculation, Mathematics holds the vulnerable prospect of succumbing to contriving and defining symbols to represent or describe "categorical relationships" that have no cause-and-effect connections to physical-universal reality — that cannot be validly proven through a research event or experimental process for objective applicability.

Sometimes, relationships between contrived symbols that attach to a concept or hypothesis may fit the system of description or the frame of reference invented by the mathematician, but the abstract conceptual symbols and the relationships they purport to describe are not necessarily representative of physical Laws of universal Reality. Conceptual symbols may connect with each other to establish "relationships" as described by the mathematician in his 'operationalization' of the concepts; however, these "relationships," though they may fit into that descriptive System, might not be true-to-reality quantifiable value relationships that exist in the Universe or in "the real World," such as applicable mathematical equations that validate physical Laws, e.g., $f = ma$.

Consequently, derivative conclusions ensuing from "solving the equation" are self-justifying and unscientific, virtual and futile, with no valid, reproducible application to "the real World."

For example, the mathematical symbols, $f = ma$, and $F_g = G \, m_1m_2/r^2$ that describe how gravitational forces affect moving objects, embody the concepts of Mass, Velocity, Force, Momentum, Distance, Gravity, and Acceleration, the sum of which, are not only consistent with "the real world" in conceptualization, definition, and "experiential operationalization," but the abstract system that symbolizes the variables and explains their "relationships" or "exchanges" with one another in equation Form, is internally consistent with the quantifiable results in true-to-reality research, experimentation, and "real-world application." These results, as obtained via the abstract equations that describe the "relationships" or "exchanges" between the foregoing variables of Classical Physics and Celestial Newtonian Mechanics, are proven to have real "connective correspondence" with what takes place "in the world," and with what takes place between these variables as represented in the equations — symbolic equation-based "relationships" have become "the true equivalents" of objective relationships that exist in Reality. These Relationships apply to factual events such as pushing a rock out of place from a mountain-top position, to the motorized horse-power that drives an automobile, or to the thrust required to lift and propel a rocket-ship into outer-Space.

"The Physics of our day" appears to have lost this type of true connectivity to real-world physical events, natural phenomena, and universal processes. Astro-evolutionists are engaged in a collective ritual that is akin to a "religious observance" as they blindly "toe the line" in feverishly "bending backwards" trying to "prove right" the theory of evolution.

Begun as a biological theory of origin rivaling Biblical Creationism, the theory of evolution slowly "graduated," from socio-biology to "cultural shaper" of views and sentiments, as it finally penetrated the real or "hard" sciences, such as Physics, Astrophysics, and Chemistry, while its adherents have aspired to make it a "totalizing worldview" and a "normal paradigm."

The theory of evolution has no real scientific solution to problems of Physics that concern current research activities, e.g., a resolution of particulate motion patterns now estimated as "waves of probability;" a unifying Frame that embodies a general theory of force; how to exceed power boundaries that prevent particulate Curvature trajectories from holding researcher-designed structures.. The complexity of the sub-atomic sphere's (if an atomic sub-structure truly exists) gravitational dynamics makes it a daunting task for physicists to formulate its "categorical Curvature Forms" within the framework of Quantum Mechanical Theory. Sub-atomic wave-particle relationships, if any does exist, might not readily fit into the conceptual scheme of Quantum Mechanics, though quantifying their Mass-Energy cycling-patterns might yield some form of mathematical formulation that remains faithful to Continuum Laws of physical-universal Reality.

Every electronic instrument and every process utilizing the electromagnetic spectrum, as even the Human Brain electrolytic colloidal structure, is limited by the Speed of Light. Because they are "functionally activated" by the electromagnetic spectrum, the research apparatus of observation, analysis, and measurement contains externalities that might affect particulate behavior. Because, fundamentally speaking, "Light-radiation-energy is chasing Light-radiation-energy," though different spectral radiations have different "electromagnetic signatures" in wavelength, frequency and photon energy units, still, they are utilized for experimental engagement "in chasing the electron."

Their combined operations, e.g. streams of electromagnetic radiation Energy generated by research equipment and particulate physical-Law-determined behavior, result in incompatible effects, from the observer's frame-of-reference, the sum of which, becoming an obstacle to deterministic results, as predicted by physical Laws, and hence, the term "Quantum Gravity."

But particulate behavior is regulated only by applicable physical Laws, and not by researchers' point-of-view. Be that as it may, faithfulness to Continuum applicability of validly proven Laws of Physics and fidelity to their genuine scientific applicability must be sustained via equations and formulae that mathematically re-translate our understanding of these natural phenomena, e.g. for Quantum Mechanics particulate "behavior."

Mathematical formulae or equations "burn the steps" to applied Physics as required for progressive mental apprehension of physical reality via understanding of conceptual relationships described by mathematical variables. For they co-determinant variables are "shortcuts" to comprehensive understanding for objective technological applications. Mathematics is thus "a language" with icons that encapsulate or compress relationships between variables, the sum of which, predetermining "equivalency outcomes." Mathematical equations or formulae constitute "a throughput embodiment" of the Input-Process-Output mechanism for yielding "quasi-immediate" operational results.

However, though mathematics is necessary for technological applications, Human Beings can understand certain natural phenomena, events, and processes, complex operational relationships of which might appear to not be amenable to "computational modeling." Is there a mathematical equation for the bio-mechanism called "metabolism"? Is there a mathematical equation for the human capacity for "pattern recognition"? Is there a mathematical equation for human species specific pattern of DNA-RNA replication?

A computer, for example, can solve a mathematical problem much faster than a Human Being. And a Human Being may utilize computerized processes without having to go through the calculative steps that would logically lead to those conclusions. It is analogous to the cashier at the grocery store who might understand the calculative steps, but, as she is equipped with an electronic calculator or cash register which computes every price respective to bought quantities, she allows these operations to be performed "at the touch of a button" on an equipment keypad.

These electronic machines are "smart" or "user friendly" but the Humans who "run them," though wholly aware of how they work, might defer to the machine's results, in totally believing them to be as "nearly infallible."

Computers do perform erroneous instructions due to "viruses," and do "break down" due to, not only Entropy, but also to defective parts. Their inherent physical capacity for efficient and fast output performance is also consequential to the fallible nature of their human designers and manufacturers.

Computerized machinery should not foster functional illiteracy regarding "the older methods" of learning "reading, writing, and arithmetic"— through procedures that extol the virtues of the "read-write-speak" methods of instruction. "Reading, writing, and arithmetic," are said to be the Fundamentals of grade school instruction. These Fundamentals of childhood development formerly included instruction in religious studies, speech, history, and the Sciences. Thus, in order to counter "push-button routines" of computerized technology, students in our technological Age whose Minds need to get enriched for deeper analytical reasoning processes, would greatly benefit from "the old methods." Computers, programmed with 0-1 cybernetic binary numeric encoding, work "at the click of a button," in ways that appears to be sufficient for developing familiarity, comfort, and utility. However, caution and prudence must be exercised so that a wide "continental divide" does not separate the few technocrats — who design-and-operate the electronic network of machines — from "the mass of consumers" who are too busy working and laboring in order "to make a living."

To remedy this "evolutional," counter-social structural "management-labor" anomaly, formative years in educational instruction, must include activities from which students are bound to benefit, such as, in moral instruction, analytical reasoning, scientific literacy, historical knowledge, and development of mathematical aptitude. This platform of instructional design for educational learning must operate within the boundaries of God-commanded principles of righteous living for lawful, peaceful, and just, "Continuum prosperity."

The more extensive became the "evolutional infiltration" of all branches of knowledge, scientific research results that purported to be accurate and valid, tended to exercise a disproportionate effect upon individual morality and public attitudes: Human Life was "devalued" to the "level of animals," and social relationships in general were viewed as "instrumental media" to be "appended" to the pursuit of wealth.

Thus, "socially-emulsified" scientific theories, processes, applications, and operations have had a tremendous impact upon our culture at large, collectively constituting a frame-of-reference into which collapse, all perceptual lenses engaged in social problem-solving.

Technology often appears to be "the first-choice-means" in conflict resolution, rather than primarily relying upon spiritual biblical applications to guiding students towards righteous inner-motivation for the exercise of moral freewill.

"Science," therefore, is not a "culturally-neutral occupation;" nor is it a "socially-inert engagement:" For, Scientists' individual belief systems and even private morality, do play decisive roles in shaping scientific direction and in molding purposive goals.

The theory of evolution should not be extended to Human behavior in the guise of Socio-Biology; nor should the Theory of Relativity be exploited to extend "equivalency" between all moral values purporting to provide spiritual frameworks for "inspiring-and-impelling" Human activities.

The Theory of Relativity was formulated to represent a descriptive pattern of physical Input-Process-Output relationships between Matter and Energy. Dr. Albert Einstein did not extend this "Relativity Theory" to all things applicable to Human life. He did not mean to extend "Relativity:" Neither to Human behavior, nor to social relations; neither to religious morality, nor to spiritual values.

The Theory of Relativity has been validated as a physical theory describing a physical process or event, and not as a theory of personal morality designed for guiding social activity.

These corrections are important for, the maintenance of absolute standards of Morality only correspond to the absolute "standard of limitations" that prevails in Physics due to the constant value of the Speed of Light in a vacuum, fixed at 186,000 miles per second.

In order that Physics expand beyond the Theory of Relativity, it must remain anchored to real mathematical Theoretics that correspond to real physical processes, events and phenomena, rather than to speculative forays that project Human society as "the equivalent" of the world of animals; or that extract social interpretations for Human beings from observation of the animal realm; or from inanimate processes that are in "relations of Mass-energy cycling" within the "Thermodynamic Frame" of the physical Universe.

For, it is not "accidental" or due to "random chance" that we, Human Beings, are both genetically endowed and spiritual predisposed to believe in One God, Creator of all Things, the sum of which, corresponding to the absolute standard which the fixed-value of the Speed of Light represents for limiting the extent-and-expanse of universal processes and natural phenomena.

Human beings are spirit-beings in mortal bodily vessels with God-endowed gifts that are potentiated with awesome creative capacities for spiritual discernment, moral living, constructive productivity, and technological innovation.

Faith in One God, Creator of all Things, is but the fore-knowledge of secured blessings already willed by divine Providence on our behalf through our Savior, Jesus Christ Messiah of all Flesh, as we righteously align our earthly living with God's commandments, including our pursuit of scientific discoveries that alleviate Human suffering and diminish the harshness of Human labor in the physical Universe.

In the same manner that Conservation counters Entropy in the physical Universe and in Nature, absolute Faith in One God, Creator of all Things, counters the vicissitudes of Human Nature and the deeper recesses of the Human tendency to moral turpitude and behavioral malfeasance.

As stated above, the "leap of faith" required to accept as true the assumption that "piggybacking" on "relic cosmic background radiation" will reveal the Universe as it was "billions of years ago," is the same "leap of faith" required to believe that "Santa Claus rides in deer-driven sleigh" that flies through the skies.

Physics, to the extent that it endeavors to prove the theory of evolution, is not operating as real Science, but as a branch of atheistic mythology for conditioning Human Beings to a social ideology of "struggle" and "violence."

The theory of evolution is nothing but a "spiritual boot-camp" for training men and women in moral depravity and for conditioning children and young people for reckless activities and destructive behaviors that result only in undermining Cultural Organization and in endangering Human civilized living.

Astro-evolutionists have marshaled their vested interests in "protecting their turf" while fanatically preventing children and youth from learning about alternative scientific approaches, and about other constructive theories that provide an explanation for the beginnings of Human life in the Universe.

But at the same time, post-Relativity Physics is at a mathematical standstill, even as new evolution-steeped "theories," such as "String Theory," only evoke "gravitational rigging," rather than logical "scientific formatting." Now go figure! A "randomly-motivated chaotic Mind" is feverishly pursuing valid physical Laws that prove the Universe to have a pre-determined Organizing Principle for ordering "intelligent operations," the sum of which, structuring "the Continuum flow" of cosmic phenomena and natural processes.

Is it then a mystery that for over a hundred years there has been no other paradigmatic theory following in the footsteps of the Theory of Relativity; and no "totalizing equation" after the example of Albert Einstein's $E = mc^2$?

We cannot make the "leap of faith" that accepts the conjecture of "big bang explosion theory," as the last word on universal beginnings.

But consistent with all the Laws of Physics, as established by our Creator, we will continue to believe in Scientific Creationism as revealed in the Holy Bible. Since God had created the Universe, and since we are created by God "in His own image and unto His own likeness," then, with the Mind he endowed us, we surely can understand physical universal Reality and its organizing Laws, the sum of which, do not contradict, but agree with, divine Creation.

We are not "struggling to prove" Biblical Scientific Creationism, but rather, we are committed to discovering its intrinsic workable laws and operational principles, as intelligently designed, and as "synchronously embedded," within the Universe by Almighty God, the Creator of all Things that exist.

"Big bang theory of universal explosion" and all doctrines trickling down from it, constitute, therefore, a pagan and "godless declaration of Origins" consistent only with the theory of evolution, wrought out, as falsely conceived socio-biological presuppositions, concerning the nature of Human Beings.

But explosive chaos does not yield organized complexity. Randomness does not yield symmetrically attuned iterative patterns of thermodynamic Mass-energy cycling!

It is inspiring to note that all the knowledge accumulated to-date in the "hard Sciences" has been obtained either by laboratory experiments that prove asserted theories, or by serendipitous discoveries "in the Fields of Living Human Existence," made possible by technological innovations from already experimentally proven theories.

The bulk of discoveries have been made through our taking advantage of the microscope and the telescope — which, again, are mechanical extensions of Electromagnetism, for making vision-based real observations, as sustained by formulated understandings, whose connections with realities described and relationships observed, had remained faithful to the Scientific Method, rather than succumbing to ideological fanaticism. In the Knowledge-field of Physics, the bulk of these true-to-Reality scientific discoveries, ended with the fabulous Einstein equation $E = mc^2$.

In addition, all major scientific discoveries in Physics had been made by Scientists whose religious sentiments and spiritual attitudes that "inform moral behavior," were anchored in the Judaeo-Christian Heritage — as previously explained, in the context of the Law, the Prophets, and Christian moral commandments of Righteousness.

It has been reported that Dr. Albert Einstein said "God does not play dice with the Universe." Einstein was not a godless evolutionist. He thought the Universe had too much order and organization to have originated from "random chance." Both Special Relativity in AD 1905 and General Relativity in AD 1915, preceded the Scopes Trials of the 1920s. Einstein died in AD

1955. The theory of evolution became part-and-parcel of Twentieth Century public schools curriculum beginning with the National Defense Education Act in AD 1958. That Act included revision of biological text books to feature evolutionist theory to the exclusion of Biblical Creationism. The emulsification of Evolutionism in our nation's public schools would then continue to the point where its godless doctrines would infiltrate the majority of social institutions.

What has happened since then, — stagnation, disorientation, and futility in post-Relativity Physics theoretics — is that, no true-to-Reality innovative Theory has emerged, to have its principles designed for our creative understanding of universal phenomena and natural processes! The catastrophic results speak for themselves!

As an ideological paradigm, the theory of evolution and the whole doctrinal appendage to its advocacy, shape not only what types of questions scientists are asking, but also how they proceed to research answers to these questions.

As ideology, the evolutionist paradigm adds not one iota of real physically-proven Knowledge to the stock of the "hard" or mathematically-driven Sciences. As such, the theory of evolution is a "Socio-biological theory of Man/Humankind, designed to engineer a specific mindset, which in turn, is expected to coax certain types of activities-and-behaviors in adherents, so as, "to prove it to be true." The evolutionist paradigm is "a closed box" where only spiritual death and mental stultification can "survive" via "the terrorism of peer discipline."

Culturally speaking, "live for today," has become a "New-Age Religion mantra" promoting "physicalism" and "temporalism," — with "no thought for the morrow," thus imperiling the constitutional imperative to our Form of Free Government: In order "to secure the blessings of liberty to ourselves and our posterity."

Astro-evolutionists engage in circular patterns of indoctrination that socially engineer our young people with a worldview fostering close-mindedness, intolerance, willful deceptiveness, and selfish recalcitrance, but for self-idolatry and eventual societal disintegration. For, denying God's existence, the facticity of His love for us, and the self-evident Truth of His caring about "the Human condition," "our Human condition," also contradicts the reality of good and evil, and the reality of right and wrong; and thus, also the reality of the Devil, and the Reality of Heaven and Hell.

Because the evolutionist paradigm is a controlling ideology in the mental apparatus of astro-evolutionists, they have been producing mathematical abstractions that are disconnected from "the real World." Unproven evolutionist assertions, such as the so-called "big bang explosion, "intermediate species," "biological missing links," and "relic cosmic background radiation," and then, "Strings Theory," engender a type of illusion-making functional preoccupation, psychological operations of which, limit the human capacity to abstract conceptual schemes that agree with extant universal processes and natural events.

The vastest chasm exists, when it comes to differentiating Human Beings from the apes, as to our divinely endowed Gift of Spiritual intelligence. Spiritual creativity and moral intelligence are awesome, miraculous, and God-created differences that separate a Human Being from a monkey!

Mathematical abstraction activity extracts from actual processes of Human understanding, a system of symbolic characters that are quantifiably amenable to calculative manipulation, not only in terms of semantic inter-exchanges or intra-relations, but also in terms of inter-conceptual or operational-definitional content of quantified variables that constitutes a valid correspondence between the mathematical equation and the realities, phenomena, events, and processes observed, the sum of which, they purport to explain.

The abstract expressions, attached conceptual descriptions, and definitional operations must match real outcomes for scientifically reproducible applicability, via technological instrumentality. Otherwise, they are in the realm of "virtual" or "imaginary reality"— that is, a pseudo-Reality Form, or "counterfeit Reality simulation," that is only symbolic, as it exists only in the Minds of mathematicians, yet, without any real material substance or physical connectivity anchored to "the Real World of Universal Reality."

Though Physics is a mathematically-driven branch of Science, without a God-inspired spiritual impetus for inspirational discovery, it is not self-sufficient in "logical theoretics" for establishing Continuum coherence and consistency in "objective re-translation" of universal Reality. For, Human Beings do possess, have, own, and live "within an undeniable DNA-need" for spiritual fulfillment and divinely connected Moral Intelligence. But, when an agenda of self-idolatry "enters the picture," only disastrous results characterize "the pursuit of profiteering." As to the true-to-Reality, Science of Physics: It amounts to "the pursuit of utter futility!"

Already proven and validated physical scientific laws and theories — such as the Law of Gravity, the laws of motion, electromagnetism, and the Theory of Relativity, — were products of thousands of years of Judaeo-Christian worldview dynamics, the climax of which, serving as a fruitful, innovative, free-spirited, "no nonsense true-to-Reality, Scientific frame-of-reference.

In sum, We the People, had developed "patterns of relating" to the Universe in "inner-ways" that validated God's omniscience and the unique, exceptional giftedness of Human Beings who are temporarily "passing through a terrestrial spiritual Journey," albeit in a mortal biological vessel, yet, within God's natural Creation, for spiritual fulfillment, scientific discovery, technological maturity, and prosperous living.

Because no evolutional proposition or doctrine is amenable to mathematical formulation while the theory of evolution has invaded Physics — still, a mathematically-driven branch of science, — Physics has become ideological and not "mathematical." Given that scientists cannot merely "sit down to twiddle their thumbs," then, these days, Astro-evolutionists have become "TV-personalities" to the example of Carl Sagan, rather than to the credit of Dr. Albert Einstein.

Yet, the theory of evolution remains a God-denying and God-opposing theory of origins, and hence, impelling it to become a pagan alternative religion that primarily competes with Scientific Biblical Creationism.

Dismantling evolutionist fallacies that have distorted the Scientific Method due to "feats of illogic" is a necessary element of restoring integrity to scientific inquiry without the ideological appendage and godless agenda of social engineering.

We cannot "travel back in time" by "piggybacking" on so-called "relic cosmic background radiation." There are no such things as "biological missing links," because such ideas contradict the Continuum Principle whose operation is uniformly applicable throughout the whole Universe and throughout Nature.

Consequently, the outcome of research that is predicated upon the theory of primeval "relic cosmic background radiation," whose measured and recorded detection is supposed to "embody the universal Past," will be inconsequential when it comes to paradigmatic innovations that must be made in order to propel theoretical physics beyond the applicable limits of the Theory of Relativity in order to "break new ground." The trickled-down theory of the "big bang explosion" is untenable and unscientific; it has remained unproven and invalid in the light of rigorous investigation by the very Laws of Physics the theory of evolution pretends to uphold. Like the theory of evolution, the theory of a "self-starting universe" must be discarded if Physics is to advance mathematically, by sustaining discovery of scientific physical Laws for proven, true-to-Reality, technological applicability.

The electromagnetic properties of Matter get "re-translated" into "structurally-bound Frame conditions" that engender gravi-metric Differentials-and-Opposites, designed for the sustenance of Continuum Gravitational Curvature Motion.

Each "re-translation" or "re-transformation" necessitates the utilization of alpha-numeric Constants to "normalize re-calibration" of "frame Differentials-and-Opposites" for System equilibrium dynamics.

In that manner, equivalents of the Gravity-force substantiate the reformulation of magnetic field motion force in order to link interdependent Frames for Continuum functional thermodynamic cycling, e.g., the atomic Frame with the molecular Frame; the planetary Frame with the solar system Frame; the interplanetary Frame with the inter-galactic Frame. Frame-bound thermodynamic cycling Differentials-and-Opposites overlap each other, in order to energize Curvature-force-motion interconnections that unify Gravitational Continuum Space-Time.

"Energy is never created nor destroyed but always transformed:" Therefore, the operational dynamics of all validly-proven, applicable, and reproducible, physical Laws, tell us that: There is a pre-deterministic iterative universal Oneness-Structure designed for "a Continuum-sustaining gravi-metric Frame," consisted of "Curvature Motion Force Differentials-and-Opposites," characterizing the time-lapsed-duration occurring between the "orbital jumps" of an electron in response to certain activities of molecular formation.

The same gravi-metric phenomena take place to trigger "the perihelion shift of planet Mercury," in response to rates-of-change in gravitational-momentum quantities and intensities of solar emissions, as well as to rates-of–change in Field-force angular momentum and Gravity-force strength-projections, as "added to" rates-of-change in overlapping interplanetary Mass-in-Motion dynamics.

****

Beyond the Applicable Limits of Post-Relativity "Hard" Physics

As examined above, star projected Light is emitted from cosmic Mass, now-existing and now-cycling within Gravitational Continuum Space-Time, and not from past processes that occurred so-called "billions of years ago."

The electromagnetic spectrum can be said to "carry information" only as engineered into mechanical devices utilized for data processing and audio-visual communication. However, Light emitted by stars or Light reflected by planets and other moving Mass frames, cannot be said to "electromagnetically encode" information and data about past universal processes. Stars are not "cosmic recording devices."

Recording devices re-play an electromagnetically supported projection of pre-registered data. As in a movie projector, a beam of Light is not the equivalent of the Source from which it is being emitted. The projected picture on a movie screen is not the equal in substance to the real object, event or process electromagnetically represented, but is an artificial substitute captured or recorded at a certain point in Time for later reproduction.

To assume that this latter audio-visual reproduction is the same as actual travel to the past Reality of the projected scene, is also to assume one could enter that projected scene simply by "piggybacking" on the projector's Light beam. The projector is a physical machine and by definition, its purpose is to project or transmit and transfer unreal images to our senses by a process that does not have any real connectedness with extant, real-Time physical events, objects or processes.

The Sun emits visible Light composed of photon particles; but the Sun's material characteristics that substantiate it as a Star composed of "excited Matter," are accountable for its "cycling nucleated plasma-Energy-Mass;" they are not the equivalent of a beam of visible Light. Stars emit the whole Electromagnetic Spectrum, not merely visible Light.

Flashlights also emit beams of Light; but we can hold the flashlight in our hands. We cannot hold the Sun in our hands. Light "carries information," but the information has no material substance for equivalence with radiating Sources.

In the movie projector, the "carried information" that is reflected towards our retina through the projector's Light beam is only an electromagnetic pictorial reproduction that encapsulates the illusion of motion which appears to give to the scene, the appearance of Reality, but not the substance of Reality. Screen-projected reality is "virtual Reality," having only a superficial resemblance to real things, events, occurrences, or processes. Thus, the Light that reaches us from cosmic distances can also be Light reflections from non-radiating sources, in the same manner that, at night, the Earth's Moon "shines" the Light it receives from the Sun.

Cosmic radiation emitted by universal Mass-frames does not "behave" as a recording device such as a movie camera. It cannot register data encoding for re-broadcasting. Astro-evolutionists are wrong in proposing that extant Light from celestial bodies is "encoded" with past universal data. Their quest to "capture" the so-called "relic cosmic background radiation" in measurable Form with the "help" of electromagnetic-spectrum-based mechanical instruments, is a pursuit in futility.

In contrast, known solar radiation or Light rays being emitted as the universal electromagnetic spectrum, is a real physical event-fact, sustained by real physical gravitational Energetics of the Sun, as projected within the "Present Curvature Time-frame;" meaning that, there is Continuum from the Past to the Present, and from the Present into the Future. Thermodynamic Time, as measuring "Solar Energy cycling," is factually being processed by cosmic bodies, such as the Earth, having their own magnetic fields, real existence of which, in true connectedness to our senses, is not "virtual," but rather, true-to-Reality representative of Energy transformation processes.

Real-time Light-vision, as electro-magnetic wave-particle rays emitted-radiated by a real object in unaltered real-connectedness to our sensory perception, does not continuously reflect an image in our eyes without being sustained by a real-Time, extant, external, ever-present "on-Source;" hence, at night-time, light bulbs with an on-and-off switch, however, substitute for day-Time solar radiation.

Universal Time-travel to the past, or "piggybacking" on pretended "relic cosmic background radiation" of a supposedly billion-of-years old "big bang explosion" is a subjectively self-induced illusion grown out of a misconceived application of Light's function in human visual processes. "Relic cosmic background radiation" is a regrettable offshoot of the theory of evolution, which is masked as a possible "vindication" of so-called "random biological evolution," the sum of which, being the paramount consummate interest of astro-evolutionists. The theory of evolution is designed to challenge the veracity of Biblical revelation of Scientific Creationism.

Though the theory of universal expansion proceeded out of Relativity extrapolations, the "big bang theory" did not originate from mathematical operationalization of the Organizing Principle's "chief-Mechanism:" INPUT-PROCESS-OUTPUT, but rather from explanatory schemes rooted in fabricated and contrived evolutionist doctrines for which there is no true scientific proof or validated evidence.

The current problem in mathematically-driven "hard Physics" is to find real connectedness between relationships-of-Force and logical mathematical themes arising there-from that will attempt to conceptualize these "Field-and-Gravity Relationships" between Mass-energy bodies "transacting" within the Space-Time Continuum through rotary centri-vectored Forms of Motion-Force as analytically framed for measurable, quantified, unified, integrated contiguous Oneness.

Field-and-Gravity Motion-Forces, from the sum of which, emerge a "tapestry of equivalent Forces," keep the Universe "running the right way," e.g., affording the Earth "to retain" its natural satellite, the Moon, in "quasi-self-sufficient Curvature Orderliness, designed-created for sustaining Earth life-support Systems, as well as for "distancing" each Space-body within "safe Relativity interactions" that afford them, respectively, each one's own "Self-identifiable Cosmic Signature," e.g., Mercury has a "perihelion Shift;" "Like charges repel; opposite charges attract:" Yet, the Hydrogen Atom only has one nucleic positively charged Proton; and even though, absent a Neutron, still, the sole negatively charged Electron revolving around Hydrogen's "singular Nucleus" does not "crash into" the nucleic center where stands the oppositely charged Proton.

As embodied in "Continuum Motion-Force Curvature," magnetic field effects upon Mass-energy bodies-in-Motion are "re-transformed," "re-translated," or "re-transmuted" into Gravity-force "equivalents" that activate thermodynamic Mass-energy cycling, beginning with the Quantum Mechanical Frame, passing through the Planetary Frame, to climax into the Solar-System Relativity frame-of-reference, in wholesome or holistic integration with the astro-cosmic solar-planetary ellipsoid plane that is constituted of a "multi-factorial Force-Frame" as conjoined with inter-galactic gravi-metric relations, the sum of which, holding the Universe together in unified Oneness Integrity.

These Frames interdependently overlap in uniform internal consistency and integrated operational dynamics as the integrated "Input-Process-Output mechanism" engenders specialized patterns of System equilibrium Energy cycling.

How do we proceed from the Law of Universal Gravitation, the Laws of Motion, Electromagnetism, Continuum Relativity Space-Time Curvature, and the Law of Energy Transformation, to enter then into "the Realm of Mathematical Theoretics" that will tangibly embody and concretely symbolize the incontrovertible unification-relationships which "connect-link-bind" all Motion-forces that work together in "concerted synergistic-symbiotic operation," even as they are "held accountable" for harnessing and executing "the thermodynamic cycling" of Mass-energy-bodies "moving" within the vast expanse of our extant Universe, into an Orderly Oneness Organizational Complex?

The Universe is fundamentally constituted of: Matter and Space, and their intrinsic properties, as they are "undergoing iterative thermodynamic changes" during measurably identifiable, computable, periods of Time!

Thus, regardless of the ways in which cosmological processes and cosmographic events are "compartmentalized" for purposes of study and analysis, we understand that from the Atom to the Sun, and from the Solar System to the Galactic Frame: The Universe has-owns-possesses integrated Oneness via uniformly transformed-transmuted "gravitational Curvature tensor-binding pressure-force dynamics," — all valid physical laws work similarly where conditions are relatively identical — "converted modes of which," are only replicated Forms of iterative Mass-energy states as impacted by projected Field-Energetics and Gravity Force Equivalents, the sum of which, re-iterating in the Form of "congealed frames," the same Form of "solar electro-magnetic, radio-gravi-metric, thermonuclear Energetics," e.g., Given that "Energy is never created nor destroyed but always transformed," then, petroleum oil is merely "another Form of Solar Power;" so is the phyto-chemical "Process of Plant Photosynthesis!"

The Sun emits "thermonuclear gravi-metric Field-projected radiation," Curvature Energetics of which, take the forms of magnetic field force, gravity force, strong force, and weak force, in order that specific Mass-in-Motion Frames, "bound-captive" in differentiated "states of Energy cycling," can "thermodynamically re-translate" via "the universal INPUT-PROCESS-OUTPUT principle, the electromagnetic properties of Matter, geared for "Frame-specific" dynamic System equilibrium, e.g., Earth ecological processes are "Solar-Radiation dependent." Still, the Earth-Moon planetary System is in "dynamic equilibrium" as a self-contained, "closed," but not an "isolated" System.

Field-forces active within the Atom and the Sun, within the Solar system and the Milky Way Galaxy, and even within Earth-gravity dynamics and Moon-orbital motion mechanics: All, partake of the same "Gravitational Curvature pressure-tensor-force Dynamics" that operate "to bind" the Proton and the Neutron together, and "to cause" Electrons to revolve around the atomic nucleus, in such quasi-deterministic ways that allow, not only (1): The operation of molecular bonding energies within respective "proximal particulate distancing;" (2): Appropriate "bonding-Force distributions" that still allow conditions conducive to "specific elemental densities," respectively; (3) But also, "neutron capture," and "electron escape," for forming heavier isotopes, and for transmuting particles to Positrons.

Thermodynamic cycling variables, Curvature pressure boundaries, Gravity force equivalents, Field force analogs, and co-determinant Relativity momentum-Motion patterns/paths/trajectories: All, "interweave," "intersect," "interface," and "intermingle," within "webs-of-overlapping-entanglements" in order to "potentiate" the strong force and "calibrate" the weak force, such that the "Electromagnetic properties of Matter" get "re-translated" into "Mass-energy structural units" that co-labor in-consort to fulfill Frame-specific functional equilibrium, e.g., Earth Life-support systems and Ecology are titivating at optimum vivification, as designed for life-saving operations that sustain the prosperity of Humankind, even as Mercury is undergoing its periodic perihelion Shift.

In applying the Continuum Principle, the Thermodynamic Principle, and the Equivalency Principle, astro-physicists and mathematicians need to begin, from "the common Curvature Fundamentals" that permeate all Mass-frames that are cycling in gravitational Continuum Space-Time. Matter-mass radiation-Energy frames, cycling in Curvature- motion, are fulfilling the "Input-Process-Output mechanism" characteristic of the Organizing Principle, geared for Frame-specific functioning, e.g., the Sun gravitationally radiates; the Earth ecologically sustains Life, just as "electromagnetically- potentiated Field-force projections" engender centri-vectored rotary Motion Forms (centripetal/centrifugal, "with-Gravity entanglements") that climax into Whole Mass-energy Systems due to operations of "the strong force" that sustain Energy Conservation for predominance over Entropy tendencies as generated by "the weak force."

Conceptualizing a comprehensive paradigm that bridges the Quantum, Newtonian, and Relativity frames-of-reference for innovative inclusion of Magnetic-field Energetics and Gravity-force Dynamics and "equivalents," would be configured via interposition of alpha-numeric Constants that "normalize" or "reconcile" strong force and weak force Differentials-and-Opposites, the sum of which, embodying "in condensed form," the "electromagnetic thermodynamics" of universal "Continuum tensing-pressure-Force power" — (the ultimate Field-gravi-force equation explaining why "things clump together") — as enhanced by "conversion factors" that assist in establishing "Relative entanglements of Equivalency."

It is self-evident that the Universe, in all the diverse richness of its "entangled overlapping structured Frames," does "stay together" as "held-unified" by a multiplicity of co-equivalent but opposite "Force-effects that exponentially factor," not only in "energizing" its functional operations, but also in "preventing" or "forestalling" its immediate dissociation or sudden disintegration.

By all observations and appearances, Conservation is a "much stronger" predominant process than the incremental inception of the "tendencies to Entropy." After all, though containing within ourselves the ultimate corporeal final terminal "process of Entropy" that will eventually climax into "death," we slowly develop, grow, and thermodynamically prosper, to then mature into "old age" until the "Wellsprings of Electro-Chemical Generation" that "feed" our "Organic Energetics" are exhausted through proverbial "wear-and-tear" and the end (death) comes.

In short, apparently, we do have a "long lifespan" as compared to other biological beings on the Earth. To put this "in context," it is a fact that many Human Beings live to be over 100 years old!

How do all electromagnetic properties of Matter get "re-translated" into "Field and Gravity Factorials" that synergistically sustain Curvature pressure-force Motion tensor-metrics in order to engender "the strong atomic force" and "the weak molecular force" that simultaneously co-labor to operate in sustenance of specialized functioning of thermodynamically-driven Frame-cycling Energetics?

Due to their interconnectedness, indivisibility, interdependence, and overlapping "range-of-influence" in unifying Continuum Space-Time: The Quantum Mechanical Frame, the Newtonian Celestial Mechanics Frame, and the Space-Curving Relativity Frame-of-reference, all, have to be integrated in "Simultaneous Synchronized Symmetry," by "A Unified Theory of Universal Motion-Force" that correlates Curvature pressure-force Motion-Forms with "Field-tensing" and "Gravity-stressing" analog-equivalents, the sum of which, accountable for "Curvature Binding Energetics," i.e., Space-bodies are "bound together" in continuous-contiguous rotary Forms of centri-vectored Motion as induced by "equivalents" of both Electromagnetic Field Force and Gravity Motion-Force, and "combinations thereof" or "composite forces thereof."

"Motion-force equivalents" allow us to walk, run, jump, or stand-at-ease upon the Earth, while our "body-Mass" is exerting skeleto-muscular force that is "greater than G-1 earth-gravity-force parameters;" yet, all these "force equivalents" are "composite forces" emerging from the "overlapping Relativity-interactions Complex" in which are "entangled" both Electromagnetic Field Force and Gravity Motion-Force, i.e., as engendered by Revolution and Rotation.

In reality, there is no "structural separation" between these Frames; and if there appears to be "compartmentalization," it is "artificiated" in the Human Mind by conceptual schemes that narrow the scope of inquiry to a reductive matrix of Evolution-based possibilities.

Is not the Motion-Force of Gravity "the analog-equivalent" of "Field Tensing Energetics" in the same manner that a Field is "an emergent property" of Electromagnetism?

Does not Gravity appear to stand as "a different category of Force" due to the absence of "particulate catalysts" accountable for generating "vectoring mechanisms" that arise from "Curvature Motion Energetics?"

In ferreting out "A Unified Theory of Universal Motion-Force," there are Fundamentals that we must keep in the foreground, even as we engage in formulating a coherent paradigmatic framework for Mathematical Physics Theoretics:

(1) "Energy is never created nor destroyed but always transformed;"

(2) Einstein discovered "the Equivalency Principle" that prevails between Matter-Mass and Energy.

(3) All proven-valid Laws of Physics governing the material Universe are equally operational in every region of Space where gravimetric conditions are similar;

(4) There is "Integrated Continuum Symmetry" between all Mass-Frames that are in Motion due to "force equivalents" characterizing the "range-of-manifestation entanglements" to which both Electromagnetic Field Force and Gravity Motion-Force are subject, the sum of which, producing "quasi-self-iterative structural operational similarity" between such Mass-Frames in-Motion, e.g., Electrons revolve around the atomic nucleus; planets revolve around the Sun; the Solar System revolves around the Milky Way Galaxy; the Milky Way Galaxy revolves around the Andromeda Galaxy.

Gravity-Force, as we experience it on the Earth is "a weak force," in the sense that it is globally permeating in effect as a "cumulative resonance" of centri-vectored Curvature pressure-force, the sum of which, re-translated as a "Frame of Dynamic Trans-sonance" that correlates all rotary motion Forms into "coherent trajectories," as applied within boundaries of our earthly physical environment, e.g., centripetal, centrifugal, curvilinear Forms of Motion and "combinations thereof" or "composites thereof," as executed on Land (walking; or driving an automobile), in the Air (flying an aircraft), and in Water (sailing on a ship; piloting a submarine under Water).

The strong nuclear force and the weak molecular force unite atomic particles in forming aggregate mass frames as emergent properties of electromagnetism. The strong proton-neutron force is structured to facilitate Conservation of Energy as the weak force is structured to allow for molecular formation in processes of energy production, accumulation, utilization, consumption, and emission.

Electromagnetism is a fundamental property of Matter. Mass in motion "re-translates" these electromagnetic properties into Forms of Force that must serve to "bind" Energy into specialized frame cycling patterns of transformation.

The Speed of Light, as a "motion vector" in direction and magnitude "embedded" within the Electromagnetic Spectrum (as wavelength, cycles per second, photon energy in electron volt units) proceeding from solar core-centric Mass, Field, and Fravity, is the limiting factor for all phenomena, events, Laws, and processes.

Gravity is not only a force but it is also a "motion vector" as it is "centri-focused." No one and nothing can breach Speed of Light boundaries; no one and nothing can breach its "exponential function operations" as a "constant" or "conversion factor" in the equation: $E = mc^2$, in accordance with which, Energy cycling has boundaries as delineated by the Mass to Energy ratio being the equivalent of c^{-2}, by way of $m = E/c^2$, yielding $m/E = 1/c^2$, or c^{-2}, which would conform to the "congealed status" of planetary frames. Thus, where "excited Mass" is negligible in a unit of Mass, so is the amount of Energy present for "conversion" from that unit of Mass.

In accordance with the ratio of Mass to Energy, given that the value of c^2 remains as a "fixed constant" as required by the Equivalency Principle established by the equation $E = mc^2$, then, results are that the Earth "embodies" very little nucleated mass energy. However, the Energy to mass ratio, being the equivalent of c^2, or $E/m = c^2$, would characterize the essential nature of the Star, in the sense that a tremendous amount of Energy can be obtained from a little unit of Mass, e.g., a one-megaton Hydrogen bomb.

Thus, in order for Energy to be produced so as to embody any Form of potency, a certain amount of mass ("m") must be present. Consequently, where Energy must be present, "m" cannot equal zero.

Mass therefore is inversely proportional to c^2, as c^2, is inversely proportional to mass. The greater the potential for nucleated chain reactions, the more "excited" the frame, e.g., a Star; the less potential for nucleated chain reactions, the more "congealed" the frame, e.g., a Planet.

The Earth is a "congealed frame" wherein atmospheric Hydrogen constitutes less than 2 percent of its volume, and oceanic Deuterium is "trapped" within sea waters that are also composed of sweet water and salt compounds.

If the Earth were "billions of years old," as astro-evolutionists presuppose, and if the earth represents "cooled cosmic nuclear big bang plasma explosion residue," then, would not Uranium and Radium have decayed to inertness within such a vast span of unfolding Time? Why would Radium and Uranium exist today within "Earth congealed frame" to radiate cosmic energy analogous to solar radiation emissions? Why is Radium radioactive, while "embedded" within an "earthly congealed frame," yet without having participated into nucleated chain reactions common to Hydrogen in the Sun? Could Uranium and Radium have originally been created so as to emit cosmic radiations while belonging to a "congealed planetary Energy frame?"

Or is there an on-going process within Earth geology, unfolding operations of which, climax unto their transformation into radio-active isotopes, in the same manner that there is a geological process giving rise to fossil fuels? Under what conditions could both processes "thrive side-by-side" within the "bowels of the Earth?"

The Sun not only has vast quantities of "catalytic particulate constituents" for transmuting Mass into nucleated plasma due to the tremendously extreme heat-temperatures pressurized by the Sun's powerful magnetic field, but these particulate constituents also are also part-and-parcel of its great Matter-mass. Extremely immense nucleated atomic radiation and plasma ionized condensate Energies are compressed by the Sun's highly potent magnetic field via exponentially powered temperature and pressure Differentials that yield numerous counter-vectored convection Forces, the sum of which, accountable for engendering gravitational Curvature force dynamics throughout "the heliosphere."

As field energy involves tensing and stressing, pushing and pulling, bending and curving, attracting and repelling, contracting and expanding within Curvature metrics of torque-power pressure force Energetics, Gravity is expressed as a "Force of attraction" between centers-of-mass that "modulates" into many "equivalents" taking the Forms of centrifugal and centrifugal variants.

As the Sun projects a potent magnetic field that causes the Earth to revolve and rotate, Gravity Motion-Force resonates as "an emergent property" of Curvature tensor-metrics causing centri-vectored Forms of motion along the projected path of Electromagnetism where develops a "matrix of variant Motion extensions" as dependent upon application of additional acceleration Force from whatever sources derived, e.g., an aircraft can, via the use of various quantities of engine thrusts, perform many feats of aero-acrobatics that "capture" the sum of all curvilinear motion-forms along a 360-degree axis of Rotation.

In the same vein, within the atomic frame, as Curvature pressure stress motion resonates into field force energetics, field tensing resonates into gravity force analogs that "re-translate" the electromagnetic properties of matter into micro-Curvature Differentials where the strong force and the weak force operate with "equivalent intensity" for maintaining electron orbital-distance fidelity. The force required to compel the electron to revolve centripetally around the atomic nucleus is relatively equivalent to the force tending to push it outwards centrifugally away from that same nucleic center-of-mass, composite operations of which, keeping the electron in its proper orbiting level.

The proton-neutron force that "solidifies" bodies-with-mass is a cumulative effect of electromagnetism as particulate charges constituting Matter interact to form gravitational attraction-repulsion dynamics. Electrons "push against" the nucleus to "compact" proton-neutron bonding into a strong force. And as they are simultaneously repelled and cannot "escape" from micro-Curvature pressure force, they engage in nucleus-centric motion.

The nucleic strong force whereby protons and neutrons are held together is a micro-encapsulation of Curvature Energetics that generates Gravity tensing and Field force stressing "equivalents," specific to the quantum mechanical frame. "Gravi-metric elasticity" of the atomic micro-orbital plane displayed as variants of range-specific vectored curvilinear Motion Form, is interpreted as "random waves of probability" because of Mass-induced motion-variants within compressed-Time micro-Energetics, the sum of which, defying instrumentally applied calculations designed for measuring Quantum Mechanical Frame parameters, such as inability to precisely determine simultaneously, both the position and momentum of a quantum mechanical particle.

Because centri-vectored rotary Motion-forms are primary, both the strong nucleic force binding protons and neutrons together and the weak force allowing electron orbital motion-variants as well as molecular change and radiation emissions, holistically "resonate" as Gravitational Curvature "binding energy," operational exertions of which, climaxing into the formation of greater mass frames.

Atoms then accumulate to form Matter via the exchange of "energy states" as driven by negatively charged electrons that revolve around atomic nuclei, due to a force that is weaker than that between protons and neutrons. Electrons jump from one energy level to another, even as molecules are formed to cumulatively aggregate into greater bodies-with-mass.

Thus, it is easier for an electron to change energy level than it is for a proton to be separated from a neutron, just as it is easier for a man to escape from the pull of Earth gravity than it is for the Moon to escape from Earth center-of-mass and center-of-gravity.

While the proton-neutron complex maintains its stable constitution within micro-Curvature-metrics that stabilize the atomic Frame, it is electrons that "jump from one energy level to another," even as molecules are being formed to accumulate and aggregate into greater mass frames.

Thus, the atom "thermodynamically cycles" as well. But because the strong proton-neutron force prevails over counter-centric forces or centrifugal forces, Motion is fundamentally centri-vectored as Conservation of Energy overcomes entropic decay.

Thermodynamic cycling "embeds" forms of resource replenishment that involve production, accumulation, abundance, utilization, consumption and restoration at the same time that Conservation and Entropy sustain one another's processes, geared for fulfilling required parametric values that climax into dynamic System equilibrium. However, there are "thermodynamically terminal products" which are considered as "waste," in that they have lost "the capacity to be transformable into Energy" that can do useful work, e.g., Styrofoam, or they are "transmuted into toxic thermodynamic Forms" that are harmful to Human life, e.g., dioxin, carbon monoxide.

Atomic nucleus forces and molecular change bonding dynamics embody both Conservation and Entropy via thermodynamic cycling of elemental components for composite formation of substances and compounds. In that manner, the Law of Conservation of Energy is sustained from element to element (via the strong force) while "electron exchange" allows for chemical reactions that trigger energy production and utilization (via the weak force), as thermodynamic cycling for consumptive Entropy processes occur within frame-specific operational conditions of specialized functionality, e.g., a combustion engine "consumes petroleum Energy" in order to produce traction that allows for vehicular mobility.

We witness within "iterative redundant forms," bounded patterns of range-specific Frames of energy transformation, even as conservation-entropy processes are cycling from frame-to-frame, e.g., the Atom is a constituent of all functioning ecological Frames on the Earth even as the Sun's nucleated fusion-fission processes are radiating gravitational Energy. As acted upon "by gravity equivalents" of electromagnetic field force, the complex entanglements of which, yielding Continuum Space-Time Curvature pressure-tensing Motion Forms, every universal Frame operationally functions in consonance with synchronized simultaneous Unified Oneness Integrity that is characteristic of Gravitational Continuum Space-Time. Though Gravity Motion-Force is an emergent property of Electromagnetism, it is Gravity that "controls" for maintaining dynamic System equilibrium in Mass-in-Motion Frames, albeit operating within their specific ranges-of-Force-manifestation or "spheres of influence," e.g., It is no accident that this Star which we call "the Sun" can only "keep" nine Planets within its own "sphere of influence" or "range of Force manifestation."

"Gravity Force equivalents" or "the body-Mass binding energies" prevailing as micro-Curvature pressure force in the atomic nucleus constitute "a greater force tensor" than that operating at the molecular frame, due to "nucleus-centric entanglements" of particulate charge polarity and Space-displacement motion forces, as induced by centers-of-mass, centers-of-field and centers-of-stress Energetics.

Thus, it is easier for an electron to change energy level during molecular formation aimed at "congealed Frame coalescence," than it is for a nucleic proton to be separated from a nucleic neutron even during application of extremely high temperatures-and-pressure Forces required by a nuclear chain-reaction event. Likewise, it is easier for a man to escape from the pull of earth Gravity by walking and running, than it is for the Moon to escape from earth center-of-mass, center-of-field, and center-of-gravity.

Therefore, within the Gravity Frame-of-Reference, the greater the Mass, the greater the Force of Gravity; but within the Electromagnetic Field Frame-of-Reference, the greater the strength of the Field-Force, the more Mass must decrease, hence, the apparent "controlling prevalence" of Gravity at the Solar System Frame and of Electromagnetism in the Quantum Mechanical Frame, i.e., cause-and-effect mechanisms, the sum of which, explaining why an electron revolves around the atomic nucleus faster than a Planet revolves around the Sun.

However, the above-described "Frame-Specific Operations Complex" belonging to every Mass-in-Motion frame constituting the Holistic Universe, co-exists and co-labors, within the same unified overarching Frame-of-Reference: The Universal Gravitational Space-Time Continuum.

Both the strong force in the nuclear frame and the weak gravity force at the planetary Frame — within the Earth and within the Earth-Moon Frame — are indivisible and inseparable within the gravitational field Curvature Reality of Continuum Space-Time.

While the strong force at the Quantum Mechanics Frame is apparently dependent upon Electromagnetism, however, due to negligible particulate micro-Mass quantities, Electromagnetism is "the equivalent" of Gravity. Nonetheless, at "the intermediate level" where Mass-frames are more distinguishable from each other as relatively independent functioning "units of thermodynamic transformation," Earth surface Gravity Force and Earth-Moon Complex Gravity force "equivalents" are apparently more Mass-frame dependent, e.g., the greater the Mass of an object upon Earth surface, the greater the centripetal force of Earth gravity exerted upon it.

Yet, respective to the Sun's exceedingly great magnetic field Curvature pressure-Force and the emergent Gravity Force it generates, the Equivalency Principle that correlates Mass to Energy, must also hold, regarding the force of Electromagnetism and the force of Gravity. Consequently, the "total Sum of Energies" required from the Sun to keep the Solar System "in operational Oneness Integrity," is relatively equal to "the Sum of all the Energies" consumed by all the Planets in their respective Frames of manifestation as they thermodynamically cycle in fulfillment of the Input-Process-Output principle of Energy Transformation. Output by all the nine Planets is relatively equal to Input from the Sun to keep them in orbital Motion upon the Solar System ellipsoid plane of Revolution.

Consequently, the interplay of proportions and ratios of Electromagnetism field-force to Gravity force at the Solar System frame-of-reference respective to amounts of Mass, would be accountable for Planet-specific Gravity-force and rotation period, e.g., If we hold Earth to have Mass:1, Gravity as G:1, and rotation period upon its own axis in 1 day as r:1, then, given that Jupiter is 317.8 times more massive than the Earth (greater Mass); it exerts Gravity equating 2.36 times that of the Earth (greater gravity); and completes 1 rotation period upon its own axis in .41

days (faster rotation), which would direct us towards concluding that indeed, all other parameters such as distance from the Sun and Revolution period held as "constants," the Equivalency Principle respective to Gravity Force and Electromagnetism Force is being observed, in the sense that the ratios and proportions to which they are exerted must correlate to the properties of Mass-in-Motion relative to all projected co-determinant Sun parameters and variables.

We would also expect that, the number of Moons consistently "held captive" by each Planet that does have Moons, respectively, would correlate to each Planet's particular Mass-in-Motion properties, relative to co-determinant parameters and variables as-above considered as "constants."

Planetary gravity, e.g., Moon geosynchronous orbital motion, is the macro-expression of micro-electromagnetic properties of Matter, e.g., electron revolution around the atomic nucleus, as structured and engineered by magnetic field Energetics that activate Mass Relativity Gravitational Curvature dynamics.

Field tensing forces causing rotary motion patterns emerge from charge polarity dynamics inherent in the electromagnetic properties of Matter. Minute or negligible micro-gravitational curvature force, micro-field force energetics and their "emergent equivalents," account for pre-determined mutually attuned elemental molecular change, e.g., H_2O.

The strong electro-motive gravi-field micro-atomic charge-forces (positively charged proton/neutron nuclei), "framed" within electron-driven centri-vectored motion forms (electron bonds), combine with negatively charged electron binding-energies as "calibrated" by atomic nuclei interactions, in order to form molecules, the sum of which, then bonding/binding together to constitute greater bodies-with-mass.

Micro-nucleic atomic forces, e.g., the strong force, the weak force, field force, and gravity equivalents, have greater binding strength than the magnitude reflected by progressive macro-molecular accumulations and macro-gravity mass expressions of force, for it is the quantum mechanical frame that constitutes the infrastructure of Energy Conservation. A piece of cloth can be torn into many pieces, while the proton and the neutron "to the last cloth atom" will remain "clung together."

Atomic nuclei preserve processes of Energy Conservation akin to the ways in which thermodynamic cycling properties of electrons "able to traverse" from one energy level to another, allow for Entropy processes to unfold as well, even as greater mass agglomerations are being formed, after which, the iterative patterns structured by operational processes of Energy Transformation begin again within frame-specific conditions.

All particles participate in sustaining the electromagnetic properties of Curvature binding Energies from frame to frame. With variables of temperature and pressure within "closed-chamber volume conditions activating chemical reactions, elements having a mutually attuned affinity for molecular bonding can exchange electrons to form other elements or compounds whose interactive dynamics do not deconstruct proton-neutron relationships, except in nucleated processes whereby the tremendous amounts of heat-pressure momentum Energetics cause nuclei fusion, dissociation, or neutron release.

Cumulatively, atomic nuclei preserve the electromagnetic properties of Matter for "transmutation entanglements" that engage recombinant frames of energy cycling. Entropy processes cycle, even as Energy is being transformed into its equivalent forms or states, such that Mass-frame Differentials sustain Continuum resource production, processing, and utilization for functional thermodynamic Conservation of Energy.

Both the strong force at nuclear level and the weak force at gravity level are indivisible and inseparable from field and Curvature binding Energetics that hold the Universe together as a whole thermodynamic Energy frame. Gravity is not only an emergent property of field-induced centri-vectored forms of motion, but it is also the macro-expression of the strong nucleic forces that unite in micro-binding energies to form molecules, such that "freed electrons" at the outer orbital levels, bond, respectively, in order to constitute greater bodies-with-mass.

Consequently, quantum and molecular frames, as well as Newtonian and Relativity frames, should not be compartmentalized by abstract mental systems, but rather should be unified in an integrative conceptual scheme that genuinely attempts to resolve the gravi-metric "cycling subtleties" that appear to differentiate them. For, it is the dynamic engendered by overlapping Frame Differentials and Opposites, complex Synthesis of which, together "re-combining" to "give unification" to the Universe, via operations and processes in order that the Universe continues to function as it does. Their thematic indivisibility and framed interconnectedness help in interfacing tensor-metrics and stress-dynamics engendered by Opposites and Differentials accrued within the Relativity Frame, as implied via Gravitational Continuum Curvature Energetics, the sum of which, to be symbolized by mathematical field equation theoretics that explore a Unified General Theory of Universal Motion-Force, positing, establishing, and operationalizing: Relative equivalency between Field-force manifestations and Gravity-force expressions.

Gravity does not display any particle or wave characteristics akin to those that are properties of the atomic frame and the solar nucleated frame. Gravity does not partake in any characteristics of electromagnetism. Gravity is an emergent property of gravitational centri-vectored forms of Motion as potentiated by magnetic field binding energies effecting Curvature pressure tensing force that causes planetary frames to move in response. The Sun's core plasma condensate activities prompt Mass-Frames "to gravitate" towards each other in ways that cause energy cycling, even as Curvature tensor-pressure-force exerts upon the Earth a concurrent Motion imperative to revolve and rotate.

Field-Forces, as summoned by relative interactions between the strong force and the weak force, are emergent analogs or "simulated replicates" of the electromagnetic properties of Matter. Electromagnetism is the constitutional foundation of Matter, as it is activated by particulate electro-motive charge displacements that are "re-translated" into field-force forms — electric fields engender magnetic fields, and vice-versa.

Gravity, however, is not present in the absence of centri-vectored rotary forms of field-generated Curvature tensor-pressure-force Motion.

The Newtonian frame, the Quantum frame, and the Relativity frame have been explored in depth. The electromagnetic spectrum, electromagnetism, and nuclear research have provided insights into workings of the Solar Frame. Accelerated particles in cyclotrons appear to have generated problems concerning the interpretation of "mass energy gains" that parallel or mimic the difficulties inhering in "re-translating" electron "behavior" as "waves of probability."

How do following Frames: Newtonian Mechanics, Relativity Dynamics, and Quantum Mechanics, interconnect, overlap, interface, and transact for Human extrapolation towards the formulation of a Universal General Theory of Motion-Force?

Problems encountered in post-Relativity Physics would seem to be research-result dependent, on the one hand, and on the other, paradigmatic in nature; that is, being a "problem of observation."

Would acquisition of "new data" cause a change in astro-evolutionist assumptions regarding universal beginnings and the essential nature of cosmic events and natural processes?

There also appears to exist, a problem in "grasping" or "taking hold" of the data necessary for a Unified Theory of Motion-Force. Yet, the data and reality to be observed and analyzed are in the Universe for us to discover. However, paradigm postulates, methods of measurement, technology being utilized, theoretical mind-frame, and the gravi-metric entanglements within environmental dynamics of research itself, appear to impinge upon observation, experimentation, data collection, and result-oriented analytical deductions. Perhaps there needs to be a "psycho-spiritual affective revolution" that will then inspire researchers in reinvigorating the hope for theoretical discovery.

The necessary spiritual transformation must come from within, as a change in the perceptual apparatus of the researcher, in order to effect a new direction and pursuit. We were born as little infants and found the Universe "as it is." Thus, we cannot "re-engineer Nature," but rather, must discover its physical laws, understand its natural processes, and decipher its cosmic phenomena.

There must be a change away from the evolutionist paradigm that attempts to annul or neutralize the intrinsic coherence of Continuum physics. The invention of a new astro-paradigm would account for the omniscient powers of Almighty God in endowing the Human Family with co-creative capacities for innovation and discovery.

All reference Frames, from the quantum mechanical to the Newtonian, from Relativity to electromagnetism, appear to have been pushed to their limits, hence, the quest for a General Theory of Force.

Must a "General Universal Unification Theory" be necessarily in terms of a general theory of Force? Must a general theory of Force be necessarily in terms of Field dynamics? If so, then how do we describe or define a "Field?"

What constellation of Energy conditions would give rise to an "Electromagnetic field" as opposed to a "Gravity field?" How does a Field emerge? What is the "constitutional foundation" or "Fundamentals" of a Field?

In what ways is the Force of Gravity differentiated from instrumentally perceived manifestation of a Field-force?

Electromagnetism from which emerge the strong force and the weak force is made manifest "in analog forms" or "as equivalents" of the Gravity Force, the sum of which, being analog forms of Curvature tensor-pressure-force Motion energetics.

All types-kinds or "Forms of Force" operating within the astro-planetary Frame, such as Quantum mechanics, Newtonian mechanics, Electromagnetism Dynamics and Relativity Energetics, will have to be taken into account in order to establish "proportional equivalency" or "ratios" between their cause-and-effect manifestations, by mathematically formulating "conversion factors" or alpha-numeric "Constants" that would "normalize" or "reconcile" respective Integrative Mass-in-Motion frame-cycling Fundamentals: Differentials, Opposites, and thermodynamic processes pertaining thereto.

Is not a line a continuous succession of individual dots? Is not a beam of Light a continuous stream of individual photons? Therefore, integration of Differentials, Opposites, and Thermodynamic Energetics is crucial to discovering the fundamental pathways taken by the Input-Process-Output principle, via the numerous "mechanisms of Energy Transformation" that re-translate the Electromagnetic Properties of Matter, even as Mass-in-Motion Frames overlap "in continuous quantum functioning" for Universal Unified Oneness.

****

ELECTROMAGNETISM AND GRAVITY AS MASS-DEPENDENT "EQUIVALENTS"

The force of Electromagnetism or the Strength of the Magnetic Field-force being projected by a unit of Mass-in-Motion, is "relatively dependent" upon:

a. The distance between the opposite bi-polar extremities of the cycling unit of Mass-in-Motion, e.g., Earth Arctic and Antarctic; the Sun's opposite magnetic poles, the sum of which, possessing opposite electric charges, respectively, and operating to cause centri-vectored Forms of rotary motion, e.g., Rotation and Revolution by the Earth as caused by the Sun's nucleated electromagnetic-gravitational radiation Energy.

b. The amount or quantity of Mass "contained" within the object, Space-body, or unit of Mass-in-Motion projecting the force of Electromagnetism, e.g., The Sun has much greater Mass than the Earth — More than one million Earth-size Planets can "fit inside the Sun."

c. The Form of Energy that is thermodynamically cycling or being transformed by the unit of Mass-in-Motion projecting the Electromagnetic Field-force, e.g., The Sun's form of Energy is nuclear Energy, and hence, its projection of a magnetic field force of such great strength as to keep nine Planets, among which is the Earth, revolving around its center-of-Mass-in-Motion.

The "power" or "torque" or "strength" exercised by said Mass-in-Motion unit yielding rotary forms of centri-vectored motion corresponds also to the Form of Energy being transformed by that Mass-in-Motion unit, e.g., the Earth is "able to keep" one natural satellite or Moon, whereas the Sun is "able to keep" nine Planets around its "sphere of influence," "sphere of manifestation," or "area-of-force projection."

Not only has the Sun greater Mass than any of the nine Planets, but the Form of Energy cycling therein is obtained from thermonuclear reactions undergoing many differentiated thermodynamic processes such as thermonuclear transformation, convection, transmutation, and compression, the sum of which, projecting gravitational field and magnetic field "torque-power" or "force-strength" of such great magnitude as to cause all the Planets to coalesce around it as they travel along its ellipsoid plane of Revolution.

However, the Force of Gravity and its "equivalents" or "g-Forces" are not a "primary force" but an "emergent force." Gravity receives its properties of motion-force from the Electromagnetic Field-force, out of operations of which, it emerges, e.g., centripetal and centrifugal curvilinear Forms of Motion-force that are vectored or "attracted" towards a center-of-Mass, and/or composite combinations thereof, e.g., a rock thrown in the air will fall back towards Earth center-of-Mass; a pebble thrown upon the surface of pond will bounce numerous times until it sinks to the bottom of the pond towards Earth center-of-Mass; or centripetal and centrifugal curvilinear Forms of Motion-force that are vectored or "repelled" away from a center-of-Mass, e.g., an aircraft can achieve an altitude or distance away from Earth center-of-Mass, which can be a centrifugally vectored motion-force climaxing to ten miles in Earth atmospheric Space.

The Force of Gravity, therefore, is present within a unit of Mass-in-Motion, e.g., the Earth, as an emergent property of Motion-force when there are centri-vectored forms of Motion-force, e.g., Rotation and Revolution, as generated, caused, or engendered by a corresponding unit of Mass-in-Motion projecting a greater electromagnetic Field-force, e.g., the Sun's great magnetic field force, acting upon the Earth, thus causing Earth Rotation and Revolution, from which emerge Earth gravity-force and its equivalents.

Why does Gravity-force emerge from Electromagnetic Field-force? Because of the fundamental constitution of Matter, Electromagnetic Field-force is primary: It precedes and supersedes Gravity-force. Without the presence of Electromagnetic Field-force, there cannot be any Gravity-force.

Matter is constituted of particles having electro-motive charges causing "space displacement Forces," — because these charges "attract" each other when opposite, and "repel" each other alike, the composite synergy of interpolated charges, respectively, causes these particles to move in rotary Forms of Motion-force, such as those animating the strong force and the weak force, which on the one hand, hold the positively charge Proton and the no-charge Neutron to "join forces" in "sharing" the atomic nucleus; and which on the other hand, impel negatively charged electrons to revolve around said atomic nucleus. Concurrently, these particulate charges engender a dynamic "area of manifestation" within which, are simultaneously projected: 1) An electro-magnetic field projection; 2) A Gravity-Motion force exertion; and 3) The emergence of "gravity-force equivalents" causing irregular effects that deviate from expected atomic patterns of Motion-force.

A gravity-force "equivalent" usually exerts "a form of influence" that deviates in vector and effect from regular Gravity-force Motion patterns. This "deviation" or "skewed Motion-force" — demonstrating a departure from regular Gravity-force Motion patterns, i.e., centripetal, centrifugal, and/or rectilinear composites thereof within the "geo-Sphere of the Planet,"— arises from the composite-combined synergy that accrues from Electromagnetic Field-force and Gravity Motion-force symbiosis. For example, Gravity-force on Earth is G:1; however, Gravity-force "equivalents" compel the Earth to "wobble" to a degree that enhances its "slightly flattened" irregular spin at the Poles, which in turn, triggers "slight bulging" at the Equator.

In the same vein, within our Solar System's "three-dimensional Curvature Energetics volume" that engenders the "bending" or "curving" of certain proximal "regions of Space," which in turn elicits planetary "responses" thereto, then, thereafter causing formation of certain "areas of influence" that impact Planet Mercury: This Mass-in-Motion dependent "gravito-magneto symbiotic synergy of field-Forces" exerts such "a dynamic" upon the Continuum of Space-Time within the vicinity of Mercury, that its complex operations temporarily synthesize the pertinent "Gravity-force equivalents" that are accountable for the periodic perihelion shift of the Planet.

Electromagnetism gives rise to "the strong force" (Proton-Neutron/Proton-Proton/Proton-Electron/Atomic Nucleus-Electrons) and "the weak force" (Electron "bonding" or "pairing" for formation of molecules and electromagnetic spectrum emissions characterizing nucleated or non-nuclear thermodynamic processes), summative operations of which, yielding Field-force variables and Gravity-force co-determinants that activate centri-vectored Forms of Curvature tensor-pressure-force Motion, from which emerge "composite-combined equivalents" of the "gravito-magneto Force," e.g., solar magnetic field climaxes into Earth Revolution (helio-centric) and Rotation (geo-centric), from which emerge, Earth-surface "equivalents" of Gravity-force (Earth core-centric) that are either consonant with G:1 ("free-falling"), or greater than G:1, such as the multi-forced and omni-vectored curvilinear acrobatic feats that can be performed by pilots with a well-designed and properly equipped aircraft.

Field-forces involve causative electromagnetic particulate charges that interact via "displacement-motion forces," as explained above. Consequently, as an emergent Force-property, Gravity does not display any particle-driven characteristics, though in Mass-in-Motion operations, its effects are analogous to "field-caused interactions," and hence, physicists oftentimes referring to Gravity, as a "Field."

A field-induced form of motion means that there exists a "theater of interactive influence" or "a sphere of transacting influence" between the Mass-in-Motion Frame projecting the Field-force and objects or Space-bodies that Field-force causes into Forms of centri-vectored rotary Motion patterns.

Gravity-force is described as "an attractive force" between centers-of-Mass; but, as noted above, "Gravity-force equivalents" can also cause "repulsion" (centrifugal) rather than "attraction" (centripetal), or composite Motion-forces thereof.

Though Gravity-force-caused Motion is analogous "in exertion" to a projected Field-force-caused Motion, — in the sense that Gravity also "acts upon Mass-frames" to a comparably similar extent as would an Electromagnetic Field, e.g., the Moon is held by the Earth in orbiting motion even as the Sun "holds" nine Planets within its ellipsoid plane of Revolution; — Gravity operations are differentiated as co-linear and co-equal Force-forms from Field operations, in that, Gravity does not originally display an electromagnetic basis, e.g., no "Gravity-particles" are being emitted.

The Earth does have "a gravito-magnetic Field" that causes Moon axial rotary motion and revolutionary motion, as encompassed by properties of Continuum Space-Time Curvature Energetics, yet the "attractive force" between the two bodies, as originally depicted by Newton, is understood to be solely center-of-Mass-dependent. However, in the Earth-Moon Relativity Frame, both magnetic Field-force and "Gravity-field" force are "Curvature-entangled" within the "rotary Motion-force action-dynamics" that are vectored towards Earth center-of-Mass and Moon center-of-Mass as conjoined together at the "barycentre;" that is, at the point where the two centers-of-Mass "intersect." That barycenter or point of intersection is located inside-within the Earth at about 1707 kilometers below its surface.

Is it possible to differentiate between the two forms of Curvature force exertion — Electromagnetism and Gravity? Moon orbital motion is due to gravitational Curvature pressure-force motion-dynamics, tensor-metrics of which, "calibrated" for dynamic Earth-Moon System Equilibrium in response to "the synergy relationship" engendered by "the all-planets gravito-magnetic fields ensemble" acting upon it in-conjunction-with solar gravitational radiation Energy "bathing" the Earth-Moon-frame Complex. As intimated above: Electromagnetism and Gravity forces are co-linear and opposite in directional vectoring within the Atomic Frame; but they are "relatively equivalent" in "exertion strength" within the Quantum mechanical frame-of-reference; and hence, the relative stability of the Atom as a "stand-alone" unit of Mass-in-Motion Energy-Cycling Frame. However, differentiation between earth magnetic Field-force strength and "equivalents of the Gravity force" is more easily effected for instrumental detection, when framed within the structural operations of Earth-surface Motion by objects "powered-thereupon" for movement away from Earth G:1 gravity-strength, e.g., an automobile traveling in curvilinear Motion from Oakland, California to Springfield, Illinois, even as Earth magnetic field force-strength and Earth Gravity:1 are held as "constants."

As a Field involves properties of Electromagnetism, Earth surface Gravity is an emergent property of centri-vectored forms of Curvature Motion-force engendered by Electromagnetic Field-force.

To reiterate, Gravity is "relative" to Mass-in-Motion Energetics, whereas the Force of Electromagnetism is "relative" to the "distance dynamics" between oppositely charged polar extremities, — oppositely charged poles of the Unit of Mass-in-Motion projecting such an Electromagnetic Field possessing "a limited range of strength-force exertion" that in turn generates "Mass-in-Space-Displacement-Motion," i.e., Opposite polarity charges will cause generation of Motion due to "attraction" between opposite charges (Positively-charged Proton and Negatively-charged Electron) and "repulsion" between like-charges (Negatively-charged Electrons), the complex-exponential mutual exertions of which, "co-labor together" to create the "Field" or "area of manifestation in Continuum Space-Time" for the "Force-exertion" within the "proximal vicinity of the field-projecting Mass-in-Motion," where objects placed therein begin to engage in rotary Forms of Motion, such as Rotation and Revolution. Electrons revolve around the atomic nucleus due to such a dynamic engendered by oppositely charged particles — namely the positively charged Proton and the negatively charged Electron — as the Neutron's presence effectively establishes "a respectable distance" that prevents these two primary particles from "merging" or "crashing" into each other.

The only Element not possessing a Neutron in its atomic nucleus is Hydrogen, and hence, reasons why it is considered "the First and lightest Element."

In Hydrogen, even in the absence of a Neutron, the sole negatively charged Electron does not "merge" or "crash" into the sole positively charged Proton, due to minute micro-nano amounts of Mass-in-Motion Differentials between the Proton and the Electron, as impacted by "indistinguishable micro-nano Force-equivalents" exerted by both Electromagnetism and Gravity.

Commonly, for other "heavier" or "more massive Elements," where the Neutron is present in the Nucleus, Mass Differentials between the Proton/Neutron Nucleic Complex and the Electron(s) — in conjunction with the Neutron's "relatively massive gravitas-presence," — hold all revolving Electrons "within a respectable distance from each other" and "within a respectable distance from the Atomic Nucleus," even as indistinguishable "equivalent Forces" exerted by both Electromagnetism and Gravity, compel all atomic particles "to properly behave" into atomic stability that is essentially geared for Dynamic Elemental System equilibrium. And, hence, the necessary "gravitas-presence" of the Neutron(s) in the Atomic Nucleus of "heavier Elements" in order to prevent negatively charged Electrons from "merging" or "crashing" into the positively charged Protons, e.g., The Element, Helium, has two Protons and two Neutrons in its atomic Nucleus and two Electrons revolving around its atomic Nucleus; it is the presence of the no-charge Proton that prevents the two oppositely charged particles — Protons and Electrons — from "clashing" with each other. Though the Neutron has no charge, and hence, is electrically inert, it plays a dynamic role in stabilizing and equilibrating the Atom for "Continuum preservation" of Energy, and by that, thwart "the advance" of Entropy by avoiding a "short-circuit," which would occur, were the revolving Electrons to "crash into" the nucleic Protons. In that capacity, the Neutron might be said to serve as "the nucleic instrument" by which the whole Atom is "mechanically grounded."

How are Gravity-force, Field-force, and their respective "equivalents," related to Curvature energetics? Given that "the strong force" and "the weak force" are specialized "analog re-translations" of the electromagnetic properties of Matter, as "embedded" within synergistic manifestation of a Field-force, — Electromagnetism has an intrinsic basis in interacting opposite electrical charges that engender Field Motion-force properties, — then, it is Gravity that must be mathematically formulated into a "tensor-like Force" or "torque pressure-force," parameters of which, would encapsulate "Curvature-stress dynamics" of Mass-Energy cycling, as units of centri-vectored Motion-force, e.g., Earth-Moon Motion-force Energetics.

Gravity does contribute to "holding Forces" that keep the Moon in its geo-centric synchronous orbital trajectory due to "attraction between centers-of-Mass;" however, it is Earth magnetic Field-force strength that engenders Moon rotary Motion patterns of axial rotation and revolution around/with the Earth.

The Moon orbits the Earth due to Curvature forces as controlled by Earth center-of-mass, center-of-gravity, and center-of-Field, cumulative Energetics of which, engendering centri-vectored Forms of rotary motion; however, solar system ellipsoidal plane "tensor Energetics" accounts for a constellation of "pressure-stress Forces" that cumulatively climax into planetary location and Moon positioning, hence, the 23-degree Earth axial tilt necessarily fixated to accommodate the Moon's Mass, which amounts to one quarter of the Earth's own Mass. And in turn, Earth axial tilt accounts for solar gravitational Field-radiation Energy to only "graze the Earth" at the poles, hence, allowing for polar icecaps formation, the sum of which, playing a crucial role in maintaining Earth regional zone temperatures within livable ranges.

Planets revolve around the Sun and the Moon orbits the Earth due to projected solar gravitational Field-force, as predicted by Newton's Law of Gravitation, and as extended by Einstein's Relativity Theories. Thus, "attraction-repulsion mechanics" and "bending-curving dynamics" go "hand-in-glove."

Though "differentiated-isolated" by methods of scientific analysis for purposes of Human Understanding, the "attractive gravitational Force" and the "Relative bending-curving of Space" occur contiguously and simultaneously — Newton and Einstein complement each other. Hence, Newton's discoveries are antecedent to discovery of the Laws of Electromagnetism by Maxwell and Faraday. Still, due to the Equivalency Principle discovered by Einstein, Gravity is also a derivative of Field Curvature Energetics emerging from emitted-projected nucleated solar gravitational radiation Energy, the complex interplay of which, activating, so to speak, the "bending-curving of Space" as formulated by Einstein's Theory of Relativity.

Consequently, in ferreting out a comprehensive unifying Theory of Motion-force that will explain how the Solar System operates, in its parts and as a whole, — the Solar System being conceptually-theoretically representative of "a micro-Universe" as an analytical-mathematical platform from which discoveries can be "extended" in application to "the Universe-proper," — Newton, Maxwell-and-Faraday, and Einstein, as complementary contributors to our understanding of numerous physical Laws of the Universe, should not be "separated-isolated," neither conceptually-theoretically; nor analytically-mathematically.

Inter-planetary-solar centers-of-Mass, centers-of-Field, and centers-of-Gravity, in totality with their "composite equivalents"— as engendered by the Sun, the Planets, and by the Solar System as a whole — also contribute to Earth and Moon locations, positions, and patterns of centri-vectored rotary forms of Motion-force, "as embedded" within the ellipsoidal revolutionary "tensor-Plane," the complex configuration of which, providing an intricate gravi-metric instrumental System-mechanism that is accountable for, not only perturbations such as Earth "wobbling," "Earth bi-polar oblation," and "Earth equatorial bulging," but also for the perihelion shift of planet Mercury as well as for the "eccentric" orbital trajectory of planet Pluto. For, the fact is that all these apparently differentiated anomalies or deviations are occurring within the same Universal Continuum Reality, but which apparently is "dissociated," only in the Human Mind.

Thus, these apparently different Forms of cosmic phenomena lay on a spectrum of correlated-connected universal structures or Frames, summative, synergistic, and symbiotic relationships of which can be the only basis for generating a theoretical-mathematical framework, from which to launch a scientific paradigm for discovering "The Unified Motion-force Theory of Universal Continuum."

Nuclear plasma radiation Energy traveling with momentum Motion-force Curvature Energetics leaves the Sun to impact Planets positioned within the ellipsoidal plane of Solar System Revolution. The Sun's "excited Mass-energy cycling state" reveals nucleated quantum dynamics of fusion and fission, compression and distention, contraction and expansion, the omni-directional centripetal-centrifugal complex configuration properties of which, climaxing in "convection dynamics" as encapsulated in the mathematical equation $E = mc^2$. Solar thermo-nuclear plasma condensate processes are sustained via reaction cycles that ensure "Continuum gravitational-Field-equivalent radiation Motion-force emissions" as emulated within the intrinsic structure the Electro-magnetic Spectrum that is indicative of all "elemental nuclear reaction-chains" holding the Solar System together. Thus, spectrographically obtained color-lines or "spectral lines" that mark the existence of certain periodic-table elements serving as "Sun nuclear fuel" are demonstrative of "element-fuel cycles" that successively enter nucleated Energy chain reactions for gravitational radiation-supported field Curvature projections-and-exertions.

Integration of all universal Frames: Quantum Mechanics, Newtonian Mechanics, Relativity Dynamics, Curvature Energetics and Continuum Properties as engendered by Electromagnetism-and-Gravity within a unifying paradigm and absent the need to append socio-biological doctrines trickling-down from the Theory of Evolution, would indeed free the Scientific Method of "excess ideological baggage."

All universal Frames display Mass-structures, respectively, as Units or Entities of thermodynamic Energy-cycling-Motion-force. Because Force and Motion are Mass-dependent, they are also indivisible, and hence, the term "Motion-force." Continuum range-specific accelerated-momentum Curvature-tensor-pressure-force Motion, as limited by the fixed value of the Speed of Light, i.e., Our Solar System can "accommodate" only nine Planets, is the relative foundation that must undergird a theoretical framework aspiring to analytically and mathematically unify all Physical Forces "embedded" within Gravity, the Electromagnetic Field, and their "composite equivalents" that animate Nature, and the known Universe.

****

MIND REVOLUTION: Apprehension of Continuum Universal Reality from Spiritual Discernment to Objective Conceptualization

We are spirit-beings in a mortal bodily vessel. Spiritual life influences scientific activity. Moral living and civic virtue impact the direction of technological applications. Rules of the Road encourage lawful and danger-free driving habits. Drunk driving costs lives and damage property.

Beneficent individual and social utilization of technological inventions flow from constructive moral values instructing conduct that affirms Human Life from God-inspired spiritual Liberty.

"Energy is never created nor destroyed but always transformed." Electronic machinery, Human electrolytic sensory modalities, and the Universe made-up of "Electromagnetic Matter-mass" (i.e., Protons, Neutrons, Electrons etc) in Continuum Space-Time, are operationally constituted of electrical impulses (Human Nervous System), or of some type of radiation Light Energy that fits right into the Electromagnetic Spectrum (e.g., Electron Microscope).

Can the Human Mind perform such a prodigious feat of symbolic abstraction in theoretical Physics, as in applied Mathematics, that will propel Science beyond the limits of post-Relativity principles, given that all research events, methods, and processes have to rely upon mechanical instrumentation that exploits the Electromagnetic Spectrum — the same Spectrum upon which Human vision and mentational processing are based? Thus, in words that "flirt with reductionism," the question being asked is: "Can Light-radiation-energy pursue and catch Light-radiation-energy?"

Given that all natural events and universal processes have "limits" that embody-display "ranges of effectiveness" within which take place the operational dynamics of the Input-Process-Output mechanism, the sum of which, animating Energy Transformation for cycles of Conservation and Entropy, then, human inventions that take advantage of the Electromagnetic Spectrum, e.g., electronic technologies of research, also possess "interface limitations" or "interference effects" that no innovative correction can amend — unless the Equivalency Principle is "operationalized" in ways that "reconcile" or "normalize" the Mass-in-Motion differentials, Motion-force opposites, and "Relativity equivalents" between Electromagnetism and Gravity that sustain Energy production and radiation propagation within the holistically integrated universal Space-Time Continuum.

Do electromagnetically-based operations of research technologies "interfere" with "electromagnetic data," being studies, e.g., in reaction chambers or particles colliders?

Because of "limits" imposed upon all Forms of Energy pertinent to "cycling Units of Mass-in-Motion," e.g., Earth-Moon Complex, the Solar System, by the fixed value of the Speed of Light, then, as for all things in Nature and the Universe, technological functionality of electronic machinery also has "ranges of effectiveness" or "boundaries of application;" hence, the proverbial need to build "a more powerful telescope," or "a more powerful microscope" having greater magnitudes of resolution, which commonly proceeds by "embedding" as many types of "energy units" with as many differentiated frequencies-and-wavelengths as possible, as provided by the Electromagnetic Spectrum, e.g., X-ray telescopes; Electron microscopes; Magnetic Resonance Imaging; Positron Emission Tomography, etc...

After all, our vision as activated by visible Light has perceptual limitations, and hence, our invention of these devices, the microscope and the telescope. When we look into the heavens, we notice that even distant stars possess "luminosity limits" or "heliospheric boundaries" from whose faint constellations we only perceive "points of Light."

In accordance with the "Thermodynamic Organizing Principle" that "holds the Universe together as One," overlapping and interdependent Mass-in-Motion differentials and oppositely vectored but co-linear Forces functioning as "equivalents" of Electromagnetism and/or Gravity, are accountable for "re-translating" the "Electromagnetic Properties of Matter" within the boundaries pre-established by the Speed of Light.

Remember Schrodinger's thought experiment with the cat! Per various interpretations, the cat could be "dead or alive," or "alive and dead" — until the researcher-observer opens the box. The purpose for mentioning "this thought experiment" is that, according to Heisenberg's Uncertainty Principle, a researcher cannot simultaneously determine a particle's position and momentum. The "Schrodinger thought experiment" embodies a "problem of Mind" or a "problem of abstract Understanding" rather than one of physically obtained data evaluation. The status of the cat would be independent of the researcher's observation, in the same sense that the Moon's rate of axial rotation would be independent of "the kinetic Frame" of the researcher on the Earth, except that the researcher must account for "Motion Differentials" and "Force Differentials" that would impinge upon experimentally-mathematically derived results.

Does the Human spiritual capacity for "intuitive apprehension" play a role in Experimental Physics analysis? Can this "psycho-affective-domain mystery" in all the depths of its abstract intangibility, "connect" or "interface" with "Universal Fundamentals," — material processes and cosmic phenomena that reach into the deepest essential physical roots of universal constitution, structure, composition, and organization?

What if the Atomic Frame (Quantum Mechanics) and the Star's Nucleated Frame (Solar System Energetics) constitute both micro-limits (the Atom) and macro-boundaries for "Mass-in-Motion in Thermodynamic Energy-Force Cycling" (the Star), — as we are "traveling" from the Atom to the Sun, our "bounded journey," just as imposed by the fixed value of this universal standard: the Speed of Light in a vacuum?

Can we escape from "Speed-of-Light boundaries" already imposed by inevitable applications of "The Thermodynamic Cycling Principle:" To all "things, phenomena, processes, and events" that physically-materially exist?"

In short, fundamentally: "What if there is "nothing after" the Atom and the Sun?"

How far can Human Knowledge go, in a Universe that has Electromagnetism for its Foundation?

Are we not only studying: "The Electro-Magnetic Properties of Matter" within Continuum Space-Time?

The Human Brain is innervated by electrolytic interactions between neurons that effect "information-processing," via synaptic cleft connectivity and receptor-specific chemical binding properties. Electrical impulses in the brain, electricity in lightning, electrical current running throughout the wiring of a house, and electro-motive dynamics of the Atom in either "congealed" or "excited" states, electrical charges animating a computer, electricity driving the nuclear dynamo of the Sun, etc . . . , are only "converted states" or "transformations" of the same Form of Electro-Magnetic Energy that is common to all universal Frames. But these Matter-mass Frames possess "re-translated" electromagnetic Differentials operationally-indicated as "Force-opposites," for examples, are "registered" by voltage, wattage, and amperage, as structured within specific Frames designed for specialized functions.

In short, the Study of Matter-Mass within the universal Space-Time Continuum is only the Study of the Structural Organization of the Electro-Magnetic Properties of Matter!

As Humans, we are Beings who find ourselves in inevitable relationships with the natural environment. Though we are spiritually intelligent Human persons in biological flesh, we drink the water that comes from the wellsprings of the Earth's water table; we breathe the very atmosphere from which rain, snow, and storm systems originate. Thus, if we see ourselves from a "Bio-thermodynamic point-of-view," we can then ask the question: Could "one Form of Energy" with creative intelligence, pursue, find, and understand "another Form of Energy," in order to discover how the essential constitutional characteristics of Electromagnetic Energy are structured and organized for universal whole-System functioning?

A "percept" is an "intuitive catch" by the Human Mind, of a certain event, process, or phenomena. From such an apprehended phenomenon, an apparent "precept" or rule, law or principle can be fashioned, for "embodiment" into a "concept." A concept is an idea or thought regarding the phenomenon's formation, structure, organization, and operation, as applied to real Forms. We proceed from a percept to a precept, the complex synthetic intersect of which, culminating into a concept.

Thus, mental abstraction can be encapsulated into formation of a percept yielding a precept to arrive at a concept. These abstract cognitive processes, though limited in prescriptive scope, are "entangled with" and "embedded within" the Human Soul as "objective categories in Human imagination" which work together to engender a pre-conscious Form of Understanding that is "fundamentally intangible in nature," called "Spiritual discernment."

"Spiritual discernment" is also a Scientific Form of Apprehension that yields the objective formation of a concrete Principle for transforming abstract perception into organized Universe-anchored conceptualization.

We are Spirit-beings in a mortal bodily vessel. Spirit has pre-eminence over mortal biological flesh just as conceptual planning precedes objective project development. For, concerning all endeavors geared for acquiring Knowledge, it is Form, Blueprint, Template, Pattern, Code, Rule, Law, or Principle that always precedes conceptual embodiment for practical or experimental application designed for Purpose and Hypothesis-testing, as well as for analytical results, and subsequent technological invention.

Abstract intuitive-mental-cognitive processes precede concrete objective embodiments as technologies. Ideas give rise to activities that initiate creative pursuits. "Intuitive discernment," as abstract thought processes towards conceptual embodiment, always precedes physical discovery of phenomena for material-physical applications.

Why do matters of "spiritual abstraction" always precede matters of "concrete physicality?"

God is Spirit! He had to first exist, in order think the Universe into being, or before He spoke it into materiality.

Spiritual abstraction precedes physical concreteness in the same manner that an experimentally validated Theory precedes a pertinent-corresponding functional technological application. Mental analysis of data is of an intellectual nature that is in character and essence different from merely witnessing, observing, and compiling data. Compiled observations must be scientifically analyzed. And this is, in essence, an abstract theoretically-anchored enterprise that climaxes into objective conceptualization of the Scientific Method for deriving conclusive results from research findings. Neither domesticated apes nor "wild beasts of burden" can do such miraculous things!

As unique Beings created by Almighty God, our mMind does possess this analytical power to integrate, differentiate, or "unify opposites," within observed data, for hypothesis formation and Scientific Method application.

Constructive objective scientific analytical capability is rooted in the status and role we ascribe to our spiritual capacity for learning, thinking, discovering, and creating. Through moral education, we are encouraged to pursue individual maturity for constructive, lawful, just, peaceful, and productive living. Contrary to biased evolutionist dogma, monkeys cannot go to church to worship God nor go to school to learn algebra! It is really a disgraceful defamation of Human character to promote "ape ancestral lineage."

In short, Spirit matters. And for us, whether acknowledged or not, "Matters of the Spirit," do predominate in all our activities. Thought processes arise from mentational operations performed upon external inputs, the sum of which, springing-forth from our inner-being.

Our Mind also participates in fulfilling requirements of the Input-Process-Output mechanism (sensory inputs/schooling), animating the Organizing Principle that constitutes and drives "the dynamics of Human Understanding" (cognitive processing/learning), in ways that edify spiritual maturity, moral creativity, and emotional intelligence (abstract activities that concretize beneficent behavioral responses to external inputs, now "internalized and processed," for operationalized planning, and applied production of technological inventions.)

Dogmatic partisanship amongst astro-evolutionists has tended to thwart openness to new ideas, and receptivity to new theories that may yield paradigmatic transformations that are so direly needed for transcending current sterility in theoretical astrophysics.

The "religion of secularism" permeates the scientific culture; and "the spirit" that "its core values" tend to engender is the "spirit of intolerance." From the outset, technological applications proceeding from "Darwinian struggle" and "Malthusian genocide" for purposes of population control appear to be "fueling the fires" of war-making for "conquest of spoils."

From "the genome project" to "fetal stem cell research," Evolutionism is the godless, yet systematic paradigm that controls formulation of procedures for scientific research, due to "turf protection" ensuing from collectively coerced partisanship, operations of which, discouraging thoughtful moral discernment and curtailing beneficent individual expression into an oppressive "esprit de corps" that stifles fact-finding dissent and suppresses truth-seeking research activities.

Issues or matters of eternal significance are brushed aside as Evolutionists cater to destructive human tendencies that instead do need the rudder of spiritual instruction and the path of moral guidance.

Temporalism or "only today matters," engenders "spiritual waywardness" or "a state of despair" that only leads to moral indifference. And "physicalism" or "only the flesh matters," corrodes inspirational urges of the Soul to aspire to "newness of Life" as the Mind succumbs to transient but deadly hedonism.

This "controlling Darwinian spirit" has orchestrated a lawless apathy to human suffering and decadent coarsening of the Culture. Godless policies that encouraged irresponsible behavior have caused the disintegration of the nuclear family, the abandonment of mothers and their children by fathers, and the pursuit of selfish hedonism.

There is also a form of "crass materialism" that glorifies spiritless flesh as "a religious ritual of carnal existentialism" encapsulated in the saying "the monkey with the most toys wins," lowering Human Worth to "spiritless instincts of apes in the jungle."

"Physicalism" and "Temporalism" eventually lead to cruel attitudes that tend to give credence to the evolutional myth that Humans descended from "cosmic dust," the amoeba and then, the apes, thus relegating Human Beings to a status deserving of no more affirmation than these animals elicit in the human heart.

The evolutional worldview is dehumanizing and demeaning. It degrades and devalues Human worth and the lives of children. Its godless doctrines bind the human spirit, destroy inspirational motivation, and consequently limit individual achievement. Evolutionism is spiritual bondage! Evolutionism is biological genocide!

And at the same time, Evolutionism imperils the moral incentive for lawful, beneficent, constructive, loving, just, and peaceful Human activity — acquisition of material things from whatever means derived, becomes an end-in-itself that justifies laziness and rationalizes criminality.

Evolutionism endangers the innate capacity of young people to develop required God-endowed sensibilities that would prompt them to mature into righteous leaders of tomorrow.

Evolutionist doctrines destroy the good Judeo-Christian spiritual impetus morally designed for scientifically driving our righteous innermost desires towards engagement in activities that improve "the human condition," thus, practically nullifying our inheritance of sacred blessings that had already empowered us, by divine Providence, to own an edifying worldview that prospers our lawful liberty and improves our moral living towards fulfilling our deepest yearnings towards forming "a more perfect Union."

Evolutionism is debilitating Humankind into anarchic hopelessness. For, succumbing to "doctrines of Temporalism" and Physicalism" endangers our spiritual capacities for moral greatness as measured by the comprehensive Good our constructive individual and social activities might achieve on behalf of ourselves and our society-at-large.

Attitudes ensuing from "Temporalism" and "Physicalism" tend to deny the eternal significance of Human freewill, because of fatalistic conclusions regarding "intractability of Human nature."

But God empowers our spirit unto changed hearts or transformed inner-being, as we are called to righteous acts that justly prosper our personal estate and peacefully support our societal welfare. Chosen action from morally-inspired given consent ought to be encouraged so as not to promote "group think" geared for "standardized responses" to ad hoc rules of social interaction and political participation.

Evolutionism engenders "mass-produced personalities" for exploitative, cynical, and divisive control. "Labeling" becomes a substitute for individual character development. While freewill choice in individual action remains the most sacred of all God-given blessings to the Family of Humankind, actions do have consequences, for good or evil, underlining the fact that the stark Reality of Human Spirituality bears eternal significance for the destiny of Humanity in the Universe.

It is true that God created us "from the dust of the ground of the earth," to which we return, after we "pass away." But, we are "more than dust." For, the Holy Bible declares we are created by Almighty God "a little lower than the heavenly beings." Likewise, the Declaration of Independence of the United States of America boldly states that it is a self-evident Truth that our Creator endowed us with certain inalienable rights, among them being the Rights to life, liberty and the pursuit of happiness. It is also self-evident that no mortal Human Being of flesh and blood can ever take them away. For, that which mortal men give away, that, they can also "take away."

Thus, given that mortal flesh-and-blood cannot establish any eternal thing that endures forever, then, only our invisible, immortal, omniscient, and immanent Creator, Almighty God, is the sole Author and Giver of our Rights. For, these inalienable God-endowed Rights can never be "made alien" or "be separated" from us, as Persons, Human Beings, and Citizens. Violation of these inalienable Rights by other Human Beings, either expressly or negligently, does not entail that they do not exist or nor means that they are not inalienable. Human Beings have a "Fallen nature" and a "fallible constitution;" and thus, they err and falter. However, so too, through the redeeming sacrifice of Christ Jesus on the Cross, God has made a way for amending and correcting these errors and mistakes. As we confess our sins, our repentance gains us forgiveness for a clean-slate beginning "in newness of Life."

Our inalienable God-given Rights are eternal in value, worth, breadth, and duration from generation to generation. Given that we have these imperishable God-endowed blessings, we ought to be inspired, motivated, incentivized, and impelled from-within, to live our temporary existence on the Earth, peaceably, justly, lawfully, righteously, morally, constructively, and joyfully, even as we partake of "the more abundant life," with comprehensive soul-prosperity that sustains us in full anticipation of our heavenly inheritance: Eternal Life in Christ Jesus.

Our culture must change paradigm. There must be an internal heartfelt inner-transformation that will renew our Minds and reinvigorate our spirits, for the rekindling of our Hope and our Trust in God, our loving Creator. We must overturn the evolution-rooted godless worldview that degrades Humanity to the spiritless depravity of wild apes — in order to restore psychic integrity and inner-coherence to the Human heart, for "re-connecting" Human life with genuine meaning, true dignity, real integrity, and beneficent purposes for living as unique creatures whom God our Creator has deliberately set apart from wild beasts and animals, from inanimate physical Nature, and from the material Universe.

We must regain our authentic selfhood as children of God by reviving our relationship with our Creator through our Faith in Christ Jesus who rose from the dead to empower us with God's Holy Spirit for prospering on the Earth as "the righteousness of God." Otherwise, the Good that we desire, the blessings that we seek, and the happiness that we yearn after, will remain ever-unattainable.

Consequently, we must renew the Human Mind in the Goodness of God and in the love of Christ Jesus, so that our freewill no longer serves as an ominous scaffold for cultural suicide, societal disintegration, and violent genocide.

Undeniably, we are both bio-physiological and spiritual in fundamental constitutional Human Nature. Thus, in true-to-Reality terms, we are both spiritual and scientific, and thus, intrinsically capable of both invisible abstraction and visible objectivity. True, we are innately flawed. But we do have the moral resources by which to elevate the fulfillment of our natural needs above irrational passions and deadly desires.

Therefore, to "know our rightful place in the Universe," we must attend to both our physiological and spiritual nature, to both our intellectual and biological characteristics, and to both our moral and behavioral fulfillment. Mind and body are indivisible, inseparable, normally and naturally unified in "entangled identity" and "joint operations:" Biological-energy wise, we are naturally thermodynamic; but because of our inner-moral constitution that activates, informs, and instructs our freewill, we are also essentially Spiritual-Beings in albeit mortal bodily vessels.

It follows, then, that, given the nature of our true natural constitution as Spiritual Beings, Evolutionism is an evil dogma of aggressive struggle that does spiritual violence to the Human Soul. Evolutionism exploits Human weaknesses and faltering tendencies in order to instill hopelessness, resignation, cynicism, misery, and despair into the Human heart for purposes that are inimical to our personal well-being and general welfare. But God empowers us by His Spirit in Christ Jesus for love, peace, joy, lawfulness, justice, creativity, productivity, and prosperity.

"Immorality in Science" or "the misuse of Science" as accompanied by exploitation of our God-given natural resources and societal institutions, has been driving public education, government policy, and social programs: From the Genome Project to Stem Cell Research; from abolition of school prayer to "legalization" of infanticide by abortion; from disintegration of the nuclear family to the promotion and propagation of sexual perversions like homosexuality and lesbianism.

Spiritual morality does matter in a big way! And it is crucial that Human Beings in decision-making positions be imbued with the moral character and spiritual maturity demanded by such awesome responsibilities. We must return to foundational Judeo-Christian spiritual principles, godly commandments, and moral values that inspired the miraculous Founding of the United States of America.

There has been fanatical attempts and a feverish movement perpetrated by astro-evolutionists and bio-evolutionists "to chain our culture" to the tenets of "Darwinian struggle for survival of the fittest," as entangled with their agenda to emulsify the evolutional paradigm regarding universal beginnings and the creation of Human Life on the Earth. In their evolutional attempts to engineer an "enterprising mob," they have sought to devise and enshrine godless government policies as "the norm" for our whole nation, while violating the liberty-securing republican principle of self-government and the federal principle of limited government, which in totality, frame the constitutional foundation of our representative, electoral, democratic republic.

In seeking a new paradigm, we need to rejuvenate our society as "a culture of life." Scientific activities must be designed "to celebrate Life" for the constructive purposes of Human edification that does not disparage the integral value of the core Judeo-Christian principles and God-given commandments, i.e., the Law and the Prophets, and Christian principles and standards of Right and Wrong and of Good and Evil, which have, since the time of our Founding in 1776 AD, affirmed our social heritage and historical development for the Good of our posterity, e.g., technologies devoted to dismembering and killing Human babies in their mothers' wombs should be abolished.

America's Founders designed our form of free government as a model for Humanity, with this prudent and cautious characteristic: decentralization, separation, and division of government's constitutionally granted powers.

This form of free government affirms scientific study and observation of "decentralized" natural processes, cosmic phenomena, and universal events that proceed in accordance with physical Laws. Those physical laws govern, as perceived in the 18th Century due to discoveries by Isaac Newton, how the Universe is "held together" in accordance with the Law of Gravitation.

The Law of Gravitation discovered by Isaac Newton has not been made void by Einstein's Theory of Relativity. Rather, they complement each other, as each is applied, as pertains to the material-physical conditions "calling for" operations of the former, or of the latter, respectively.

Atoms retain their integrity as Electrons bond to form molecules. Elements and compounds take forms that can be analytically detailed in their constituent components. The Solar System comprises nine Planets and a Star unified-in-whole, via an ellipsoidal plane of Revolution.

Each planet also thrives as "a Whole Energy Sub-System" with its own cycling patterns that fulfill the Input-Process-Output mechanism geared to sustain its own Mass-Frame Differentials and Opposites in dynamic System equilibrium. In the same way that it is uniquely created God-endowed individual Persons who form the greater social family called the Nation, individual Planets form that to which we refer as the Solar System.

The Earth is the only life-planet in the whole solar system. Valid physical laws do not annul but complement each other in cumulative, overlapping, and consistent application. As a unifying astrophysical paradigm is being sought, it does not imply the "melting away," or disintegration of Mass-Frame boundaries and their overlapping "spheres of influence" or "spheres of manifestation." Therefore, Quantum Mechanics, Newtonian Mechanics, Relativity Dynamics, and Electromagnetism-Gravity Curvature-Field-Energetics will retain their applicable validity, even as a new theoretical framework is conceptualized towards Universal Motion-Force Unification.

The Atom is still the smallest indivisible quantum or unit of Matter, though Atoms are necessary in the formation of molecules and greater agglomerations of Mass; the Law of Universal Gravitation is still operating even as Relativity Dynamics engender Curvature-tensing Motion-force in Mass-Frames engaged in thermodynamic Energy cycling.

Thus, all experimentally proven-valid physical laws inter-relate, as Energy-cycling Frames intersect and interface in Continuum Space-Time for unified equilibrium.

The evolutionist "mind-cast" has a great influence upon the research environment. In formulating a new paradigm in theoretical physics that will integrate the Quantum Scale with the Solar-System Frame, consistent with the Theory of Relativity in unifying all fundamental forces, the Human Mind, self-consciousness, religious sensibilities, historical understanding, and Bio-Physics will have to be taken into account. For, we know that "atheistic Evolutionism" is no less than a "ritualistic religion:" The religion of Secular (godless) Humanism.

How do evolutionist doctrines factor into experimental designs? Paleontology, Biology, Physics, "Psychology," and "revisionist History," are all suffused with evolutionary assumptions and unfounded conclusions that impact personal beliefs and social advocacy. For example, dinosaur bones are ritually worshipped in museums, with the same zeal that the apes are promoted as "distant ancestors" of Humanity.

The so-called "big bang explosion," "relic cosmic background radiation," and/or "missing biological links," agitate, instruct, inform, and quicken theoretical approaches to scientific understanding of universal phenomena and of Human history. Through attempts at "tunneling Human perception" within the godless materialistic framework of Evolutionism, astrophysicists have transmogrified Relativity Theory beyond recognition while conjuring such things as "String Theory," "Eleven-Dimension Universe Theory," "Holographic Universe Theory," as well as existence of "Black Holes," "Dark Matter," "Bosons," "Quarks," "Gluons," or what-have-you, etc…, all of which, unproven by valid mathematical equations and/or true and factual experimental results. Rather than faithfully abiding by the strict experimental research practices prescribed by the Scientific Method, astrophysicists rely upon "the Peer Review Method" that ultimately yields to "approving" or "accepting" unfounded and unproven propositions made by their physicist colleagues. Thus, Evolutionism confines Humankind to a pre-deterministic destiny or to a pre-conceived system of self-destructive personal experiences and social behaviors that, even they, concede, will eventually lead to "Terminal Species Entropy:" That one way or another, Humankind will self-annihilate!

We know that Human life starts at conception whereby the DNA of each parent, selectively reproduces by RNA replication as they combine to make up the Human baby. DNA is constituted of genes — the complex operations of which, constituting the micro-structure for macro-organic growth and development, until biological-physiological maturity of the Human Person is achieved.

Micro-structure processes, e.g., DNA-RNA replication, upon which macro-organs or macro-processes are erected, are deterministic of Energy forms utilized by Human metabolism, e.g., respiration by Oxygen, and proceed according to "sensitive dependence on initial conditions," e.g., The First Law of Bio-Genesis states that Human Beings are reproduced from other Human Beings who are neither fish nor monkey.

Though we begin with DNA-carrying genes, genes combine to make up cells; cells combine to make up organs. Each organ has a specific chemo-electrolytic physiological-colloidal environment encompassing its specialized biological functions and processes. And each organ has a specific function akin to its own kind. For example, the heart performs pumping action that circulates Oxygenated blood; we breathe through the lungs, operations of which, oxygenating Human blood; the Human Brain is constituted of neurons that are specialized cells regulating physio-metabolic-biologic processes as well as organ functioning, via "command-and-control mechanisms" that appear to be analogous to a computer's binary encoding operations designed for executing "lines of instruction," e.g., "It's either All or None," is an expression often evoked regarding how the Human Brain operates that refers to the fact that Brain neurons either fire or do not fire at all.

However, no "specialized gene" has been discovered to specifically account for the Human Capacity or Predisposition for Abstract Creative Intelligence.

Though bio-genetic micro-structures, i.e., DNA, are responsible for the Forms of Energy utilized by the Human Body, they do not predetermine the Forms or products of Human creative expressions. Nor do they pre-determine a Person's "level of intelligence" or "how intelligent" a Person is. "Intelligence Tests" rather reveal "the performance quotient" of a particular method, e.g., "multiple-choice examination instruments," devised for "assessing Human Intelligence," but operations of which, resulting only in extracting certain Forms of acquired knowledge or types of learned skills from a Human Person, as "experienced" by that particular Person throughout his or her lifespan during specific periods of time.

Thus, given that it is not possible to establish definite differentiation between "innate/born-with intelligence" versus "learned knowledge/acquired intelligence," then it is "the pertinent effectiveness" of the "examination instrument" itself that is being assessed rather than the specific Person's "level of intelligence."

The Human capacity or pre-disposition to have self-consciousness might be said to have a genetic basis, from the incontrovertible fact that, Bio-genetically, Human life begins at the point of Species-specific genetic conception. However, the genetic foundation or genetic basis, does not pre-determine the content, organization, expression, purpose, and direction of individual Human thought processes. Human attributes of character, apparently, do not have a genetic basis, or "specific physical genetic address" akin to the computer's binary encoding method, i.e., 0 and 1, that requires a "physical address."

Respiration and metabolism, sleeping and eating, digesting and eliminating, are "biological pre-dispositions" that can be said to have "a genetic basis." However, the possession of spiritual and moral attributes, and of abstract and analytical qualities that climax into personality traits of individual-personal character, are "not physical gene-location dependent," nor are they amenable to a "physiologically-emplaced DNA-address."

Evolutionism as a philosophical doctrine, promotes "the image and likeness of the beast," i.e., the ape/the monkey, instead of "the image and likeness of our Creator," Almighty God.

But the qualities, attributes, traits, and characteristics ingrained, demonstrated, and displayed in Human Beings, do indubitably point to "spiritualized abstraction" rather than to immutable, pre-wired, genetic-instinctual, pre-determined, functional pre-dispositions, i.e., Bees cannot "decide" not to make honey, but Human Beings can decide not to learn how to ride a horse. Hence the self-evident Truth: That, because we possess the benefits and advantages that inhere in "freedom of choice," we are indeed "created unto the image and likeness of God."

God is invisible Spirit! He has imparted to us a portion of His own character: We possess and own spiritual pre-dispositions for moral abstraction and for creative understanding.

But the ape-monkey is soulless and spiritless animal-flesh or a wild beast devoid of the pre-disposition to intangibly-spiritually think or perform mental functions of symbolic abstraction.

God gives moral standards of right and wrong and the knowledge of good and evil to Human Beings. And from this factual self-evident Truth, flows the sine-qua-non necessity that God also requires us to have personal responsibility and individual accountability for consequences ensuing from our willfully chosen actions.

Monkeys live in a jungle where wild animal-predators devour each other as food-prey, while Human Beings live in civilized Society that is regulated by constitutional laws based upon God-inspired and spiritually derived expressly given consent.

Conclusively, then, it is not too difficult to understand the godlessness and destructiveness of evolutionist goals and purposes. The strident advocacy by astro-evolutionists and other materialistic philosophers-doctrinaires that they must be allowed to inculcate children and young people with ape-ancestry-based worldviews that devalue Human Worth to the beastly level of jungle animals, aim only to exploit their youthful energies for promoting such deadly ungodly activities as abortion, infanticide, pornography, illicit drug legalization, prostitution, homosexuality, lesbianism, Nature-worshipping, Satanism, and so-called "animal rights," the sum of which, making the fulfillment of Human spiritual and biological needs subservient to "instinctual animalism," from which derive lawlessness, degradation, violence, suicide, and ultimately, the extinction of the Human Species.

We must jettison "the evolutionist-ape mentality" that engenders the falsehood that Human personality and character can be "mass-produced" for acquiring self-destructive traits and anti-social attributes that result in hijacking Human life for "the pursuit of exploitative profiteering," e.g., drug-addiction peddling for corrupt gain, sex trafficking, slavery, infanticide, genocidal warfare, resource expropriation, etc…

But each Human Being is uniquely, specially, and exceptionally created by God, to be endowed with innate capacities and intrinsic predispositions for spiritual wisdom, intellectual creativity, and moral righteousness, the sum of which, allowing for just, peaceful, loving, and lawful personal prosperity, as well as for civilized societal edification.

Alligators and apes also have "biology" yet they have no spiritual and moral self-consciousness for changing or altering "their pre-wired" animal existence. Nor can Human Beings teach or "domesticate" alligators and apes on how to "utilize their instincts" or on how to take advantage of their "pre-wired genetic predispositions" so as to result in their building a rocket ship designed for a Moon-landing.

Thus, "creative consciousness" and "intelligent self-awareness" cannot be simply an "emergent property" of biological structure and physiological functioning. If it were so, we would not be the only Species with abstract imagination and symbolic thinking!

Undoubtedly, then, Spiritual-moral Self-consciousness is not a character-quality, trait or attribute of all biological Species: But they are exceptionally peculiar only to Human Beings! Why? It is because God created us "in his own image and unto his own likeness."

There is physical creation and there is spiritual creation or rebirth. Animals can only have physical birth. However, Human Beings not only have physical birth, but they can also renew their Minds through spiritual rebirth!

Human Beings are born as mortal sinners inheriting the sinfulness of Adam and Eve, our biological ancestors, to then embark into a difficult and perilous journey, during which, God prepares spiritual and practical provisions for their redemption, so that they might be re-empowered for true Liberty and real prosperity, by way of spiritual rebirth in Christ Jesus Messiah, our Lord and Savior.

We are spiritually reborn by our faith in God whose Spirit we inherit when we accept Christ as Lord and Savior, to believe on His name for our very well-being and welfare. Because Christ redeemed us from sin, death, and Hell, we are sealed as heirs of God's Kingdom who become forgiven, justified, sanctified, and glorified sons and daughters of God. The will to repent in the God-endowed Liberty wherewith we were created brings forgiveness unto the spiritual power with which we were predestined to exercise our freedom for moral living.

Evolutionists are wrong in equating Human Beings with the ape Species. "Gene expression" alone cannot account for the great differences that separate the Human Family from unthinking animal hordes. "Gene expression," as specialized integration and organic differentiation, is present in all species: Bees utilize flower-nectar to produce honey; birds utilize twigs and grass-stems to build nests.

"Gene expression" is insufficient in fully explaining "Species differentiation." The Human Family and all other Forms of Life are separated by numerous "genotypic differentials:" Differentiation of intelligence-in-kind (the Form of Human Intelligence) and differentiation of intelligence-in-degree (the level of Human Intelligence).

And the fact of "being set apart" from the natural and animal realms, is made manifest, not only in "differentiated Human intelligence," but also in differentiated "essential Human Species constitution." God created us "in his own image and unto his own likeness:" We are totally different from animals in spiritual and moral self-consciousness; in creativity, productivity, and destiny. Thus, we, Human Beings, are pre-destined to be victorious over godlessness by our faith in Christ Jesus, due to our knowledge of Good and Evil and our innate capacity for differentiating between the two.

Spiritual self-consciousness for righteous living is a unique God-given endowment peculiar only to Human Beings. We are the only Beings who have been empowered by God with the capacity to transform our own selves, while we simultaneously engage in transforming Energy into its diverse differentiated utilizable states or Forms. As we spiritually aspire to eternal significance, we touch Heaven with righteous works that glorify God and edify the Human Family.

Scientific Creationism is the only paradigm that is consistent with all known physical laws. Faithful believers in God are not "struggling" to prove his existence; nor are they engaged in "conquest" of Nature in order to ferret out its physical laws. We accept God's powers of divine Creation and only seek to discover the laws He designed therein. No psychological conflicts arise from this type of scientific enterprise.

We are intelligent Beings engaged in the pursuit of scientific knowledge, engaged in studying and evaluating an objective thing that is neither self-conscious nor sentient. The Universe is inanimate and unthinking, and therefore cannot be said to have "intelligence." The "intelligent design" witnessed in Nature and in the Universe is consistent with our Creator's character-qualities: God structured the environment for us to live in and thus imparted to its operation a "property of organization," the Organizing Principle, for our own benefit.

The evolutionist concept of a "self-starting universe" is inconsistent with the logic of spiritual creativity: That there is redundancy of Form in iterative patterns of Creation. God not only created the Universe but he is also engaged in preserving it, as he is interacting with the lives of Human Beings. When we pray, spiritual communion replaces doubt-induced despair, for faith overpowers uncertainty by providing victory over external circumstances.

The heavenly realm is the archetype from which the physical realm obtains its substance. Spirit speaks abstract conceptual objectivity into concrete physical materiality. God, in whose image and likeness we are created, said "Let there be light, and there was light." Consequently, Man said "let there be the light bulb." This pattern of spiritual creativity is repeated in human life as we productively transform and build upon the natural stock of resources already created for our benefit by our omniscient and omnipotent Creator, Almighty God.

A watch found on a beach attests to the existence of its creator — the Human inventor. In the same vein, the Universe affirms the existence of its Creator, God. Every thing that physically exists in three-dimensional Continuum Space-Time, must partake of an ever-present "on-source." Thus state the Laws of Thermodynamics, the first being Conservation of Energy, the second being Entropy. And Almighty God is the only eternal Source for embedding the "Organizing Principle" observed in Nature and the Universe.

God had "set us apart" from the inanimate world from the beginning of Creation, in order that our relationships with the environment might not develop on a "religious basis." God gave "dominion" or "management" of the Creation to Humankind, so that Human ownership of parts of the Universe might be consistent with "responsible stewardship" to our constructive benefit and life-affirming advantage.

A "self-starting Universe" is inconsistent with the Laws of Motion, the Law of Gravity, the Laws of Thermodynamics — the Law of Energy Conservation and Entropy, — the Organizing Principle, and the Continuum Principle.

Only Scientific Creationism provides us with a conceptual theoretical Frame that can unify the apparently disparate mysterious phenomena, events, processes, principles, and laws that govern universal Reality.

Since we must take ourselves into account, — Human consciousness as an added "apparatus of dimensionality" — post-Relativity theoretical integration must encompass physics, chemistry, biology, psychology, philosophy, history, religion and spirituality, so that Human personality can "relate to" the complex interplay of parameters and determinants that factor into our utilization of the Electromagnetic Radiation-Energy Spectrum which our brain, our vision, our research equipment and analytical resources utilize within the Gravitational Curvature of Continuum Space-Time.

The "psycho-Physics of Human experimenters" appears to have already been included as "part of the equation" that factors in scientific research, even as astro-evolutionists pontificate their Darwinian religion in defiant denial of God's existence. As they are pursuing justification for "the big bang explosion," astro-evolutionists are only exposing the fallacy of their self-starting, ex nihilo inanimate, material Universe, the sum of which, immersing Science under the rubric of prideful ignorance: When astro-evolutionists don't know, they fabricate and contrive impossible scenarios, e.g., String Theory, that contradict the very Laws of Physics they purport to adhere to.

In sum, we must redeem scientific activity by anchoring its research environment to the "peaceful spirit of discovery" that makes technology benevolent, rather than to the "violent spirit of struggle" that transmogrifies it into a "malevolent enemy."

The worthwhile enterprise of scientific research must sustain internal consistency and contextual coherence with the Continuum Principle, by unifying all things in the Universe, i.e., the Laws of the Universe with their real range-specific functional operations, yet without participating in "the pagan worship of Nature," and without engrossing ourselves into its "flowing streams of unconsciousness," e.g., That inanimate, unthinking, and insentient trees "have spirits."

An integrated, unified Universe can be perceived and conceptualized only by persons who cultivate an integrated psycho-spiritual constitution that is impartially and independently set apart from Nature's stochastic cycling mechanisms. Integrated intuitive-analytical "psychographics" will empower "the Language of Mathematics" with the operational connectivity and relevancy that it presently needs, in astro-evolutionist attempts to define, describe, analyze, and resolve the Universe, wherein must subsist no inner conflicts due to divisions of mind, soul, heart and spirit.

Otherwise, in the absence of such mentational-spiritual integration, "psycho-interferometrics" will arise from internal battles generated by conflicts of interests that will consume the raging Human soul. Astro-evolutionists, in trying "to prove the big bang explosion"— are only found to be opposing God, fighting Human nature, degrading Humankind as "descendants of apes, and pretending to "conquer Earth-ecological Nature," all at the same time. Hence, the convoluted farfetched explanations brandished in "finessing the stark contradictions" characterizing evolutionist Socio-biology philosophy.

But, due to the way in which evolutionists approach the problem of discovering a mega-theory of unification for the physical Universe: It is apparent that Biology does fit into the Continuum-Scale of universal Reality, in the sense that:

(1) We live on the Earth and must interact with its ecological environment;

(2) Biology is the climax of all scientific laws that govern the integration of Consciousness with a "Unified Theoretical Resolution of the Universe" that can then be substantiated by Mathematics.

However, the "natural tension" between the Human mind and the Human body; between the Human spirit and carnal Human flesh-nature; and "between the head and the heart" can be best addressed only via a spirit-based resolution that puts all things into moral perspective, including accepting the mortality of Man, the Reality of his Sinful Nature, and the eternal existence of God, our Creator.

This "existential tension" is thus resolved in a way that empowers us for psychic health, perceptual soberness, spiritual intelligence, and emotional maturity — for, we can only worship God, and not the natural environment. And only in worshipping God and not ourselves can we live in psychical integrity and emotional health. Only in worshipping God and not Nature or the environment can we find rest for our souls and peace for our hearts.

We must accept Human Beings as Spirit-Beings, in albeit, mortal bodily vessels; but as Beings who are empowered with temporal lives while possessing unique endowments, such as awesome creative spiritual powers of abstraction and inventive creativity —constituting "the psycho-bio-physics" of Scientific Inquiry, the sum of which, pursued for constructive-beneficent life-giving technological applications.

It is then that sober moral judgments are transformed into "intuitive embodiments" that are spiritually-guided to yield "a form of encompassing universal resolution" that prevents Human Species self-annihilation and averts Human Species self-extinction.

From our proper "Theoretical Resolution of Universal Reality" as substantiated by valid mathematical formulations, will emerge just, peaceful, lawful, and life-affirming applications of scientific discoveries, aimed at fulfilling applied technologies that are soul-satisfying —

rendering no such afflictions as "Man v. Machine," or "Human v. Robot;" and thus, precluding any grievous heart-wrenching regrets or prescriptive doomsday lamentations.

We need to assume a psychically unified spiritual position or stance, as inspired by coherent mental logic that is empowered with the enriched capacity to rediscover and apply the Scientific Method in a way that provides Reality-checks to our perceptual creations via a Mathematics that substantively results in scientifically applicable equations and formulations.

We can now perceive that insertion of the "Uncertainty Principle" within the pseudo-scientific paradigm of Evolutionism, causes the emergence of "Psycho-bio interferometrics," — presumptive Human inner-attitudes and preconceived operational beliefs that lead to false analytical conclusions, i.e., surmising a randomly-driven, chance-activated, probabilistically-performing, accidentally-existing, self-starting ex-nihilo Universe, — fail to derive an accurate mathematical Theoretics that yields valid formulations and proven equations corresponding to Real universal processes, Real cosmic phenomena, and Real natural events, e.g., In consonance with Evolutionism, NASA (National Aeronautics and Space Administration) pontificated that life on Earth originated from Mars; astro-evolutionists bound to the federal-research-grant agenda conjured up "black holes" and "dark matter;" "String Theory" and an "Eleven-Dimension Universe," all of which unsubstantiated by reproducible scientific experiments or proven-valid mathematical equations that yield technological applications.

In short, $E = mc^2$ remains "the last word" on "universal Physics." Yet incessant astro-evolutionist activities continue to prevail in misleading not only the so-called "scientific establishment" but also students, teachers, and other people in the world-at-large.

Yet, as affected or "coloured" by homeostatic conditions of physiological energy cycling that are subsumed under the rubric of "Biological Thermodynamics," "Psycho-bio-Interferometrics" is directly amenable to inevitable operations of the Second Law of Energy Transformation — the Law of Entropy, from which not even "the most intelligent Species on the Earth," Humankind, can escape.

For example, the tension between abstract conceptualization and operational objectification embodies simulated-analogs of "bio-Entropy events" that also correspond to our sinful spiritual-moral nature, as stated in the Holy Bible: In the same manner that we are prone to errors in judgment and decision-making, when we get tired, we sleep, eat or rest, as needed for resource replenishment, stamina restoration, and energy reinvigoration.

Thus, in a way, it is "normal" that difficulties arise in "re-translating" conceptual apprehensions into concrete objectification. For, primarily, we are "Spirit-Beings" to whom God imparted a mortal biological body answering to the Laws of Thermodynamic Transformation of Energy. Thus, it is conclusively apparent that the "natural tension" that existentially inheres between theoretical abstraction and objective understanding of observed universal phenomena, will always persist.

Neither naivety nor cynicism should predominate in our seeking a life-affirming understanding of "the Human condition" for "scientific re-translation" or "logical interfacing" with cognitive discovery of natural phenomena, cosmic events, and universal processes. Only a motive of well-placed Hope should govern our approach to universal understanding. Given that "spiritual things are spiritually discerned," God's Spirit helps to guide our scientific judgments towards beneficent moral thinking, in applicable ways that "commune" with "the Mind of our own spirit." As correlated with activities of the Human intellect that are designed for finding solutions to objective problems, inner-peace and spiritual rest conform to a "thought-life" that perseveres in bringing our faith to fruitful resolution of physical matters that might "overtax our fleshly-carnal capacities." (Romans 8:1-11; 26-28, Holy Bible, KJV).

Redundant recurrence of iterative quasi-similar structural Forms of Organized Order permeates Nature and the Universe, at the same time that Energy Conservation for resource production is countered by consumptive thermodynamic processes of Energy-cycling that accentuate the effects of Entropy, e.g., as a flower has its own respective blossoming duration or "life-span," or as a ripened fruit endures towards degradation and returns to seed, a Human infant continues to grow towards full biological maturity until his/her physiological and mental development climaxes into geriatric aging that incrementally approaches eventual death.

As the Conservation-Entropy Complex is on-going via resource utilization in fulfillment of the Input-Process-Output mechanism, even our thought processes in the exercise of our freewill, inspire "sharpening," for more proficient correspondence to extant Reality as well as for more beneficent result-oriented activities. Our "inner thought-life" must be spiritually nourished through communion with our Heavenly Father. As the Holy Spirit of Christ Jesus Messiah informs our decisions, our temporary lives are accounted and reckoned for developing plans and establishing purposes that are designed for manifesting worthwhile accomplishments and meaningful achievements that accord with affirmation of our God-ordained mandate "to secure the blessings" of Life, Liberty and pursuit of Happiness.

The conceptual framework of "interferometrics" operates as we acknowledge sinfulness in Human nature (our spiritual condition) as well as the tendency to decay (Entropy) that prevails in natural processes and physical events, the complex of which, engendering an enduring tension within Human inner-being, even as we attempt to apprehend and discern "how things work."

This "Interactive Interface Complex" does not amount to "struggle for survival of the fittest" as evolutionists would have us believe. But it rather demonstrates exacerbation of "affective conditions" governing "human negotiation" of universal Reality, as encountered in willful and deliberate efforts at "navigating through the pitfalls of Human existence." This "psycho-material Interface" carves-out a difficult road, due to inherent spiritual-moral factors that impinge upon the physical scientific process of fashioning theoretical correspondence or thematic fitness to discovering the observed-witnessed "output-producing relationships" that are intrinsic to "the Universe-as-it-is."

Within the physiological-biological Frame, cells display duality in "firing-cycling form" that takes "excitatory" and "inhibitory" patterns of operational processing. Certain genes will undergo growth during predetermined time-periods until the specific organ they are designed to build has matured for functional specificity, after which these genes will stop growing and thus become "inhibitory." In the human brain, certain neurons do not "fire" all the time; sometimes, they are "excitatory," and other times, they are "inhibitory."

For example, in the optical sciences, "interferometers" are utilized to separate beams of light via mirror positioning, and light-splitting lenses. In Chemistry, where temperature and pressure are controlling variables, certain elements will react for specific time-periods and only with certain elements and not with others. Gas expansion or contraction, solubility or liquefaction, suspension or saturation, all have naturally limited time-spans and conditions of reaction that can be contrived in the laboratory only under specific parametric conditions. Even in the Sun, hot-particle "excited" plasma gases reach a specific maximum temperature to then "undergo cooling" as they release radiation energy during which they return to the solar core and repeat the process. The sun's inner-core, convection zone, radioactive zone, and coronal region interact under extreme magnetic field forces that cause nucleated reactions to fast-cycle in "compressed Time." Gravitational nuclear convection Energetics comprise controlled plasma core-compression, expansion to the convective zone, dilation to the radio-active zone, expulsion to the corona, and ejection into the helio-sphere, after which the energy cycling process reverses towards the inner-core.

This "entangled duality" of Conservation-and-Entropy Cycling, brings forth light and radiation and heat Energy, as well as projection of Curvature binding dynamics, in the form of magnetic Field and Gravity force equivalents, within the structurally-framed interplay of the variegated manifestations of Solar Power, the sum of which, being the very wellsprings of Earth life-support systems.

No earthly Life would prosper without continuous Solar Inputs that feed-maintain-sustain the whole ensemble of Earth ecological processes, atmospheric-meteorological-oceanic phenomena, and hydro-geological events.

The element, Hydrogen, cycles from Deuterium and Tritium to stop the nucleated reactions at Helium. It is conceivable that other nuclear fuel cycles, e.g., Nitrogen cycle, Carbon cycle, Iron cycle, etc . . ., might also have chain-reaction boundaries. Solar particles "excitatory" states and "inhibitory" states — in fusion, fission, convection, compression, dilation, contraction, expansion, heating-and-cooling transmutations, etc . . .— accompany each other in cyclic patterns from which Field and Gravity forces derive, as emergent properties that co-exist with determinant Curvature variables that effect Earth revolution and rotation.

As solar Energy cycles due to Differentials-and-Opposites that co-exist in gradients of temperature, pressure, gravi-metric variables, and magnetic Field parameters, Earth seasons also cycle in recurring periods, as an integrated, complexly entangled ensemble of structured Curvature-motion patterns, the sum of which, correlating with conservation of solar system Mass angular momentum that is consistently sustained within Continuum Space-Time.

According to the Christian worldview, Human Beings are nurtured as individuals to become healthily "well-rounded Persons" inspired to moral maturity for accomplishing great things. Christians are exhorted to live by faith even as they judge righteously with wisdom, compassion, and charity, with calmness of spirit, tranquility of Mind, and peacefulness of heart. God has empowered us to be representative of his divine Person on the Earth, as "vessels" or "embodiments" of His own Spirit. In knowing his commandments and in faithfully applying them to human living, we bring to Earth the living manifestation of heavenly will for ourselves, in the lives of our neighbors, and in our society-at-large.

The evolutionist worldview aims at "mass conditioning" of people-groups for reacting to artificially orchestrated social conditions of struggle and scarcity, within the framework of which, Evolutionism attempts to standardize human conduct, as falsely induced from animal-behavior based observations. From such unregulated asocial observations of animal behavior that is apparently estimated as random and probabilistic, evolutionists contrive the primacy of so-called "instinctual needs" over cause-and-effect ratiocination, which they thereafter transfer to the Human Species for inevitable applicability. Given that Human Beings are spirit-beings inherently imbued with a moral nature, these contradictions give rise to conflicting pathways of need-fulfillment, the complex interplay of which, sabotaging the innate Human capacity for emotional intelligence and psychological maturity. Spirituality is thereby shunned and morality becomes instrumental to desired ends.

The theory of evolution destroys the special uniqueness of individual Personhood to engender a "spirit of collectivism" aimed at mass-producing Human personality for purposes of manipulative control and exhaustive exploitation.

The theory of evolution severs our spiritual communion with God by making void the redeeming love of Christ whose sacrifice on the Cross and resurrection from the dead already freed us from the bondage of sin, debilitation from emotional despair, and degradation from moral helplessness. While our Judeo-Christian heritage emphasizes a spiritual Continuum between God and Humankind through the empowering love of Christ, the theory of evolution separates, isolates, and divides people into a kind of "atomistic animalism" that degenerates into struggling to survive, but only in cynicism, pessimism, resignation, apathy, and indifference.

Rather, it is necessary to re-integrate Human personality by restoring "the Principle of Spiritual Continuum" between our Creator and ourselves, so that, in our approach to post-Relativity Theoretics, our scientific activities are inspired by God and activated by the knowledge of his everlasting love for us, the sum of which, to direct our application of the Scientific Method towards discovering the true-to-reality, unified essence, of universal Oneness-Integrity. Human personality should no longer be divided for conquest due to ingrained conditioning to "instinctual animalism" that dishonors spiritual fulfillment for enabling Human character to the crass pursuit of materialism. For, the dissection of Human personality corresponds to our current inability to discover a unification theory that correlates with the Oneness of extant physical Reality. The conflicted Human inner-world cannot truly interface with outer external objective Reality, for apprehending universal Oneness.

We must free ourselves from the pattern of disconnectedness that has resulted from applying the Scientific Method in analytical terms that divide, isolate, dissect, and compartmentalize all categories of real things to which we relate, including Human Beings, as "subject to conquest." But it is indeed true that we are in the world, but not of it. Non-biologic Matter is as unthinking as it is unspiritual. Yet it is divisible in terms of atomic particles or radiated rays that are also expressed or made manifest as "waves." This "entanglement" in duality of Form within Curvature Continuum Space-Time encounters pressure-force motion Energetics that complicates particulate pathways and trajectories, the sum of which, being rife with difficulties ensuing from attempts at simultaneously measuring both position and momentum.

Due to great near-Speed-of-Light velocities and micro-mass quantifies: Field, Gravity and Curvature tensing-forces within the atomic frame thrive as "momentum Motion-pressure" in "compressed cycling Time." The electron moves vertiginously fast when compared to the speed at which our planet Earth revolves around the Sun — a journey that takes the Earth more than 365 days or one whole year!

In regards to re-translating the electromagnetic properties of Matter into Mass-in-Motion that yields equivalents of the Gravity Force, the lighter or less massive an object, the faster it can move within Curvature Continuum Space-Time, i.e., Light, constituted of nearly mass-less Photons, moves in vacuum Space, at a Speed equal to 186,000 miles per second.

At the atomic level, the positively charged "massive" proton is "attracted" to the negatively charged electron, while both are simultaneously "repelled" by each other. The oppositely charged Electron in relation to the positively charged Proton does not result in a "crash of sparks," due to the "grounding effects" of the charge-less but comparatively "massive" Neutron at the atomic nucleus. This complex dynamic allows Electrons to revolve at a "safe distance" from the atomic nucleus, without "short-circuiting."

In turn, "this dynamic" engenders Electron trajectories that take the forms of centri-vectored motion, patterned as "composite embodiments" or "combined expressions" of rotary, rectilinear, and curvilinear paths, hence, interpretation of Electron-behavior by astrophysicists as "waves of probability." But, "probability" rather refers to Human incapability to fathom, decipher, or understand the physical cause-and-effect relationships that yield Electron-behavior.

For, in the material Universe where physical Laws scientifically govern and control, no event is "probabilistic" in occurrence, and no phenomenon happens "by accident" or "random chance." The deficiency is not in Nature or the Universe, but apparently in the limits of Human understanding, or knowledge, as "attached to" the limits of extant electronic technologies being utilized in research-and-experimentation activities.

In Quantum Mechanics, tensing, stressing, pushing, pulling, contracting, expanding, compressing, transforming, transmuting, and dilating Forces are "in entangled relationships" within micro-Curvature dynamics that differentiate particulate motion from planetary motion.

The theory of evolution advocated the isolation of Man in "struggle for survival of the fittest," as "an ape-like creature" devoid of spirit and morality, merely instinctually breathing-and-moving through activities of "environmental adaptation" to contrived circumstances of hostility and forced situations of violence, due to "inevitable competition for scarce resources," outcomes of which, being amenable to "natural selection." "Natural selection" actually evokes "the law of the strongest," idiomatically coined as "Might makes right."

"Might makes right" has become a colloquial cultural icon commonly denoting a specifically unknowable yet consciously driven social organizational cadre and/or societal structural order predicated upon "the natural expectation of mutual conquest."

Since Charles Darwin's (1809-1882 AD) evolutionist philosophy got "emulsified" in the professional university-world of socio-economic political theory, various "mental categories" have been invented to dissect the individual Human Person into "chunks of matter" for purposes of selective exploitation in fulfillment of evolutionist prophecy:

(1)   Herbert Spencer (1820-1903 AD), in applying Darwinian Theory to economic society, proposed that certain people were "fitter" than others

due to pre-ordained personal genetic determinism that caused them to be more effective than the others in "struggling for conquest."

(2)   Karl Marx (1818-1883 AD) conceptualized the system of thinking referred to as "dialectical materialism" involving the interplay of three logical elements, analogous to "The Trinity:" Thesis and Antithesis (contradictions that clash) producing a Synthesis (a unity of opposites), the sum of which, to operate in the individual's mind and in society-at-large to denote a process resulting in a different but optimal new level or "state of society" or "social condition:" The Socialist State intermediately; and the Communist State, finally.

(3)   Frederick Taylor (1856-1915 AD) clothed autocratic control of workers by managers in a conceptual scheme called "scientific management."

(4)   William G. Sumner (1840-1910 AD) declared the rich to be "the fittest" in his socio-biological theory of Social Darwinism.

(5)   Sigmund Freud (1856-1939 AD), via methods of so-called psycho-analysis, also undertook to divide the individual Human Person in terms of the "Id (irrational unconscious or instinctual passions); the Ego (conscious and subconscious rationalizations); and the Superego (personal conscience and religious convictions),"— the sum of which, not unlike "Marxian dialectics," constituting a tripartite division analogous to atomic particulate characteristics such as the Proton, the Neutron, and the Electron. Such a tripartite appellation is also interpreted as a secular rendition of the Christian concept of "The Trinity" that signifies the essential character of God's existence in Three Separate Spirit-Persons: The Father (God); the Son (Jesus Christ Messiah); and the Holy Spirit.

(6)   B.F. Skinner (1904-1990 AD) concocted "behaviorism" by electrocuting rats in a cage as "punishment," in order to "motivate" them to drink water as a "reward"— this system of "behavior modification" whereby electric shock was given as "random" or "scheduled" reinforcement is called "operant conditioning."

This pattern of applicable socio-biological extension of evolutionist doctrines to be active in Human spiritual morality and Human conduct as attuned to respective social expectations, engendered the most cruel forms of exploitation, the sum of which, utilized to justify slavery, war, plunder, pillage, aggressive militarism, violent conquest, ethnic racism, xenophobia, and lawless expropriation of natural and Human resources for unjust purposes of corrupt wealth-building.

The perceptual lens or overarching paradigm that crafted Human relations from the worldview of Evolutionism became the perennial source from which scientific problems are investigated. The Scientific Method has come to suffer under the aegis of profiteering from "turf protection" as practiced by university Physics departments, as Human Beings continued to travail from inhumane attempts at dissecting Human personality and Human character for isolated exploitation.

The Biblical worldview, which however focuses on integrating personal character for emotional intelligence, spiritual growth, emotional maturity, intellectual development, and creative productivity, has been discarded as mythologies or products of Human imagination. But, truly, in reality, it is in the knowledge of Christ that the world would have had come to understand the true meaning of peace, grace, compassion, charity, and love. For, Christian principles of repentance, forgiveness, redemption, sanctification, and justification would become "the weapons of righteousness" that Christian missionaries, to the example of the Apostles, will have had continued to bring "to the four corners of the earth" for conscientious implementation in Human social organization and orderly nationhood. Relationships of love and peace would then re-empower Human Beings for compassionate commitment to selfless service for reciprocating neighborly relationships that inspire, motivate, exhort, and encourage them towards improvement of social conditions in which to live for mutual justice, equality, security, protection, and prosperity.

How come then that the Mind of the Human Person is internally driven to divide, dissect, isolate, and compartmentalize for purposes of study, learning, and analysis, or even of exploitation, while the real Universe itself is One in unified integrity and gravitational operation? What are the reasons for this lack of correspondence between Human inner-being and external universal Reality?

The Universe is an integrated Reality in Continuum Gravitational Space-Time, just as there is Continuum between Human spirituality and the fact of divine universal Creation.

Dissection and division, isolation and compartmentalization, are methods of study for purposes of categorization, classification, integration, differentiation, comparative analysis and problem solving. However, the comprehensive wholeness of all things that exist within the same Space-Time Continuum necessitates perceptual unification for purposive comprehensive all-encompassing Human understanding.

The Human Person, though a sinner by nature, is redeemable when he aligns his or her fallible constitution with the Will of God to thereby be saved. There is glorious and sublime Hope in the Christian faith for all who believe in Christ Jesus Messiah as Lord and Savior. It is a revolution in self-perception that is predicated upon abandonment of the "ape mentality" that has crippled individual self-esteem, social relationships, and personal self-concept.

Since it is God who created the Universe, we are not engaged in "proving it," for it is self-evident; then the raging contradictions that inhere in internalizing a "theory of conquest" while facing mortality disappear to make way for a mature self-fulfilled personality no longer seeking "struggle for fitness."

The fact is, through God's eyes and due to equality of original sinfulness condition, no Human Being is ever conceived or born "unfit." "Biological perfection" is not possible due to the Laws of Thermodynamics, the climaxing result of which, ultimately being terminal mortality.

However, there are no limits to spiritual knowledge, moral development, emotional maturity, and creative productivity.

Christianity put an end to the barbaric primitivism of human sacrifice to pagan gods, mirrored after natural forces over which Humans imagined they had no control. Moral conscience, mediated by an absolute standard of right and wrong, activates the understanding that, spiritually, or by the practice of Good rather than evil, we are becoming children of God with unique capacities for moral behavior and creative endowments for fruitful productivity.

We cannot "split" Man into "psychological quanta," teach him that values are "relative" and expect his inner-being to be in healthy, mature wholeness. These aberrations are the basis of "pysho-spiritual interferometrics" that sabotages the Human capacity for integrative universal knowledge.

Could cultural values that are incompatible with a spirit of unification be responsible for dissociative patterns intrinsic to post-Relativity theoretical Physics? Christianity unites people-groups through oneness of Spirit whereas Evolutionism divides them through aggressive self-centered isolationism.

A social order of constructive beneficent Liberty for just and lawful civilized living depends upon spiritually fulfilled and morally mature members of Society who appreciate the necessity for willfully given consent, individual responsibility, and personal accountability. Only a healthy, spiritually integrated person perceives the world as a morally unified life-giving environment, without worshipping its natural phenomena. Spiritually fulfilled Human Beings have constructive-beneficent self-perception for integrating the apparently disconnected and disjointed Reality of the Universe. A Continuum between spiritual consciousness and healthy understanding of Nature engenders qualitative discoveries in the study of physical processes for unification purposes.

Quantum Mechanics and Gravity force, Relativity and Electromagnetism, are already integrated and unified in the universal Space-Time Continuum where cycling Mass-in-motion-frames function as initially designed to operate.

Given that all things that exist are constituted of Atoms that reactively bond to form molecules in building up greater bodies-with-Mass such as planets and stars in the same manner that a line is a succession of dots or that Light-radiation propagates as "wave-particle units of Energy," then, extant Reality can only confirm the integrated Oneness of the Universe, and by that, corroborate the sine qua non necessity that the Human Mind must be perceptually integrated, as researchers and experimenters adhere to a paradigm of Science that conforms also to the integrated Oneness of physical Universal Reality.

Some astro-evolutionists are conscious of the problems that arise from contradictions that inhere in the evolutionary paradigm, especially as applied to Physics Theoretics wherein must factor the intrinsic properties of Human consciousness in regards to religious spirituality.

For example, Dr. Roger Penrose, as mentioned above, wrote "The Emperor's New Mind" (1989 AD) and "Shadows of the Mind" (1994 AD). Dr. Penrose maintained his evolutionary worldview as he created a theoretical framework that discarded Scientific Creationism, in spite of the fact that he did consider it as a valid alternative. He thought Human life and the Universe had a greater purpose than mere accidental randomness.

Another astro-evolutionist, Dr. Frank Tippler, in "The Physics of Immortality (1994 AD), in contrast with Penrose, went in the opposite direction. Dr. Tippler declared that he was atheistic, and materialist-evolutionist in outlook. Regardless of the fact he did not believe in God, he went so far as quoting Scriptures from the Holy Bible.

Dr. Tippler "piggybacked" on scriptural text to extract a pagan interpretation of Physics therein explained, to reach for a thesis that ultimately climaxed into a conclusion, called "the Omega Point." Dr. Tippler by using the word "Omega" as in Christ being "the Alpha and the Omega" proposed a "physicalist distortion" of the Holy Scriptures. He utilized Biblical terms as 'bait" for the believer to then fill them with pagan content.

According to Dr. Tippler, "God" is a pantheistic animistic Being" who is devoid of all spiritual characteristics that make for real Personhood. Dr. Tippler's "God" is at times a "he," and at others, a "she," and still, at other times, an "it;" "God" is sometimes a process, at other times an event.

Dr. Tippler only finessed all the pagan doctrines that trickled down from the theory of evolution into a "new age religion" cloaked in the aura of scientific inquiry. Dr. Tippler's work is a "Trojan horse" loaded with manipulative usage of Scriptures as a plan to de-personalize God as we know Him from the Holy Bible. As such, it is an unsuccessful attempt to "secularize" biblical themes into a materialistic framework of concepts that depersonalize the Spirit-character of God who is, in reality, the only omniscient, omnipotent, and immanent Creator of all things that exist.

As these two astrophysicists have proven, Scientists like Dr. Penrose and Dr. Tippler, are not ignorant of the larger spiritual problems that loom over our culture's research activities — That there exists a deep sense of cognitive dissonance, as a strong apprehension of the disintegration of "the inner-core of what it is to be a Human Being," made manifest by his disconnectedness from his true self, essential identity, and authentic personhood.

The more time people spend in "fighting against God," the more incapacitated their efforts at unifying the physical Universe. Astro-evolutionists glory in "worshipping" dinosaur bones, apes, or reptiles, even as they "put their hopes" in the futile pursuit of unearthing "biological missing links" with such feverish energies that consume the living passion out of their souls.

Inner-being dissociation due to evolutional doctrines is comparable to polytheism whereby human psychic energies are exhausted in worshipping many different pagan deities that compete for their loyalty. But, worshipping only one true God not only unifies the Human spirit but also integrates the Human psyche for inner-understanding and external apprehension, as carried out, in a state of self-evident perceptual wholeness.

Not only are there "psycho-interferometrics," there are also "technological interferometrics" that impinge upon astro-evolutionist approaches to proving "the big bang theory" via detection of "relic cosmic background radiation." Experiments performed in particulate physics by numerous research scientists have proven that particles do not transact with Gravitational Continuum Space-Time Curvature Energetics, as predicted by their esoteric formulation of "Quantum Theory."

"Particulate behavior" appears indeterminate, and hence, their reference to such observations, as "waves of probability." Some astro-evolutionists have proposed a composite theory called "Quantum Gravity Theory" in attempts to merge Gravity-force with apparent "particulate quantum behavior." Location and energy charge, position and momentum, apparently cannot be assessed simultaneously, hence, the difficulty in formulating proven-valid equations for particulate trajectories, gravity-force vectoring, and field interactions that correspond to quantum mechanical reality, or "der ding an sich."

"Technological interferometrics" inhere in the apparatus of research and its methods of measurement. But more revealing is how these effects are connected to psychic dimensions of experimenter participation. Complexities arise from the fact that the electromagnetic spectrum is the controlling factor in the analytical dynamics via which human vision, cerebral processes, and research technologies are engaged. As stated previously, in a way, "light is chasing light."

The visible spectrum is yet another form of the same electromagnetic energy utilized by research machinery in attempts to detect hypothetical sub-atomic particles. Particles can be manipulated and studied without being seen as in objective human vision; however, they can be detected with computerized technology in remote forms of experimentation. Whenever a quantity is increased in one form, it must be met by a commensurate change in another parameter — increases in the strength of electric fields and magnetic fields change particulate trajectories away from desired Curvature motion vectors, even as particles emit cosmic radiation.

Thus, a vital psycho-technological link is operating in conceptualization of particle relationships, if any do exist. Sterility in theoretical mathematics foretells an interruption in applicable Physics. There appears to be a connection or correlation between scientists' ability to theoretically concretize apprehensions into objective abstractions and the extant state of research technologies.

How do electronic instruments interface with the electrolytic properties of the Human Brain? How does the electromagnetic machinery of research interact with the apparent platform of electrical impulses that frames operations of Human consciousness in the study of particulate behavior?

The search for scientific truth ought to override fanatical adherence to a mythical evolutional conjecture predicated upon unscientific assumptions of "time-travel to the past," based upon spectral temperature readings of pretended "relic cosmic background radiation."

Theories addressing the problems that inhere in integrating Space, time, the quantum frame, electromagnetism, gravity, and field forces appear to have climaxed into esoteric consensus-based prescriptions requiring partisan allegiance, i.e., "peer review," rather than marshalling scientific evidence for their validity, e.g., "string theory," proposing a universe made up of strings in motion within a flat ten-dimension space-time.

This is the state of theoretical physics to date:

"Let's begin by recapping exactly what we know about string theory. There is, first of all, no complete formulation of it. There is no accepted proposal for what the basic principles of string theory are, or for what the main equations of the theory should be. Nor is there proof that such a complete formulation exists. What we know of string theory consists mostly of approximate results and conjectures . . ." (Lee Smolin, "The Trouble with Physics," Houghton Mifflin Company, New York, 2007).

It was hoped that the philosophical framework upon which "string theory" would result in advancing the world towards a comprehensive, unified, integrative, all-encompassing theory of the Universe for purposes of mathematical validation and scientific application. However, that hope proved to be misplaced.

Usually, when an impasse or obstacle is reached in a prevailing scientific paradigm whereby an apparent aberration is not addressed or answered by its Fundamentals, then Scientists proceed to engage in additional data collection, categorization, classification, and comparative analysis, the sum of which, leading to theoretical conceptualizations that would climax into explaining observed patterns in data and variables falling outside of that dominant paradigmatic framework. Perhaps post-Relativity Physics is at that stage.

Given that validated gravitational law, laws of motion and Relativity Curvature dynamics governing our solar system are equally applicable everywhere else in the Universe where conditions are similar, any distant "Star-planets complex" constituting a Solar System, of whatever scale, would demonstrate the same parameters of thermodynamic Energy transformation, though not necessarily arranged or structured similarly. Much Space-bending or Space-curving of Light would therein be likewise observed; however, given that astro-evolutionists have ascribed such "bending" or "curving" to the existence of so-called "black holes," that has given rise to a host of other conjectural phenomena for which the evidence appears to be only mathematical constructs devoid of objective proof. There are "black hole" astro-evolutionists, "dark matter" astro-evolutionists, "dark energy" astro-evolutionists, "twelve-dimension universe" astro-evolutionists, "one-dimension universe" astro-evolutionists, "string theory" astro-evolutionists, and "six-manifold universe" astro-evolutionists. If these "new energy Forms" really exist, then, absent proven evidence for their manifestation in the Universe, that would imply that they also must be factored into the discovery of a unification theory or that there are "intermediate steps" yet to be taken before a theory of universal unification can be formalized.

The task that apparently remains is to unify Continuum Universe Physics into one consistently integrated Mega-theory that accounts for the micro-world of Quantum Mechanics, as complemented by the intermediate sphere of Star-planetary Curvature dynamics. In addition, Relativity determinants must be re-configured in terms of the macro-plane of Solar system projected nucleated-reactions-driven magnetic Field Energetics via which cycling planetary Mass-frames overlap each other in sustenance of dynamic System equilibrium.

The so-called "strong force" and "weak force" are only emergent equivalents of field dynamics made manifest as encapsulated analogs of forms-of-Force engendered by Fundamentals of Electromagnetism, the complex transmutations of which, cycling in distinct patterns as structured by specific frame-bound tensor-metric Differentials and Opposites co-transacting in fulfillment of the Input-Process-Output mechanism.

But Gravity is an emergent Force-property, climaxing out of the complex interplay of co-determinant dynamics arising from Continuum Space-Time Curvature Energetics and Field tensing-pressure Forces, thereby effecting rotary, centri-vectored, rectilinear and curvilinear

Forms of Motion, e.g., Solar magnetic Field tensor-metrics within Continuum Space-Time effect gravitational Curvature Energetics yielding Earth Revolution around the Sun and self-reflexive Earth axial Rotation.

Since only the Human Mind can accomplish scientific conceptualization and computational mathematicization of A Theory of Universal Unification, then the integration of the Human inner-Person is an a priori first-part of this multifaceted, multi-factorial, and multi-sourced complex endeavor. Undoubtedly, major sources of "interferometrics," both within the Human Mind and within research and analytical technologies, will have to be identified and overcome.

And, in order to explore the conceptualization of a general theory of force comprising all qualities, values, or characteristics that would comprehensively constitute the Fundamentals of Universal Reality, there must occur, the development of a mathematically validated paradigmatic framework particularly addressing the formulation of A Unified Theory of Continuum Curvature Tensor-Pressure-Force Motion in terms of "simultaneously potentiated" functional operations of both Electromagnetism and Gravity.

"Curvature pressure-force tensor metrics" in Continuum Space-Time are emergent properties of centri-vectored rotary Forms of Motion as engendered by simultaneous operations of both Electromagnetism and Gravity, in as many analogous quasi-similar equivalent Forms as "re-translate" or "re-normalize" or "reconcile" the Electro-motive-Magnetic properties of Matter. As embodied in Electromagnetism, particulate charge momentum attraction-repulsion displacement affinities interact and transact to yield the so-called "strong force" and "weak force," respectively.

Electromagnetism, or "the union" of electricity-and-magnetism, incorporates essential ingredients or factors producing both "the strong force" and "the weak force," in that, all three Forms of Force ensue from interplay of charged particles that act upon each other, within "spheres of influence" or "areas of manifestation" that are analogous to "Field-force projections" that then effect the cause-and-effect relationships that yield "the Motion-force complex."

Consequently, given that a "Field," as in "an Electromagnetic Field," is "a region or theater of event manifestations," then, the so-called "strong force" and "weak force" are, all, indicative of "Electro-motive-Magnetic charge-displacement entanglements," the complex interplay of which, engaged in projecting "a sphere or area of interactive Force manifestations," akin to an identifiably, quasi-similar, "Field-force."

Likewise, Gravity projects "a sphere, area, or region" of "interactive Force-influence" akin to, so to speak, "a Gravity Field," the complex manifestation of which, behaving as a "magnetic Field Motion-force equivalent," (but devoid of entangled particulate charge-displacement dynamics). The projected "Gravity-field" presents co-determinant variables possessing numerical values that can be analyzed, as "quantifiable units of Force" that can then be measured, due to their acting upon Bodies-with-Mass, such as a planet or an aircraft, e.g., causing centri-vectored, rotary, rectilinear, curvilinear, or composite-combined Motion-patterns thereof; hence, the "constant" $g = 9.88$ meters per second per second for the acceleration of an object due to or within Earth Gravity-Force, or due to or within G-1 frame-of-influence.

Thus, Fundamental Forces such as Electromagnetism and Gravity constitute the analytical components of Physics Theoretics that can be formalized into a mathematical paradigm for quantifying "A General Theory of Motion-Force," the complex equations of which, to be conceptualized as "overlapping interactive co-determinant spheres or fields-of-influence."

However, though Gravity behaves as "a Field-force," that does not imply that it possesses particulate charge characteristics that would categorize it as an electromagnetically-induced type of Force. It has not been scientifically or mathematically demonstrated that Gravity has a so-called "graviton particle," as there is "no Gravity compass" to detect projection of a Gravity-force equivalent. Gravity is known only indirectly, via the behavior of cycling Mass-frames, e.g., planetary Motion Forms within a Solar System, in the presence of an external Force causing it to alter, modify, or change its inertial mode to a kinetic Mode-of-Motion, e.g., an aircraft launched into Earth atmosphere via mechanical-jet engine thrust or propeller-driven engine thrust.

****

"MOVEMENTS-OF-THOUGHT" IN MAKING SCIENTIFIC BREAKTHROUGHS

"Energy is never created nor destroyed but always transformed;" an electric field accompanies a magnetic field and a magnetic field accompanies an electric field. Conservation intermediately prevails over Entropy as "the strong force" sustains centripetal Forms of motion. Matter-mass embodies variegated Forms of "re-translated" radiation-energy that cycle in Curvature-motion within Continuum Space-Time. Though production and utilization of Energy in a closed system yields unusable Energy residuals that cannot produce constructive work anymore, the Universe has been intelligently designed for dynamic System equilibrium via

Continuum Energy cycling possessing iterative patterns of structuring wile fulfilling the task of resource replenishment, e.g., Earth geo-atmospheric events and hydrosphere phenomena could not continue in sustaining its life-support systems were it not for relatively constant iterative periods of solar radiation-Energy bombardments.

Conservation and Entropy are simultaneously entangled in all cycling-Mass frames. The Continuum Principle, the Law of Transformation of Energy, and the Equivalency Principle climax into redundant iterative patterns of periodic Mass-frame cycling-Motion, so that fulfillment of the Input-Process-Output mechanism engenders the dynamic interplay of centripetal forces, centrifugal forces, system equilibrium, and relative gravitational-electromagnetic Field equivalency. Energy Conservation is sustained as "the strong force" preserves universal Continuum, thus, allowing "the weak force" to trigger Entropy processes, the complex dynamics of which, imparting functional stability to all cycling Mass-frames.

Where a magnetic Field exists, it is accompanied by an electric Field; and an electric Field is accompanied by a magnetic Field. This "field duality" or "electromagnetic unity" engenders electric-Field and magnetic-Field characteristics from which emerges "Curvature Energetics" that causes patterns of centri-vectored rotary Forms of motion, yielding equilibrium variables that respond to Gravity inputs from Differentials and Opposites originating from cycling Mass-frames, e.g., the Earth "wobbles," and has "bi-polar oblation;" planet Mercury displays a perihelion shift.

Every category of Force replicates in analog Form "the sphere of influence" or "theater of manifestation" projected by solar gravitational radiation Energy, synergies of which, impelling Space Curvature, magnetic Field force and Gravity force "equivalents." Every category of motion-Force duplicates in analog Form the helio-centric Motion-force engendered by solar gravitation and magnetic Field projections.

Inferentially, therefore, every category or type of Energy on the Earth, e.g., dry wood stick; charcoal; petroleum; Propane, etc….: is "Solar Power," so to speak, or as transformed into its "congealed" or "excited" derivatives.

Every manifestation of "Field-force relationships" — whether Electromagnetic Field Force, or Gravity Field Force, or any other type or category of "Force" causing objects to engage in Motion, or to "attract" or "repel" each other while simultaneously inducing centri-vectored rotary Forms of Motion, " e.g. Quantum Mechanics, Newtonian Mechanics, Relativity Curvature Dynamics, Electromagnetism-Gravity Force Energetics, etc…, — is an analog Force replicating in iterative quasi-similar Form, Solar gravitational Curvature Tensor-pressure-Force that operates to yield center-directed or centri-vectored rotary Forms of Motion-force, e.g., Earth rotation upon its own 23-degree tilted axis; Earth revolution around the Sun within/upon/along the ellipsoid Solar-System plane of Revolution.

The Sun emits radiation Energy of great magnitude and tremendous strength, at the same time that it projects an extremely strong magnetic Field from which emerges a powerful "Gravity-Field equivalent," the sum of which, engendering Tensor-pressure-force metrics, born of "Curvature Energetics," as followed by the Planets, which, in consequence, follow rotary Forms of centri-vectored Motion.

"Rotary Motion-Force Curvature Energetics," as enhanced by its own electro-conductive hot iron molten magma core, empowers the Earth as a cycling-Mass-Frame with its own bi-polar magnetic Field; its own Center-of-Mass; its own Core-centric Gravity-force; and its own planetary-satellite complex, fulcrum of which, being "Earth-Moon barycentre" at 1,100 miles below Earth surface, exerting "Mass-equilibrium Force Dynamics" that holds the Moon as a natural satellite for Earth-centric, geo-synchronous, rotary Forms of Motion.

Given that the whole Universe is primarily constituted of Matter-Mass in Continuum Space-Time, then, Curvature Motion pressure-tensor-Force Energetics are iterative properties of Electromagnetically-derived cycling Matter-Mass-Frames that "get re-translated" into quasi-similarly framed structural patterns of organization. Electromagnetism gives to Matter its Mass-properties of Motion. Matter's "natural properties" are organized for orderly functioning, in "differentiated fulfillment" of the universal "Input-Process-Output mechanism," e.g., Planet Earth receives solar gravitational radiation Energy inputs for ecological processing, the complex sum of which, sustaining its life-support systems for "Dynamic Whole-Entity Equilibrium," i.e., in spite of hurricanes, cyclones, typhoons, tsunamis, tornadoes, earthquakes, or volcanic eruptions, the Earth remains faithful to dynamic System equilibrium as a Life-Planet where seasonal cycles predominate for providing abundant agricultural harvests that nourish Human Life and livestock farming.

Consequently, discoveries in scientific Physics yielding mathematical knowledge are embedded within present-current-extant "workings of the Universe" as we witness it in its operational functioning. Thus, no evolutionary "biological missing link presuppositions" or "big bang explosion assumptions" need be posited in order to arrive at a scientific goal aiming to mathematically unify universal understanding in terms of a General Unification Theory of Motion-Force.

Given that it is the extant Universe that is under study, and not its conjectured evolutionary past, then current-present-extant phenomena, processes, and events ought to be sufficient, in paradigmatic principle and experimental practice, for "fielding" or yielding all the material-physical cause-and-effect data necessary for arriving at a General Universal Unification Theory of Motion-Force.

The "past Universe" which astro-evolutionists "worship" as a catalytic agent for effecting the Present, can never be reproduced or replicated in the first place, not only because there are no relics or remnants of so-called "intermediate structures" that have already disappeared, but also because the Past is presupposed to have "mutated into the Present," current Forms of which, embodying in their very operational functioning the evolutionary process itself, e.g., many bio-evolutionists falsely presuppose that the Human baby in utero assumes "fish-like gills" for Oxygen extraction in order to survive in his mother's womb. Fish do extract Oxygen as water passes through their externally located gills. The in-utero Human baby, however, never develops fish-like gills, but only "ingests-and-expels" the amniotic fluid through his mouth, and not through fish-like gills, to there-from extract Oxygen.

The only thing concerning the in-utero Human baby to which can be attributed a similarity with "fish condition," is that as the fish is under water, the Human baby lives within amniotic fluid. But there, also ends, the comparison.

Because the baby's lungs are still undeveloped, he ingests amniotic fluid to extract Oxygen there-from. As they are only suitable for breathing Earth atmospheric content, in-utero, lungs are unsuitable for extracting Oxygen directly from fluids. The Human baby, living within the amniotic fluid sac, must therefore extract Oxygen from the amniotic fluid for "lung processing" occurring within his developing body, but not by breathing the fluid as he would breathe Earth atmosphere. Rather, by ingesting it through his mouth, after which the amniotic fluid is then expelled there-from for new replenishment.

No one has ever seen a fish come out of a woman's womb! The Human baby in-utero has never been a fish! Therefore, at no time does the Human baby ever assume fish-like physiological Form or anatomical structures. From biological conception beginning at the fertilized union of a Man's sperm and a Woman's ovum, to in-utero organic maturity and then birth, the developing infant in the pregnant Woman's womb is always a Human Being!

No evolutional phenomenon has ever been witnessed on the Earth since the dawn of universal Creation. Bio-evolutionists and astro-evolutionists have no cause-and-effect explanation to validate the Theory of Evolution as scientific; nor do they provide valid answers as to why "inter-species evolution" that produced the so-called "biological missing links" would suddenly stop while the humanoid ancestral genotype (the apes) still exist, during the same Species-lifespan as its presumed "descendants" (Human Beings), — e.g., apes still exist within the same time-frame as Human Beings, or during the same Human Species lifespan. In other words: (1) Apes are still with us — "the link" is not missing; (2) The presupposed "ape-ancestor biological missing link" has not been unearthed from the so-called "fossil record," because it simply never existed.   .

The so-called "universal Past" as presupposed by astro-evolutionists, and which, detection of the "relic cosmic background radiation" is supposed to reveal, is not necessary as a mathematical construct, factor, parameter, determinant, or variable to our discovery of the universal unification theory.

For, current knowledge or present understanding confirms that extant Forms of Energy have ever-present "cause-and-effect On-sources" within Continuum Space-Time Curvature, e.g., Sun-light-radiation-Energy in our Solar System comes from a Star (the Sun) that is and has always been there, relationships or inter-exchanges of which, can be understood within the framework of already discovered and proven-valid laws of Physics: such as, the law of Gravity, the laws of Motion, Electromagnetism/Field-force, Quantum Mechanics, Newtonian Mechanics, and Relativity Curvature Energetics.

Quantum Mechanics address the atomic Frame; Newtonian Mechanics address the solar-planetary Frame; and Relativity Curvature Energetics address thermodynamic cycling of Mass-frames, respectively, for Frame-specific Transformation of Energy in Continuum Space-Time.

At the same time that astro-evolutionists are consumed and obsessed with activities geared to "proving right" the so-called "big bang explosion theory," they also propose that the fundamental forces of the Universe that need unification are: Gravity, Electromagnetism, "the strong force" and "the weak force."

Given that both "the strong force" and "the weak force" are apparently emergent properties of Mass ensuing from functional operations of the Fundamentals of Electromagnetism, treating them separately mathematically would only invite unnecessary redundancy that does not effect formulation of, or result in yielding, proven-valid equations.

Thus, "the strong force" and "the weak force" are not Fundamental Forces; and they ought to be treated mathematically, within the functioning framework of Quantum Mechanics, operations of which, giving rise to the Electromagnetic Properties of Matter.

However, only Gravity lacks the particulate characteristics that the other forms of Force embody. "The strong force" and "the weak force" engage "field force Forms" that "re-translate" the properties of electromagnetic charge displacement. All Force forms, from field force to the strong force, from Curvature tensor-pressure-force Energetics to "the weak force," only "re-translate" the electromagnetic properties of Matter; — They all involve particulate-charge Motion-in-Space displacement actions; except Gravity.

What can we say about the Force of Gravity? Gravity is an emergent property of Force arising from operations of Electromagnetism, functioning of which, yielding Field-force that then engenders centri-vectored rotary Forms of Motion, i.e., rotation and revolution, and/or combined-composite Motion-force actions there-from. It is then that we can speak of Gravity, as emerging from effects consequential to actions upon Matter-mass in Motion, as exerted by the Fundamentals of Electromagnetic Field-force.

Gravity emerges from effects engendered by both rotation and revolution, that are, in Force-effects, co-linear, co-equal, and opposite — Gravity emerges from centri-vectored, particles-based, Field-forms of rotary centri-vectored Motion-force, e.g., rotation and revolution. Rotation and revolution are activated by solar magnetic Field binding-Energies, as projected from the Sun, to engineer tensor-Curvature Motion pressure-Force, e.g., The Solar System ellipsoid plane of Motion-force, whereupon revolve all Planets around the Sun.

The paradigm that is most suited for purposeful unification of the Universe is Biblical Scientific Creationism, as it is the only Frame-of-Reference that empowers Physicists with the capability to discover and integrate the "Continuum Motion-Force algorithm" that God has put within the inner-workings of the Universe.

This integrative "Continuum Force algorithm" has to do with the Fundamentals of Electromagnetism, operations of which, permeating how properties of Matter-Mass Frames-in-Motion are made manifest.

Electromagnetic Properties of Matter are God's endowments of Intelligent Design to the Universe, as engineered for keeping the Gravitational Space-Time Continuum working together, within the framework of Field-Curvature Motion-force.

Discovery of such an "algorithm" cannot be accomplished while astro-evolutionists are imbued with a mindset, or rather "mind-cast," steeped in "moral relativism" or "value-free" inner-spiritual activity. Things are "relative" in accordance with a super-arching Standard: The Speed of Light, which has a fixed value, standing constant at 186,000 miles per second in a vacuum.

As a final, integrative, unified, Continuum Theory of Universal Motion-force, its mathematical equivalent will have "Relativity properties" of being induced by both Electromagnetism and Gravity, the sum of which, being the Universe's Fundamental Forces. This "algorithm" will then have the quality of dispensing an overarching, absolute, controlling paradigmatic impact, upon all the Forms of Energy, and upon all cycling mass-Frames that are made manifest by cause-and-effect relations in the Universe.

The "mental-force Energy" or "Creative Force-of-Intellect" that prevails in Human Beings is anchored in inner-being spiritual inspiration that quickens moral capacity for lawful enjoyment of constructive liberty, in peace, justice, and equanimity. It activates conscientious inquiry into God's natural Creation as "the source of original Capital," not in terms of "Economic Capital" alone, but rather, as the initial wellspring of all the Good we can ever pursue during the lifespan of our temporary existence on the Earth.

A God-inspired scientific impetus works as a "spiritual Force-field," so to speak, which is intrinsic to our psychic constitution for exhorting us towards a natural thirst for external discovery. Our hearts are no longer blind to God's love. We come to understand that Christ's forgiving power frees us for peaceful and righteous enjoyment of God's blessings.

Thus, "Spiritual centers of Human creative force" are continuously interfacing with physical centers of material Energy, in our pursuit of scientific discernment of the "intelligent design" already embedded within gravitational operations of universal Continuum Space-Time.

Newton and Einstein complement each other, as Centers-of-mass "attract" centers-of-mass even as cosmic bodies simultaneously "bend" or "curve" each other's space. The Solar-planetary System "recapitulates" in analogous quasi-similar Form, the atomic Motion-force-Energy complex. As we are factoring differentiated gravi-metric variables of Mass-cycling within Curvature pressure-force-tensor Energetics, respectively to each System, we realize that in the same manner that we are social by constitutional nature, redundancy of Form permeates natural phenomena, cosmic events, and universal processes: Atoms congregate to form molecules for enlargement to greater structures of Mass, as Human Beings assemble to form nuclear families for enlargement to greater institutions and organizations of Human society.

It is no accident that there is an affinity between our intellectual apprehension of external Reality for corresponding mathematical formulation (cosmology) and the physical organizational nature of the structure of the Universe (cosmography) — Conceptual schemes constituting frameworks of knowledge and understanding accompany corresponding objective application of structured discoveries.

This synchronous relationship between the Human Mind and externally apprehended Reality, climaxes as symbolic-abstract knowledge, the concretized encapsulation of which, can be representatively embodied in mathematically formulated derivative equations.

Human perception and understanding play a crucial indispensable role in the discovery of physical truths that factually describe perceived external Reality — detecting iterative organizational patterns in Matter-Mass Energy relationships, as made functionally manifest in structured cycling Mass-in-Motion frames, gravi-metric Differentials of which, overlapping for thermodynamic fulfillment of dynamic System equilibrium in execution of the Law of Transformation of Energy.

Force-fields that move cosmic bodies or Mass-structures in predictable thermodynamic cycles of Energy Transformation are iterative Forms of Curvature pressure-tensor-force Energetics, giving rise to equivalents of the Gravity-force. Matter can be transformed into "excited Energy" that then yields field-forces that activate "planes of universal Motion."

Relationships animating Matter and Energy as they interact within these cycling planes of Curvature motion can be mathematically codified as Human consciousness acts upon them for unified understanding. For example, $f = ma$; $F_g = G\,(m_1\,{}_x\,m_2/r^2)$; and $E = mc^2$ have led to the launching of rockets and spaceships, as well as to the creation of deadly atomic bombs (nuclear fission bombs) and Hydrogen bombs (nuclear fusion bombs).

Thus, for good or ill, the Human Being is a vital, self-conscious part, and creative necessity, in the scientific formulation of "the equation of physical Reality."

The controlling frame-of-reference, worldview, or paradigm through which these realities are "filtered," or understood and formulated for scientific application, is a necessary element of the psycho-emotional dynamic that is so essential for activating our capacity to integrate symbolic representation of physical laws with the organizational structures that govern unified natural phenomena, universal processes, and gravitational events.

Our "Spiritual optics" or "our perceptual matrix" operating to climax into how a Human sees himself and/or how he relates to his social and physical environment, constitutes "the psychic lens" through which he interprets phenomena that impinge upon his acquisition of knowledge as an experimenter-observer of material correlations embedded within universal Reality.

Is it possible that personal Human inner-vision or intellectual inner-perceptual platform for external apprehension "might interfere with" detection or discovery of the relative substance of micro-Physics, in the same manner that Human sight is also limited in the macro-sphere as regards the structure of Galaxies or in the micro-sphere as relates to atomic structure?

Is it possible that Human "perceptual optics" is determining "Quantum optics" — That "Quantum Gravity" or any other unification theory posited by astro-evolutionists is a product of their evolutionist worldview — ensuing from their belief in an accidentally appearing Universe existing due to only random chance probability, which then only leads to a probabilistic-random paradigm as to the nature of atomic sub-structure — that spiritual frames-of-reference, e.g., Evolutionism, might interfere with true estimations of the material structure of physical relationships between Mass-frames that are moving within Field-activated Gravitational Continuum Space-Time — not only in theoretical terms but also from experimental and mathematical standpoints?

Is it possible that Human ocular vision, even when aided by mechanical technology that transcends "differential magnitudes of Scale," might be unsuitable for discovering "der ding an sich," or "the true essential nature of things-in-themselves," because, like the technology, Human perception utilizes the Electromagnetic Spectrum?

Society also presents its own "field of influence" or "Social Interferometrics," which acts upon personalities; and within which, participating members choose appropriate courses of action. "The social sphere" is a real variable that affects opportunities for either "group think" or distinctive individual expression.

It is possible that "collective consciousness" is also part of the intrinsic Continuum algorithm — in the sense that a specifically sustained worldview that persists within "the scientific establishment," could influence what the experimenter might research, apprehend, understand, and profess.

It has also been posited that the mere presence of Human Self-Consciousness itself during the field of experimental operation might influence what is observed, how it is observed, how it is interpreted, and how particular results are analytically obtained.

The Scientific Method was established to facilitate the exclusion of irrelevant factors or extraneous variables while collecting experimental data. Is it also possible that there is a relationship between "collective psychic disintegration" and the apparent disorientation in the Sciences, such as in theoretical mathematics and continuum physics?

We already have, through applications of the Equivalency Principle, an understanding of Continuum between Matter and Energy, and between Field and Curvature, as activated by solar radiation spectrum Energetics. How about between Gravity and Solar radiation spectrum Energetics? How about between Gravity dynamics and Curvature Energetics? How about relationships between Field-force and Gravity-force, in effecting "gravity-equivalent Differentials?" (e.g., G-7, or Gravity-force equivalents away from G-1 Earth Gravity-force.)

Field dynamics engenders various derivative analog forms of Force, due to Differentials and Opposites vying for predominance within functional operations of interdependent, cycling Mass-frames. "Field-Derivatives" such as "the strong force" and "the weak force," are "re-translations" or "re-negotiations" of the Electromagnetic properties of Matter. For, where overlapping cycling Mass-frames intersect, they share "areas of turbulence," e.g., Perihelion Shift of Planet Mercury.

The quest for the Unified Theory of Continuum Curvature Tensor-Pressure Motion-Force, that would include Field-force, Gravity-force, Force-derivatives, and Force-equivalents, would purpose to integrate all "thermodynamically-moving/operating/functioning" Mass-frames in their cycling Differentials and Opposites, with utilization of alpha-numeric Constants that would "renormalize" or "reconcile" overlapping regions or "areas of turbulence," in order to "re-integrate shared perturbations" for dynamic System equilibrium, e.g., the Moon's matrix of phases, engendering forces causing oceanic tide surges; Earth "wobbling;" "Earth bi-polar oblation;" perihelion shift of planet Mercury; particulate mass-energy gains "re-translated" as cosmic radiation; particulate behavior interpreted as "waves of probability."

All universal events and natural processes operate within "cycling-frame ranges" whose boundaries are limited by "the common arresting Standard:" The "Speed of Light in a vacuum; and hence, the prevalence of cause-and-effect mechanisms that prompt those Mass-frames to thermodynamically cycle (— having a beginning point, a mid-point, and an end point) within Field-activated Gravitational Continuum Space-Time, in periodically occurring stochastic iterative patterns of Energy Transformation, e.g., During the 24-hour day period, relatively half or one hemisphere of Planet Earth is "bathed" in Sunlight while the other hemisphere experiences "Night-time darkness" or "absence of Sunlight."

Thermodynamic Cycling requires recurring patterns in "temporal time" (non-continuous Time or Eternal Time, as marked by iterative recurring patterns-of-operation) with a beginning, intermediate stages, and an end. Curvature Energetics circumscribes Planets as cycling Mass-Frames engaged in rotary centri-vectored Forms of motion, e.g., Earth Rotation within a period of 24 hours; Revolution in $365^{1/4}$ days or lasting one whole year.

Scientific Creationism helps us "interface" Human life-consciousness with gravitationally unified Continuum Field-binding Energies, in a consistent and healthy manner, because there are no inner-conflicts deriving from contradictory doctrines such as "struggle for survival of the fittest," that assault the incontrovertible "Reality of Human Mortality," the sum of which, nullifying unrealistic attempts at planning experiments for "conquest of nature." Ecological Nature continues while experimenters and researchers are honorably buried in the cemetery of their choice.

Thus, more realistic and true-to-Nature, is our faith in Almighty God, our Creator, who lovingly endowed us with temporal mortal lives on life-planet Earth where we are already empowered with God's Spirit, even as our Human nature is redeemed, by the love of Christ Jesus Messiah, who forgave us our sins, thus freeing us for spiritual rebirth, newness of Life, rekindled Hope, and rejuvenated and restored creativity.

****

PRACTICAL RESOLUTION OF APPARENT UNIVERSAL ABERRATIONS

When contradictions appear within a paradigm and its explanatory frame has been extended to the limits of its applicable frontier, a new conceptual-theoretical framework is created to resolve the apparent anomalies.

Before Isaac Newton (1643-1727 AD) had undertaken to explore gravitation for scientific discovery, important conceptual theoretical activity had taken place in Western culture to establish the foundation for the Scientific Method, such as the works of Nicolaus Copernicus (1473-1543 AD), Galileo Galilei (1564-1642 AD), and Francis Bacon (1561-1626 AD) that instituted a platform by which to evaluate a theoretical framework as to objective inquiry, scientific methodology, experimental procedures, and validation process.

Between Newton (1643-1727 AD) and Einstein (1879-1955 AD), — from Gravitation to Relativity — there is the discovery, understanding, operationalization, industrialization, and mechanization of Electricity, as well as the resourceful exploitation of Petroleum and its by-products, as to functional utilization, research and experimentation, and technological innovation, as marked by the invention of the Telegraph, on the one hand, and of the steam-powered engine for moving railway cars, on the other.

Productive research activity interfaced with socially engaging technology to yield discoveries in Electromagnetism by James Clerk Maxwell (1831-1879 AD), and Michael Faraday (1791-1867 AD), that paved the way for the "technologization" of the electromagnetic spectrum with devices, appliances, machines, and instruments that ranged from radio and television, to the electron microscope, and telecommunications satellites.

Scientific breakthroughs are made, when integration of inner-perceptual understanding of external patterns of physical inter-exchange between constituent elements of observed phenomena, is formulated into a theoretical framework that explains their cause-and-effect mechanisms. For example, Newton, after having analyzed what he observed to be true concerning falling objects, discovered the paradigm for gravitational attraction between Centers-of-Mass, or between moving-bodies or moving objects in Space.

As an explanatory system, Newtonian mechanics became a scientific paradigm in which real interactive exchanges as represented in mathematical equations and formulae, proceed, in reality, as embodied in corresponding symbolized relationships. Conceptual equation relationships between Force, Mass, Acceleration, Momentum, etc…, proved to be internally consistent with the scalar Differentials pertinent to each frame's specific properties, when operationalized in application.

Proportions for each variable as symbolized within the equation are attuned to the proportions that exist in the real world, in specific applications that "re-translate" their relationships in adherence to the overarching paradigm as an explanatory Frame of reference, e.g., a rocket ship launched into space, a rock pushed out of place by a bulldozer, the Moon orbiting the earth.

All Frames remain faithful to gravi-metric parameters that unify Continuum Space-Time as a whole-Energy entity in dynamic System equilibrium. The symbolic values for variables remain faithful to the real proportional relationships as embodied in real quantities for which the equation is a true representation in abstract form.

This faithfulness of symbolic representation to true-to-reality proportions is also transferable to valid applications, such as driving an automobile, flying an airplane, or the revolution of a planet around the Sun.

The need for a new paradigm arises when the explanatory scheme of the prevailing frame of reference does not explain the cause-and-effect mechanisms of newly observed phenomena in a manner that fits into its mathematical theoretics, research-and-experiments framework, and technological applications. When there are phenomena that fall outside of a paradigm's predictions, e.g., As Newtonian Mechanics had no specific explanation for the Perihelion Shift of Planet Mercury, the particular case was addressed, by formulating a new paradigm that provides a pertinent resolution, i.e., General Relativity Theory.

The new paradigm, e.g., General Relativity Theory, must retain "Scientific Continuum," experimental applicability, and reproducibility-integrity, relative to the already-proven old paradigm, e.g., Newtonian Mechanics; for, as valid scientific theories overlap each other, they share the same necessary universal Reality for a mathematical foundation: Moving Centers-of-Mass in Gravitational Continuum Space-Time Curvature.

Where mass-Frame cycling operations overlap, "regions of perturbations" arise, which must be "reconciled" or "renormalized" via alpha-numeric Constants that "renegotiate" prevailing relations and transactions between Energy, Tensor-pressure-force, temperature, Momentum, and Motion Differentials as Centers-of-Mass compete for sustaining their particular fulfillment of dynamic System equilibrium.

Consequently, both Newtonian Mechanics and Relativity Theory are internally consistent with Continuum universal Reality, as operating within the framework of Gravitational Interactions of Centers-of-Mass in Field Curvature Tensor-pressure-force Motion.

Both Newtonian Mechanics and Relativity Theory would have scientific validity within a mathematical formulation of a General Theory of Motion-Force that unifies Electromagnetism (with strong force and weak force as appendages) and Gravity, within the frame of Curvature Energetics for universal dynamic System equilibrium, as "reconciled" with Field-force tensing-pressure Motion, that impels cycling Frames or Centers-of-Mass, to functionally operate for thermodynamic Energy Transformation as they satisfy the Input-Process-Output principle of universal Continuum.

This "General Continuum Universal Motion-Force Theory" would work as an "algorithm" that frames all valid paradigms for complementary applications, not only through analytical research-and-experiment operations, but also through "real-world results" that do not contradict each other.

Given the iterative nature of structural arrangements espoused by Centers-of-Mass in-Motion within a Universe that has Oneness Integrity, then, the context for uniformity exists: As all validly proven physical laws that address specific phenomena, events and processes, will ubiquitously apply within all environments of the same kind, where physical-material conditions are quasi-similar.

Newton's Law of Universal Gravitation and Laws of Motion will operate in all solar systems where planets revolve around a certain Star. In the same vein, the Theory of Relativity will apply where stars and planets interact to "bend" or "curve" each other's space. Whence originates the need for a new paradigm? Because Newtonian Mechanics, though applying to phenomena of the same kind where conditions are similar, could not provide a valid explanation for the perihelion shift of planet Mercury; thus, this phenomenon necessitated a new paradigm whose explanatory scheme would present a mathematical frame that would formulate the cause-and-effect relationships accountable for the mercurial shift.

The "new explanation," brought about by the Theory of Relativity, however, does not invalidate or cancel Newtonian gravitational theory, but only provides cause-and-effect mechanisms for the "new conditions" that fell outside of the explanatory limitations of its reference frame.

The Theory of Relativity adds to the integrative Continuum that permeates scientific activity of the Human Mind for uniform understanding of universal events, processes and phenomena, as embodied respectively, in mathematical theoretics and applied sciences. Since the Universe is held together by physical laws that are internally consistent with the "Continuum Motion-force algorithm," the Theory of Relativity "takes off" where Newtonian mechanics had reached its frame-of-reference boundaries. And to the extent that the Theory of Relativity will have reached its applicable explanatory limits, a "new paradigm" will be needed in order to formulate the equivalent of a General Theory of Motion-force. This Unified General Theory of Motion-force would have to "take off" where the Theory of Relativity had ceased to provide cause-and-effect explanations for natural phenomena and cosmic events.

The Unified Field Theory of Continuum Curvature-Motion Tensor-Pressure Force would primarily treat both Electromagnetism and Gravity as "fields," or "areas of manifestation," signifying co-equal, co-linear, but oppositely vectored Forms of Motion-force. This "mathematical treatment" of both Electromagnetism and Gravity as "fields," would also mean the integration of Quantum Mechanics with Newtonian Mechanics, as well as the integration of Relativity Dynamics with Mass-in-Motion Curvature Energetics, all of which, being "re-translated" Forms of Electromagnetism for: (1) Emergence of Gravity-force; and (2) Causation of centri-vectored patterns of rotary Motion, namely rotation and revolution, and composite combinations thereof as embodied in rectilinear and curvilinear Forms of Motion.

The above-described frame-of-reference designed for paradigm formation inheres in the way that mathematically-driven scientific progress is achieved for Continuum inner-conceptual consistency as demonstrated in both operational Theoretics and technological applications. Valid new theories build on proven old ones, for they must remain consistent with the integral "Continuum Motion-force algorithm" accountable for aggregating the Universe as a whole-Energy system.

Theoretical frames proven factually valid will yield corollary laws and ground-breaking new applications emerging from innovations born out of systematic and consistent adherence to their scientific principles as attuned to the Scientific Method. Newtonian Mechanics had reached its explanatory limits, thus will the Theory of Relativity; and a new approach will come to unify and integrate all physical laws within the Frame of "Continuum Space-Time Curvature Energetics."

It is self-evident that "interactive Force Energetics," e.g., as field force, gravity force, the strong force, or the weak force, operate as "controlling factors" in all Mass-in-Motion cycling frames and at all levels of complexity, such that their operations fulfill the universal "Input-Process-Output thermodynamic mechanism," as Mass-in-Motion Frames, respectively, impact each other, in the quasi-iterative likeness of "re-translated Forms" of the electromagnetic properties of Matter.

"Energy is never created nor destroyed but always transformed." Force is a property of Mass-in-Motion; and it acts upon and transacts with other Forms of Mass-in-Motion, which themselves are displaying such a property in other Forms of Force. Electromagnetic Force and Gravity Force are properties of Mass-in-Motion. And given that Space is intrinsically void unless occupied by Matter cycling in "Thermodynamic Time," then, Mass can only act upon Mass. Though it is said that according to Relativity Theory: "Mass curves or bends Space," this property is not "discovered," until another Form of Mass "is thrown into such Space." The "bending effect" or "curving effect" is observed only upon Mass that is "moving" within such Space, and not as a property of the void-Space itself.

These "interactive Force Energetics" arising from gravitational Curvature tensor-pressure-Force motion-Forms, appear in many analogous patterns evoking "field tensing mechanisms" and equivalents of "gravity stress dynamics." Whether as field, as the strong force or as the weak force, "interactive Force Energetics," embody "Gravity-force equivalents" taking the Form of "re-translated Electromagnetism-force," whose electro-motive charge characteristics commonly cause "displacements" in Continuum Space-Time — like charges repel; opposite charges attract, hence, triggering rotary Motion-force around centers-of-Mass.

Electromagnetic projections of Curvature pressure force tensing engender gravitational binding energies that get expressed as field force and equivalents of the gravity force, whose complex interplay climaxes into centri-vectored forms of rotary motion, i.e., revolution and rotation.

In the physical sciences, from quantum mechanics to Newtonian mechanics, from electromagnetism to Relativity dynamics, from thermodynamics to computer physics— knowledge in each respective branch appears to have been pushed to its apparent theoretical limits.

Applications of theoretical research frameworks in Physics seem to be concerned with the foregone conclusion that Scientists have to prove the so-called "big bang theory of universal origin." In chemistry, discovery of "laws of molecular bonding" have been maximally exploited while reaction-chamber applications continue, as many elements are "re-combined" via outer-orbit electron energy bonding, respectively, under variables of specific reaction-emperature and pressure.

Chemical reaction chambers operate, as experiments on elemental inter-exchange are carried out, in order to determine how gases, liquids and/or solids will combine to form compounds. Approaches to Chemistry analytical research and experimental formats differ greatly from evolutional approaches to Physics, in that, results obtained from chemical reactions answer to chemo-physical laws upon which socio-biological doctrines have no bearing. Thus, when it comes to Chemistry research and experimentation, Scientists are free to discover processes yielding new products or compounds, without ideological impediments from Evolution Theory that would impel them to pre-construe basic analytical frameworks that distort observed data, disparage the Scientific Method, and falsify experimental results obtained therefrom.

Consequently, universal Cosmogony in accordance with Scientific Creationism as laid-out in the Holy Bible, is consistent with proven-valid universal Cosmology as well as observed and known universal Cosmography (— the study of universal origin; its form, content, organization and structure; and its general description in whole or in part, in terms of geology, geography and astronomy).

For, the Biblical account of Creation in Genesis, regarding the Universe and the origin of Human life, proceeds from divine revelation by God's Spirit. As revealed in the Biblical worldview, the Account of the beginning of the Universe and of the origin of Life on the Earth is consistent with known scientific facts that are the building blocks of real scientific knowledge, the sum of which, ensuing from true-to-reality applications of the Scientific Method in framing research-and-experimentation.

Thus, methodical study of the Universe and analytical knowledge there-from, do support Biblical Creationism, in that, no proven scientific theory has ever contradicted the textual content of the Holy Bible.

How could the Biblical Genesis Account contradict the Law of Gravity and the Theory of Relativity when the letter and spirit of the text succinctly and clearly declare that a perfect, righteous, omnipotent, and omniscient God, is the Author and Creator of all things that exist?

The Law of Gravitation states that mass "attracts" mass — Nowhere does biblical content contradict this Law. Likewise, the Theory of Relativity explains how mass "bends" or "curves" the Space upon which or within which it is acting, as the Speed of Light remains the standard limiting factor at 186,000 miles/second in a vacuum.

Primary evolutionist complaints underline the fact that it cannot be proven that God created the Universe, Human life, and all that exists. But, in the same vein, it cannot be proven that all Reality, including Human life, began with an initial so-called "big bang singularity," from explosion" of which, ensued or emerged the extant Universe, as an entity of such "organized complexity" as anchored in intrinsic corresponding physical Laws, that only a supernatural Being of limitless intelligence could have designed it.

Consequently, evolutionist pronouncements are equivalent to "fables" that can never be scientifically substantiated by corroborating evidence, such as to be obtained, as required by the Scientific Method, from either laboratory, or "field-boots-on-the-ground" research and experimentation.

Astro-evolutionists concoct objections to the Biblical Account of universal origin, from assumptions and presuppositions that are purposely contrived to compete with the Biblical worldview, the sum of which, constituting a godless philosophical framework of atheistic "random-chance principles of meaninglessness, that mask not only "enlightened self-interests" but also deliberate obfuscation and deception. Rather than elucidating a scientific paradigm, astro-evolutionists confabulate and confound the masses via farfetched scenarios that directly oppose the Scientific Method and its sine qua non cause-and-effect mechanisms or processes as required for theoretical validation and data verification.

Therefore, what astro-evolutionists require from faithful Christians, they, themselves, cannot deliver! Astro-evolutionists demand a theistic proof of universal beginnings, while they themselves cannot provide proof for universal beginnings from their own godless " big bang singularity explosion" theory. They question the veracity of the Biblical Account of universal Genesis yet are intolerant of well-founded constructive dissent from opponents of Evolutionism who also demand experimental or "on-the-grounds" proof for their probabilistic assertions.

Due to their inability to disprove the existence of God, astro-evolutionists' rejection of the Biblical Account is directly connected to their unbelief and refusal to recognize God as their only Creator. Thus, they choose to vest their faith in a presupposed random-chance probabilistically self-created Universe, which they never witnessed, and for which they can offer no verifiable or reproducible experimental proof.

Biblical Creationism, therefore, is not "on equal footing" with the Theory of Evolution. No proof needs to be "created" for verifying Biblical Creationism, but rather, God's glory as Creator, is proven within the scientific discoveries as well as within the physical Laws that we, Human Beings, understand as to be productively sustaining universal workings and operations. All in all, Scientific Creationism is consistent with both scientific laws and material discoveries that together reveal to Human Beings how the Universe really works, as a physical system qua physical system.

The Universe is always in unified Oneness! However, compartmentalized dissections and divisions exist only in the Human Mind for purposes of discovery, analysis, understanding, and replication, e.g., Quantum Mechanics and Relativity Dynamics co-exist and overlap, transact and interface, within the same universal Continuum Space-Time! Thus, how can there not be "a general unification theory" for the whole-oneness Universe as we know it!

The origin of Man as revealed in the Biblical worldview is also consistent with Human nature or spiritual Human constitution. Only in the Christian worldview as predicated upon revelations from the Old Testament and from the New Testament that The Human Person is endowed by God with a spirit, a body and a soul (mind, heart and will.)

Thus, the Judeo-Christian worldview is internally consistent with both the Human psyche or inner-world and the physical environment within which the Human lives: God created the Universe, after which, God created Human Beings "in His own image and unto His own likeness" to live and prosper within that same previously created Universe. Thus, God exists; the Universe exists; and the Human Being exists!

For life in the Universe: Every thing "fits well together" — the Creator, Human beings, and the physical-material and/or ecological environment.

There is "fine-tuned consonance" between all things created by God for purposive Human existence. "Let there be Light; and there was Light!" encapsulates a "spiritual-scientific correspondence" as well as a "conceptual-objective correlation" between our Creator, us, and the environment created for us to live in.

There is physical-biological Life; and there is spiritual-soul Life! Faithful adherence to God's commandments of righteous living on the Earth, in communion with his divine Will, constitutes, therefore, our "spiritual life-support system."

The Human body "embeds" life-support systems that climax into signifying certain "vital signs," such as heart rate and temperature, just as the Earth is built with inherent ecological life-support systems that seasonally cycle for the benefit of fauna and flora. However, concerning Human Beings, biological and ecological life-support systems must be complemented by spiritual life-support systems for inner-fulfillment and beneficent environmental interface.

Spiritual life-support systems, biological life-support systems, and ecological life-support systems, must thrive together in fine-tuned harmony for prosperous and more abundant earthly living. There is "Spiritual Continuum" between God and us for constructive moral living, which corresponds, to the "Organic Continuum" that prevails, between the ecological earthly environment and our Human body, as well as with the life we live, for good and faithful stewardship of the Earth, the sum of which, being a wellspring of natural life-giving and life-sustaining resources.

Human life within the gravitational Space-Time Continuum is consistent with the psychic integrity that faith in God brings to the Human person. Faith in God establishes communion with our Creator, such that, as Heaven and Earth are united within the sovereign exercise of Continuum divine Will, our own temporal earthly existence as spirit-Beings, develops, grows, and matures, according to the "Christ-like principles of Being-ness" that God has established for us, Human Beings, to live by, until such a time that final or terminal Entropy be fulfilled.

Mortality or final Entropy, or, Death or terminal Entropy occurs both spiritually and naturally: Spiritually, due to unrepentant sinfulness, and biologically, due to intrinsic processes of decay that inhere in all carbon-based creatures.

Great increases in Knowledge of the Universe are not incompatible with mortality, but Entropy does limit our command-and-control of Knowledge to a range of understanding, boundaries of which, continue to expand and extend, as advances from the works of previous generations become new foundations of Knowledge that are preserved, for sustaining cumulative Good ensuing thereof, towards creative innovations.

Within the Frame of our bio-organic constitution, Conservation prevails over Entropy, until such a time all our bio-systems have attained fullness of maturity, after which "the pangs of Entropy" begin to assert dominance over processes of Conservation, thus resulting in aging, systemic fragility, pathological predispositions, and eventually, in death.

In the Universe, "the strong force," as indicative of a particular framing of the Electromagnetic properties of Matter, temporarily prevails over "the weak force," in order to sustain centri-vectored Forms of Motion-force that cause Continuum cycling Energy patterns in Mass-frames to intersect within overlapping "spheres of influence" or "areas of manifestation" — otherwise termed as, "fields," in Physics.

We notice redundancy of analogous Forms within iterative patterns of energy cycling. For example, "things revolve around a center of gravitational activity," e.g. in the Atom, in the Solar System. The prevalence of overlapping reference-Frames, operational ranges for cyclical events, processes and phenomena, and interactive relationships of gravitational force in temporal Time-Space, all, establish fulfillment of the Input-Process-Output mechanism in sustenance of Continuum equilibrium. Balance is static; equilibrium is dynamic.

For human beings, earthly life ends in the fulfillment of "Terminal Entropy" which is death of the physical body; however, spiritually, final Entropy is everlasting Hell, whereas final Conservation is eternal life in Heaven.

Scientific Creationism is consistent with the kind of created beings we are — its principles, commandments, and laws are universally applicable in faithful fulfillment of spiritual, "heaven-begun representations of reality," that are materialized in Continuum Space-Time "as embodied in earthly Forms" made manifest by the will of God, e.g., the laws of Heaven will not permit Lucifer's evil spirit to perpetuate wickedness in the spiritual realm; human laws proscribe murder, theft, and perjury, for mutual security in society.

Terrestrial and cosmic processes have their template or blueprint in "the heavenly realm" where the Invisible is the archetypal Source or Form for every concrete objective physical material Form. What is spiritual is primary; what is physical is secondary. The Spiritual precedes the physical; the Invisible precedes the Visible; abstract concept precedes material embodiment.

In the same vein, paradigmatic innovations come with conscious awareness of "new things" that cause transformation within our inner-being for a new interface with external Reality via "renewing of the Mind" for spiritual discovery, conceptual innovation, theoretical initiative, formative analysis, and creative experimentation.

When we are "born again," a complex spiritual transformation takes place in our inner-being for moral maturity, spiritual growth, and scientific discovery. Inner-peace flows with external synchrony. Calmness of spirit interfaces with the Organizing Principle. For, apprehension of Reality via study of God's Word empowers us with a "spiritual-scientific perspective," as the living Mind "engages in Thought-motion" for self-renewal and transformative understanding.

Thoughts of Faith and thoughts of Hope bring forth "constructive interface" between the Human soul and the external world, of which a loving God gave us responsible and accountable Stewardship. "Movements of Mind" within blessed provisions of God's righteous Will, climax into renewed perspectives of creative discovery that bear fruit for God's kingdom and individual Human spiritual prosperity on the Earth and in the Universe. (Isaiah 43:18-19; Romans 12:2, NASB, Holy Bible).

****

GRAVITATIONAL CURVATURE "SPHERES OF BINDING ENERGY" FROM TENSOR-STRESS-PRESSURE-FORCE ENERGETICS

The Principle of Continuum is germane to integrated understanding of the Universe that will yield theoretical formulation for a grand-unifying mathematical equation.

All validly proven frames of reference overlap each other in theoretical dimensionality regardless of the specific applications to which they render utility. All valid physical laws have uniform application within equivalent frames where conditions are similar. Interconnected frames of "energy-matter-mass-in-motion," overlap as "structures of entanglement." Their cycling Differentials and Opposites are "negotiated" for integrative wholeness, via "hidden variables," acting as "reconciling Constants," respective to each frame complex striving for dynamic System equilibrium.

These mass-in-motion frames are "steady-state entities" — in fulfillment of dynamic System equilibrium — with definite patterns of energy cycling whose "Relativity ripples" effect Time-measured gravi-metric reverberations that climax as perturbations throughout Continuum Time-Space, e.g., the Earth "wobbles" and is "slightly flattened" at each Pole; planet Mercury displays a Perihelion Shift; the Moon's variable orbital trajectory causes oceanic tide surges on the Earth; particles accelerated to a fraction of the Speed of Light "shed their mass" by emitting radiation even as they gain increases in "mass momentum-Energy."

Quantum Mechanics, Newtonian Mechanics, Relativity Dynamics, — solar gravitational Field-force tensing-Energies, solar ellipsoidal plane stressor-pressure-force Energies, and the galactic-Star complex, — all, are held together via "Curvature Binding Energetics" incorporating centri-vectored patterns of iterative cycling Motion within the Frame of "Continuum Angular Momentum Synchronization and Preservation."

Each frame-of-reference or Mass-in-Motion cycling-Frame, is manifest with gravitational "Input-Process-Output Differentials and Opposites," — e.g., pressure, temperature, velocity, mass, electromagnetic signature, revolution duration and rotation periods, gravity quotient, radiation absorption rates, geo-atmospheric transformative structures, etc . . ., — that necessitate the "catalytic action" of alpha-numeric Constants to "renormalize" or "reconcile" their Relativity interactions, due to "spheres of relationship entanglements" being exerted, emitted, effected, and projected by co-transacting Centers-of-Mass, Centers-of-Gravity, and Centers-of-Field that delineate the dynamic of "Curvature motion-force Energetics" — which in totality, engrosses all Curvature pressure-tensor-stress Motion-force Energetics that comprise all composite-combined "rotary Force equivalents" generated by both Electromagnetic Field-force and Gravity-field force, and by their curvilinear and rectilinear "quasi-equivalents."

Is the strong force that binds protons and neutrons due to Field-force activity, or due to Gravity Force exertion?

At the Quantum Mechanical frame-of-reference, Field and Gravity are "entangled" within the Frame or "areas of manifestation" of electro-motive charge interactions that climax into rotary nucleus-centric micro-Curvature pressure-tensor-stress-Force energetics. Opposite charges attract; like-charges repel. Particulate charge-and-mass, counter-centric tensor and pro-centric stress metrics, "operate to compress" neutrons-and-protons into an "attractive pressure-Force dynamic" even as the attraction-force between protons and electrons (opposite charges) is weaker than the repulsive force between the nucleus and electrons due to Neutron activity — and hence, affording molecular change, as well as potent effects causing Electrons to always remain "at a respectable distance" from the atomic nucleus — even in the Hydrogen atom, in spite of absence of neutrons.

In Quantum Mechanics, Gravity is a stabilizing "nucleus-counter-centric Force" — "answering to nano-Mass-in-Motion quantities" — while the Electromagnetic Field engenders a "nucleus-centric attractive Force," the totality of whose inter-transactions, giving rise to Forms of motion-Force as embodied in "the dynamic" of electro-motive particulate charge displacements.

This "Force Dynamic," in accordance with Newton's Third Law of Motion, involving co-linear, co-equal, yet, opposite Forms of Motion-force, — namely, Electromagnetism and Gravity — engenders counter-vectored tensor-pressure-stress Energies Differentials, all of which composing a "theater-of-phenomena" climaxing into "an Atomic-frame Curvature-plane Dynamic" — quasi-similar to the Solar ellipsoid Plane-of-Revolution — but which is interpreted as "waves of probability," e.g., comprised of composite-recombinant Forms of rotary, rectilinear, and/or curvilinear motion due to near-Speed-of-Light velocities by micro-Masses-in-Motion as affected-impacted by Electromagnetism-based research-and-experimentation technologies.

What causes Electrons to revolve around the atomic nucleus — to neither "crash into it," nor "escape from it"? The positively charged proton dominates the nucleus as it projects a strong attraction force towards the negatively charged electron which is simultaneously repelled by the nucleus due to the massive charge-less neutron's "grounding-insulating action."

Curvature pressure-force Energetics predominate in "re-translating" the electromagnetic properties of Matter, as field-induced Motion and as motion-Forms arising from "equivalents of the Gravity-force."

In the atomic frame, Curvature Energetics, Gravity-force and Field-force dynamics are "in entangled-equivalency of exertion" encapsulating "the tensing-stressing climax" of electro-motive-charge displacements, in conjunction with "gravi-metric cycling Differentials and Opposites," caused by centers-of-Mass, centers-of-Field and centers-of-Gravity.

Electrons revolving around the atomic nucleus partake of all variant Forms of centri-vectored Motion composed of rotary, rectilinear, and curvilinear patterns that makeup a Complex of composite-recombinant pro-centric (centripetal) and counter-centric nucleus-vectored trajectories (centrifugal), as they are simultaneously "attracted-and-repelled" by "the strong force" that is generated by the "proton-neutron entity" at the nucleic atomic center.

There is Conservation of Energy due to "the proton-neutron strong force" overcoming "dissociative-disintegrative tendencies" contained-embedded within apparently indeterminate electron trajectory Forms. But, due to the "gravitational tensility" or "range-of-Motion variability" intrinsic to the micro-plane of Curvature of the Atom, as necessary for molecular change, periodic-table chemical elements always possess Electrons that engage in nucleus-centric motion patterns for the preservation of atomic integrity.

Molecular change is neither dissociative nor disintegrative, but rather answers to each particular Element's respective affinity for reacting with other similarly attuned Elements in order to form organic compounds and inorganic substances, e.g., H_2O (water) and H_2SO_4 (sulfuric acid).

Forces, Motion patterns and Energy cycling Differentials and Opposites are in "interactions of relative equivalency" as mediated by "conversion factors," and as "re-normalized" or "reconciled" by "hidden-unknown variables" that are mathematically "negotiated" as alpha-numeric Constants.

However, due to cycling Differentials and Opposites, "interactions of relative equivalency" do not amount to similarity of thermodynamic condition or equality of quantified proportionality; hence, the existence of "areas of turbulence" where "mass-energy-motion-force frames" overlap, intersect or interface, as they undergo range-specific changes while fulfilling the Input-Process-Output mechanism for dynamic System equilibrium.

"Energy is never created nor destroyed but always transformed" — Thus, there can be only differentiated iteratively patterned structured-organized Forms of the same Energy template! The Sun, due to its essential Energy-form and its great Mass, engenders a multi-planetary ellipsoidal plane-of-Revolution that is tensing-stressing-and-pressuring Solar-System inner-Space, with Curvature-sourced, electromagnetically-based, radiation binding Energetics, while the Earth, as a type of "ground-State" or "congealed" or "non-nuclear Energy-form," holds only one natural satellite, the Moon, which displays geo-synchronous Revolution.

For purposes of illustration, let us agree that Earth gravitational activities will be differentiated between "external" and "internal." The Earth, as a cycling mass-Frame, displays "external" magnetic-field rotary motion-Forms, e.g., helio-centric Revolution and geo-centric Rotation, while at the same time, imposing Earth-centric, phase-cycling, geo-synchronous, rotary Motion-force patterns upon the Moon.

Internally, helio-centric Revolution is countered by geo-centric self-reflexive Rotation, the complex total operations of which, climaxing into Gravity-force Dynamics upon Earth-surface, within Earth-atmosphere, and within Earth-oceans and waterways.

These Gravity-force Dynamics engender Earth core-centric Forms of motion, in rotary, rectilinear, and curvilinear patterns that follow Earth line-of-radius, e.g., the rectilinear-radial motion of an apple falling towards Earth center-of-Gravity and center-of-Mass ; a plane flying in Earth atmosphere in rotary-curvilinear motion patterns; a vehicle traveling on Earth surface in a composite trajectory, constituted of rotary, curvilinear, and rectilinear patterns as it "negotiates obstacles" in its path.

The Earth is a spherical planet-Mass in geometrical Form. A bicycle wheel with metal pokes displaying vectored intersections that cross-end at the center-of-the-wheel is not geometrically spherical, but were it spherical, it would represent a befitting illustration of how Gravity-force Motion on Earth, like those metal pokes, takes the Form of centri-vectored patterns that are always directed towards Earth center-of-Mass or center-of-Gravity.

Earth external rotary motion forms, e.g., Revolution and Rotation, are electromagnetically- induced by "Solar Gravitational-and-Field Curvature Energetics."

In response, Earth Mass-cycling dynamics, climax into "fields of influence" or "areas of exerted manifestation" that "capture the Moon" "or keep the Moon," for geo-synchronous rotary-Revolutionary Motion around the Earth. Then, as "transmuted" or "re-translated" by Earth center-of-Gravity, center-of-Mass, center-of-Field, and centers-of-Motion-Force, these "fields of influence" or "theaters of operation" simultaneously engender within the Earth-Mon frame, "Gravity-force equivalents" that, upon the Earth Surface, in its Oceans, and in its Atmosphere, operate, — at 9.88 meters per second squared, (g = 9.88 m/s^2), respective to each unit of Mass "traveling therein," — as an "Emergent Force-property" or "Emergent Property-of-Force," from external-to-Earth centri-vectored rotary forms of Earth-Mass-in-Motion, i.e., Earth Revolution around the Sun and Earth Rotation upon its own 23.88-degree-tilt axis.

Gravity-Force is thus an emergent cause-and-effect property of Mass-in-Motion (Mass whose "Motion" or "Movements" are engendered by the "Motion-triggering" Electromagnetic Field-force — the greater the Mass of an object, e.g., a spherical Planet, the greater the Gravity-force exerted thereupon: On its surface (landmass), over its surface (atmosphere), and within and upon its ocean-waters, if any (its seas,") — Said Planet-specific and range-specific Gravity-force, e.g., Earth G-1 at 9.88 m/s^2 acceleration rate, being exerted upon objects "moving" thereupon, that are "moving-and-falling" within the G-1-specific-range of "its area of manifestation" or "theater of influence," e.g., The closer to Earth surface is an object, relative to its Mass, at g = 9.88 m/s^2 in standard "G-1" acceleration rate, the greater a Force necessary to "overcome" or "break-free" from Earth "G-1" Gravity-force, e.g., When launching a 5-ton Space-rocket into outer-Space from Earth-surface, it requires greater engine thrust-Force, as opposed to when throwing an apple in the air on Earth-surface with "the thrust-force" or "throw-force" of one's hand; hence, cause-and-effect reasons-interactions why, around Earth "orbital bandwidth," "orbital rim" or "orbital area" wherein are satellites revolving, there, exists a Gravity-force that is less than "G-1," or otherwise referred to as: "micro-Gravity."

In "total void-vacuum Space," or "outside of Earth "orbital bandwidth," Gravity would be "Zero Gravity," as opposed to "micro-Gravity" characteristic of Earth "orbital rim area" of less-than-G-1 Gravity-force manifestation.

And, by the same token, given that the Moon is "less massive" than the Earth, whose "Gravity-standard" is "G-1," then we can logically surmise that the Moon presents, upon its surface, a gravity-Motion-force that is "below G-1," or "less than G-1." From our above-analysis, we conclude then:

(a) Earth: G-1 Gravity-motion force;

(b) Moon: Less than G-1 Gravity-motion force;

(c) Earth Orbital Brand-width or Orbital Rim, boundaries of which, prevailing within the circum-sphere area between the Earth and the Moon: Micro-Gravity;

(d) And Deep Void-Vacuum Space: Zero Gravity.

All universal Forces have intrinsic Mass-in-Motion-specific limits attached to the distance, geographical range, or perimeter-of-influence within which they effectively and sufficiently operate, as well as limits to the degree-of-intensity or degree-of-strength projected by their exerted "Force-pressure-tensor-stress dynamics," e.g., The Star in our Solar System, the

Sun, can only "keep" nine planets; our Earth can only "keep" one natural satellite or Moon whereas Jupiter, of much greater Mass-in-Motion, "keeps" more than one Moon, or more precisely 67 Moons.

All universal Forces, e.g. Gravity, Electromagnetism, are exerted within their specific "geo-spherical-region" as-to "range of applicable Force-strength-intensity", the sum of which, operating as "projected Curvature Energetics," respectively, which pertain to the particular "Unit of Mass-in-Motion," e.g., the Sun, from which they originate; thus, denoting encapsulated "spheres of influence," "theaters of operation," or "areas of manifestation," or simply-put, "Fields," — which is a more familiar term often expressed within "the Physics Establishment" to describe such "areas of Force-activity.

This "Field" or "area of Force-activity," possesses characteristics that are peculiar or unique only to the Unit-of-Mass-in-Motion or to the particular object's "resident-standard Gravity-force" as above-described, e.g., Earth G-1 Gravity-force "keeps" only one natural satellite, the Moon: That is its peculiar "Gravimetric Signature;" the Mass of that single Moon, being only one quarter of Earth-Mass. But Planet Jupiter has 67 Moons characterizing its own peculiar planetary "Gravimetric Signature."

The farther an object is to stray, move, or travel away from Earth centers of G-1 Gravity-force "theaters of operation," respectively, e.g., Center-of-Earth Mass or Earth "Core-center-of-Gravity," that object always requires an "external Force," e.g., engine thrust, the sum of which, must be greater than the "resident standard Gravity-force" (or G-1) of said particular Mass-in-Motion, e.g., It requires "greater external Force" for launching a 5-ton Space-rocket from Earth-surface into outer-Space than for lifting a small 2,000-pound plane into Earth Atmosphere; thus, the 5-ton Space rocket needing much greater thrust-Force than Earth "resident standard Gravity-force" at "G-1," relative to its Mass-in-Motion, in order to totally "escape" from Earth "Core center-of-Gravity" or Earth-core Center-of-Mass, for catapulting it into outer-Space.

There are cycling Differentials and cycling Opposites that espouse many thermodynamic Forms, e.g., of mass, motion, force, energy, angular momentum, planetary rotational vector, angle of inclination, revolutionary duration, Gravity-force strength, Field-force intensity, etc . . . , the complex relations and interactions of which, establishing distinctions or peculiarities between overlapping frames of Mass-in-Motion where analogous or quasi-similar "quantifiable properties" operate in iterative patterns of Energy processing. For example, as the magnetic Field of the Sun proceeds from an "excited Mass-frame," (nucleated-Energy Mass-frame) and the magnetic Field of the Earth proceeds from a "congealed-Energy frame" or "non-nucleated Mass-frame," ratios and proportions of quantifiable properties involved in Earth Revolution around the Sun, are different-in-nature and/or values from ratios and proportions of quantifiable properties involved in Moon geo-synchronous revolution around the Earth, or in Jupiter's revolution around the Sun.

Therein resides the problem in mathematically formulating "relative equivalency" between the "Gravity-field-of-influence" and "the Electromagnetic-field-of-influence," in the light of overlapping Differentials and Opposites, within the context of a "General Force Theory" that must account for "re-translations" or "transmutations" of the Electromagnetic Properties of Matter.

In what way can Electromagnetism be differentiated from "the strong force" and "the weak force" when they all partake of particulate charge displacement Motion-force Dynamics?

"The strong force" and "the weak force" are "sub-categories of Forces" that also partake in the Electromagnetic Properties of Matter functioning as Mass-in-Motion, but which cannot operate apart from the greater-atom-wide operations of the Atomic-Frame Electromagnetic Field-Force and of the Atomic-Frame Gravity Field-force; the latter itself, being engendered by the Atom-wide Electromagnetic Field-Force.

In the same vein, Gravity-force on the Earth (G-1) cannot exist or operate apart from the greater-Solar-System-wide Field-and-Gravity-Forces being projected from the Sun.

Externally the Earth displays rotary Forms of Motion elicited by gravitational field-Curvature forces that partake of Electromagnetism, whereas internally, within the Earth, Gravity-force upon objects is an emergent property of Mass-centric rotary Forms of Motion-force that cause them to move in recombinant motion patterns, along earth line-of-radius as they seek radial alignment with Earth Core center-of-Gravity and center-of-Mass.

Gravity "attracts" and Electromagnetic Field "curves" as they simultaneously thrive in "entangled Force exertions," the sum of which, being very difficult to separate or divide in essential functional operations, but verifiable only in effects.

Direct earth "external" motions are rotary in Form, e.g., revolution and rotation, whereas within the spherical earth, body-mass motions are core-centric in Form along earth line-of-radius, e.g., linear, rectilinear, curvilinear, and "rotary-arc" composite patterns of motion, e.g., An aircraft can participate in aerial acrobatics by performing all these patterns of Motion, to "spiral down" towards Earth Core center-of-Mass and center-of-Gravity along its line-of-radius.

Thus, as objects "arc along" its line-of-radius, in whatever Form of vectored Motion, Earth "internal" motion-Force Differentials and Opposites are in "entangled simultaneity" with Earth "external" rotational and revolutionary cycles, e.g., Given that the Earth is a three-dimensional Sphere, an aircraft flying from Los Angeles-USA to Moscow-Russia in an easterly direction, must continuously "make course corrections" to its flight-path patterns in order to remain within Earth spherical-volume or Earth circum-sphere periphery, while en route towards its plotted-mapped Earth-surface physical-geographical destination. Otherwise, without these "course corrections, the aircraft would eventually "pierce" Earth outer-atmosphere, to be propelled thereafter into deep-vacuum Space.

In addition, "an Electromagnetic Field-of-influence" has alternating electro-motive charge polarity displacement properties which "the Gravity field-of-influence" does not possess, e.g., a magnetic compass "detects" Earth magnetic field by pointing towards the North Pole; no machine exists to "seek out" Gravity "force indicators." Nonesuch exists! For Gravity-force is Mass-in-Motion dependent — the greater the Mass of the object, the greater the Force of Gravity exerted upon/within its Frame.

While an Electromagnetic Field displays electro-magnetic particles-induced charge-motive characteristics, a Gravity-field is an emergent property of Revolution and Rotation, both of which being centri-vectored rotary Motion-Forms that are engendered-induced by projected Solar Gravity-and-Field Curvature Energetics, e.g., there is no such thing as a "Gravity compass" to detect it as a "force" or as a "field." There is no such thing as a so-called "Graviton" to be detected by a machine as a "particle" or "wave."

As to the "Gravity-field," in order that "Gravity effects" are determined, the Observer must assess the behavior-properties of cycling Units-of-Mass "operating" within its range-of-influence.

Field and Gravity are "analogous" or "quasi-similar quantifiable properties" denoting "Force exertion," in that they both project "spheres of interactive influence" upon Matter-mass, causing it to react to that "projected influence property": A pattern of Motion will result, e.g., a Field will cause rotary Forms of Motion; Gravity will cause Motion in all its variegated Forms "along radial lines-of-attraction" towards or away from an Object's Center-of-Mass.

No Force exists without Motion, just as no Magnetic Field exists without being accompanied by an Electric Field. Motion is to Force what "thermodynamic cycling" is to Field Energetics. Any object or body that is operating where a "Gravity-field" is present, must engage in vector-directed Motion.

To the same extent that a Magnetic Field is always conjoined with an Electric Field, wherever a "Gravity-field" exists to cause "bending" or "curving" of Space, there will also exist an Electromagnetic Field causing objects to engage in rotary Forms of centri-vectored Motion around a Center-of-Mass.

And Continuum Space-Time Curvature engenders "binding-bonding Energy" in the same manner that Electro-motive charges cause in-Space-displacement as triggered by Electromagnetism, e.g., An electric fan blade assembly belonging to a fan connected to an electrical outlet is "driven to revolve" while situated at the Center-of-Mass of a copper-coil generating a Magnetic Field. Thus, no molecular bonding-reactions can take place in the absence of "Gravity-field Force-Motion as generated by the Sun;" and no molecular binding-reactions can take place in total Zero-gravity deep-void vacuum-Space. All functional universal operations take place within the boundaries of Mass-in-Motion Frames.

Planet Earth is "always on the move!" On Earth, molecular reactions are articulated while the Planet is completely "showered" with Gravity-and-Field Force-driven tensor-pressure-stress Motion, i.e., Rotation and Revolution, which in turn, engenders Mass-in-Motion Forms, within terrestrial G-1 Gravity-field Force.

Thus, as these "quantifiable properties or categories" are being considered, e.g., force, electromagnetism, gravity, motion, field, energy etc . . ., it is "the Mass-energy with Motion-force Complex, as a "Gravimetric Frame-of-Reference," that is being evaluated, in order that its constituent components are "relatively reconciled" or "relatively re-normalized" for proportional values equivalency. "The Equivalency Principle" is in operation, relative to the fixed constant value of the Speed of Light in a vacuum, not only between Matter-mass-in-Motion and Energy, but also between the Gravity-field and the Electromagnetic Field.

Energy Transformation in operational cycles within "temporal Time" or "Thermodynamic Time," transmutes Mass-frame Motion into iterative Relativity-relationships that apparently climax as functional steady-state Systems in contiguous-continuous Space.

Consistent patterns of periodic cycling allow "functioning Mass-in-Motion-Energy-Force Frames" to establish dynamic System equilibrium for Continuum wholeness, which qualifies them as "steady states." Even the Human Body is said to be in "homeostatic equilibrium."

These Relativity-relationships develop into redundant patterns of dynamic cycling that animate the application of physical laws whose operations "transubstantiate force-into-motion and motion-into-force," e.g., 9.8 meters per second per second denote a "gravity force exertion equivalency," as the exponential function, c^2, that is equal to $(3.0 \times 10^8$ meters per second$)(3.0 \times 10^8$ meters per second$)$ embody a "momentum radiation energy pressure force exertion equivalency," hence Continuum Curvature tensing-and-stressing pressure-force Energetics emerging from Solar Gravitational-and-Field radiation-Energy projections-and-emissions. It is this "transubstantiation" of momentum-Force into Motion and of momentum-Motion into Force that accounts for the iterative, quasi-similar movements of Galaxies in relation to each other, of Planets in relation to each other, and of our Solar System in relation to other Galaxies.

Cycling mass properties embody the respective gravimetric Differentials and Opposites that constitute the tensing-stressing Energies of gravitational-and-field "Solar Radiation Motion-Force Pressure," in accordance with specialized functional fulfillment of the "Universal Thermodynamic Input-Process-Output Mechanism-of-Continuum," as encoded within the Organizing Principle of Universal Operational Oneness for sustenance and propagation of dynamic System equilibrium.

Emitted solar projections constitute the equivalents of the total gravitational-field Energy Strength necessary to sustain-and-maintain the ellipsoidal "Plane of Revolution Motion-Force Dynamics" within the Total-Sum-Frame of Sun-inter-planetary interactions.

It is conceivable that: Every thing that possesses some kind of purposive operational functioning from activation of the Input-Process-Output mechanism-of-Continuum, e.g., Earth ecological life-support Systems throughout all Seasons, will also possess unique Forms of "Thermodynamic Processing" that are only "embodiment-re-translations" or "transubstantiations" of Solar Power in its numerous, variegated, differentiated Forms," e.g., Plant photosynthesis; Human respiration; the Combustion Engine.

Radiation (Electromagnetic Energy), Mass (thermodynamically cycling Matter), and Motion (Continuum Field-and-Gravity Curvature Force Energetics), make up the fundamental Complex embodying the constituents of universal Motion-Forces, as "overlapping Frames" that are diffusely represented within and throughout the greater Atom-Planet-Star Continuum-Frame-of-Reference, e.g., *Quantum Mechanics* whereby Electrons revolve around the Atomic Nucleus; *Newtonian Mechanics* whereby Planets revolve around the Solar System's Star, the Sun; *Relativity Dynamics* whereby the "Space bending-curving" Motion-force-strength of the Sun and of Planets-with-great-Mass "triggering" Planet Mercury's Perihelion Shift.

As all other intermediate objects or bodies-with-Mass "replicate" in specialized functional patterns the complex fulfillment of the Universal Input-Process-Output Mechanism-of-Continuum, "in converted forms" of the same Mass-in-Motion-Energy-Force variables and co-determinants, "Cycling Functionality" constitutes the embodiment of "re-translated" Properties of Electromagnetism, that engender gravimetric frame Differentials and Opposites geared for Energy Conservation, e.g., the Earth is a "congealed cycling frame," (non-nucleated cycling Frame) but the Sun is an "excited cycling frame" (nucleated cycling Frame).

Thermodynamic "Mass-Energy Cycling," in Curvature Space-Time, in fulfillment of the Organizing Principle, climaxes into Continuum Motion-force projection-exertion, for dynamic Frame/System equilibrium; hence, periodic processing of quasi-similar Inputs for apparently iterative patterns of functional operation, e.g., The Earth "processes" solar Inputs daily, during every Season, as one-half of its Eco-Sphere is "bathed" in Solar Radiation, giving rise to Day and Night simultaneously "appearing" daily on the Planet as a Whole — (When it's "daylight" in the United States of America, Japan experiences "Night-time."

Is it possible that Time could also be "curved or bent" due to its "gravimetric entanglements" with infinite Space, even as physical Matter-mass dimensions give definition to Continuum Space-Time?

How would "Time curvature" factor into particulate behavior at the atomic frame — Quantum Mechanics Frame?

How would quantum mechanical micro-Field-and-Gravity Curvature "interface with" the Solar System macro-Curvature, as represented by its ellipsoidal tensing-stressing plane of Revolutionary motion-pressure Force that puts planetary events and phenomena into effect — constituting a Planet's Gravimetric Signature?

The "physical dimensionality" of Time matters only when computed as "Thermodynamic Time" during which "rates-of-change" are being measured in a System. This "physicality attachment" to Time is triggered even when there is a change in "cycling mode," e.g., from inertial-Mass energy to kinetic-Mass energy. Kinetic-energy potentiality contained in an "inertial cycling mode" also triggers Time-Differentials (Thermodynamic Cycling) and Dimensional-Opposites (variables change in values), because matter-Mass is always in a state of Motion due to its electromagnetic constitution, i.e., Atoms are always in Motion as negatively charged Electrons continuously revolve around their positively charged Nucleus.

Regardless of the cycling-Mass Frame-of-application, Time is always uni-linear or uni-directional — it only goes forward, from Past, to Present, to Future. Processes always have a beginning (Inputs), an "in-between-point" ("processing") and an ending (Outputs), even as micro-atomic-Frame particulate-cycling, geared for "keeping" dynamic Atom-System Equilibrium.

The only difference is that, within the micro atomic frame, Space-Time coordinates "instantaneously compress and contract" into "nano-seconds" of temporal Time or "Thermodynamic Time."

Atomic frame properties do participate in Energy cycling as do Planets and Stars and Galaxies. As "a line is a succession of dots-points," it is also through iterative stochastic (quantized) Energy-cycling patterns, at all levels and depths of cosmic Reality, that there is "overlapping connectivity interface" between Eternity (Ever-present "non-clocked" Time) and Temporal Time ("Thermodynamic Time") conducive to their integration into Continuum Space-Time for Curvature "Field" Motion-Force Unification — for "putting together" both Electromagnetic-Field and Gravity-Field Motion-forces.

Because Gravity is an emergent Force-property arising from Motion-events caused by Electromagnetic-Field operations, there is 'Motion-choice" or "Movement-decision" only in the Gravity-Field frame-of-reference, e.g., a Person can decide whether to sit down or to walk, but not in the Electromagnetic-Field Frame-of-operation.

The Electromagnetic-Field Frame-of-operation causes Objects having an affinity for responding to an Electromagnetic Field, e.g., Planets, to engage in rotary Forms of Mass-in-Motion Force, such as Revolution and/or Rotation, around a Center-of-Mass, e.g., Solar System where Planets revolve around the Sun; Electric fan where blade-assembly revolves (in rpm) within its copper-coil-magnetic-Field-activated blade-stem-well; Atomic Frame of Quantum Mechanics where Electrons revolve around the Nucleus.

"Temporal cycling" or Thermodynamic Energy-cycling is "the equivalent" of Space-Time Motion-force, e.g., Thermodynamic Energy-cycling is a property of Mass-in-Motion. There must be Motion in order for Thermodynamic Events and Processes to take place. Thus, Thermodynamics occurs only under "conditions of Motion, e.g., A Planet-in-Motion like the Earth, has many phenomena that occur, only due to the Movements pertaining to each of its constituents or to their combined effects; Human stomach muscles continually contract in facilitation of food digestion; Oxygenated blood from Lungs that breathe Earth atmosphere circulates throughout the whole Human Organism, as pumped by the Heart muscle. Seasonal cycles trigger "ecological movements" peculiar to each specific Season that impact flora and fauna in differentiated ways. Bears hibernate in Winter after fattening themselves for "the long haul" as trees shed their multi-colored leaves in Autumn, in preparation to undergo "Winter-sleep."

Iterative redundancy of Energy-cycling patterns in fulfillment of the universal Input-Process-Output mechanism-of-Thermodynamics constitutes the embodiment of Continuum in the pursuit of dynamic System equilibrium.

As such, "Temporality" that allows for observing "Thermodynamic rates-of-change in Time" as pertains to an operational System, is a "physical unit of measurement" that "clocks" the duration of "the Unfolding of a process," such that, from "from point A - to - point B," variables and co-determinants are varying within their peculiar range-of-values — Change in quantities and Change in quality are also occurring, — even as matter-Mass-Frames engage in Energy-cycling transformations engendered by Curvature tensor-stress pressure-Force exerted-to-cause centri-vectored rotary Forms of Motion.

Thus, a vehicle travels in "Temporal Time" from original point of departure A to destination point of arrival B, in the same manner that a chemical reaction takes place from a point in Temporal Time as "Energy-state A" to arrive at its end-point destination into "Energy-state B," whether the chemical reaction is taking place within Earth "congealed" environment, (non-nucleated Environment), within an atomic frame, within Solar System ellipsoidal plane of Revolution, or within the Sun's "excited" nucleated plasma condensate reactions-structure.

Cosmic processes, universal events, and ecological phenomena are all Energy-cycling, Mass-in-Motion, and Thermodynamic-Time dependent!

The forward-vectoring of Time, as Eternity or as Thermodynamic Time, cannot be "curved or bent" towards the past, while cycling-Mass variables and co-determinants are undergoing rates-of-change within the ever-present Frame-of-Time that has "a Motion-Force vector" uniformly proceeding from Past, to Present, to Future. In our physical Universe, the march of Entropy follows a one-direction path: A broken drinking glass will not re-assemble itself given infinite Time and infinite random chance-probability! It's "brokenness into thousands of little pieces" — rate-of-change in Thermodynamic Time — is terminal and cannot be reversed! Positive Entropy is a dead-end! Mortality results in eventual Death!

Energy-Cycling changes undergone by Mass-in-Motion occur as Time is elapsing or during the proceeding of Time — but the Mode-of-Time itself cannot be changed! For the cycling-changes are in Mass-properties, the sum of which can be accounted for in their modified physical proportions, values, and ratios,; but not in the nature-of-Time per se — phenomena are occurring while possessing duration-in-Time, but Time itself is not occurring or changing within these natural processes. Time does not stop because a clock-device has stopped working —Time is always "forward-flowing," never stopping. When an event or process is "paused," Time continues to flow unabated.

How are "particulate Movements" to be differentiated from "planetary Movements" in light of Gravity-and-Field Curvature pressure-force tensor-stress Metrics?

Events and processes can have straight-forward path from beginning to end, or "convoluted paths" from beginning to end marked by "process interruptions," "changes in trajectories," "parallel simultaneities," etc . . . Linear, parallel and simultaneous processes still proceed through Time's forward-vectored march from Past to Present to Future.

It is the proceeding of the phenomenon that is "convoluted," e.g., advancing, stopping, or regressing, but not the march of Time itself. The speed of Light in a vacuum is 3.0×10^8 meters per second, or 186,000 miles per second.

But "The speed of Time itself" is independent of Mass-cycling activities that proceed in "Time-frames" characterized by different rates-of-Time; units-of-Time whether as a Second, a Minute, or a Nanosecond, only mark "duration of Time" as a period elapsing from one point-in-Time to another.

But Time itself continues its forward march from Past to Present to Future at the same essential constitutional foundational rate, independent of whether it is marked by a traditional hand-clock-device, by an atomic clock, or by a "number-counting device" that has no physical-material coordinates such as Earth Rotation and Earth Revolution.

Mass-Energy, cycles at different rates, in periods of Time, while Time itself keeps its foundational essential rate-of-proceeding, e.g., At the atomic Frame, Energy Transformation processes occur very fast or fast-cycle in "compressed-contracted instantaneous periods-of-Time," such as pico-picoseconds.

"Temporal Time" or "Thermodynamic Time" is therefore active in all Frames, regardless of the difficulty encountered within micro-Curvature environments where Curvature Energetics born of "interactive-Relative Force dynamics" effected by particulate electro-motive charge displacements climax into "tensing-stressing pressure-force trajectories" that appear to defy theoretical analysis and practical-experimental measurements.

There are no such things as "Time particles." There are no such things as "Space particles." There are no such things as "Gravity particles."

Time, Space and Gravity intersect via Mass-cycling properties whose physical dimensions and coordinates allow for quantifiable and measurable categories, such as Force, Energy, and Motion. Due to Curvature pressure-Force Energetics that triggers centri-vectored motion Forms, such as Revolution and Rotation: Time, Gravity and Space are quantifiable in terms of "measurable categories" that allow the conceptualization of specialized properties that are defined as force, motion, pressure, volume, surface, depth, height, width, length, temperature, and energy, etc....

Energy is transformable because it is a property of electromagnetic Matter-Mass in Motion. Time, per se, has no physical properties. Space properties are defined by Matter-Mass dimensions occupying it.

Gravity is not an electromagnetically-induced Force like "an Electromagnetic Field" but is an emergent Motion-force property of Curvature Energetics engendered by a Magnetic Field, the complex totality of which, climaxing into centri-vectored rotary Forms of motion.

Thus, Space-Time displays Continuum properties due to matter-Mass Frames that thermodynamically cycle in Energy transformation patterns that are consistent with Curvature pressure-force Motion Differentials-and-Opposites, intersecting to cause "re-translations" of the Electromagnetic Properties of Matter. Matter-Mass is "moving" within Continuum-Space-Time as its "re-translated" or "transubstantiated" Electromagnetic Properties undergo thermodynamic "rates-of-change" during functional operation.

Correct understanding of Time for conceptual-and-mathematical formalization, will usher the preservation of "Continuum Curvature" for universal Motion-force Unification.

Time, as expressed in minutes, hours, months, years, or light years, is a consciously experienced phenomenon, acknowledged and recognized as an achieved "distance-movement Concept" for structuring an event or process, within a "Frame-of-Duration," e.g., Mental movement of thought-processes from one idea to another within the Mind; Motion of a traveling automobile upon earth surface from a departing location to an arrival destination.

"Temporal Time," or "Thermodynamic Time" whereby "rates-of-change measurements" are being taken, is a "means-of-Unfolding" attached to a process during a "quantized-quantifiable period," as a "physical unit of measurement" superimposed over the Gravitational Universal Space-Time Continuum, for in situ understanding. Events, phenomena, and process in the material Universe have operational footprints necessitating comprehensible deciphering in the Human Mind, which is the Medium of External Apprehension par excellence! But the Human Mind possesses no physical coordinates or location. It operates by ascribing-encoding and decoding Meaning from symbolic characters that constitute the 'body of our literary language."

Thus, the Human Mind is engaged in "debunking" or "decoding" its own Systems of Apprehension, but in terms that concretize-and-objectify "its Relative interface" with external Reality — Time cannot remain an abstract concept-of-Measurement, but must be "operationalized" with physical landmarks and coordinates to which are attached definite meanings such as duration periods or Units-of-Elapsed Time. From remembering or placing events as occurring from Sunrise to Sunset, the Human Mind now can "operationalize" the course-of-events in terms of days, weeks, months, and years, because of the invention of the "clock-device" that marks Time in terms of Seconds, Minutes, and Hours! The abstract concept of Time in the Human Mind has come to gain its external concretized physical counterpart by attaching duration-periods to the Unfolding of processes, events, and phenomena from Begining to End!

"Quantum Time" as displayed by particulate trajectories in micro-Curvature Motion characteristic of Quantum Mechanics, is also the "distance-in-Time measured unfolding" of a "fast-cycling event," estimated as "waves of probability." Because of the vertiginous velocities involved, neither "the Human Mind's Eye" nor invented electronic instruments and equipment can "keep up with sub-atomic particles."

The Human Mind engages in Motion in Time as it proceeds, for example, from one thought to the other, from one idea to the other. Time proceeds, therefore, as "a certain specific condition" undergoes change from one state to another or as from point A to point B, e.g., Mind-motion; Process-motion; Energy-state motion; Cycling-mode transformation change, etc . . .

The abstract Mind requires concretized external "anchors-of-Measurement" for Understanding universal Reality in all its complex variegated Forms. A "congealed-Energy condition" yielding "a nucleated-Energy condition," — e.g., $E = mc^2$, as in a series of nuclear-bomb chain-reactions — constitutes "Movements-of-the-Mind" as Human Understanding proceeds to analyze such processes and make scientific deductions from observed results.

Ever-present Time, Time per se, or "Non-Thermodynamic Time," has no beginning or end, as its abstract immateriality is embedded in Eternity — it has no intrinsic physical indicators, coordinates, anchors, or landmarks.

Time per se is as abstract as the Human Mind. Thus, "Clock operations" and "Digital meanings" must be learned from Infancy and skills at "telling Time" must be honed unto proficient utilization throughout the whole Human lifespan.

However, "Temporal Time," or "Thermodynamic Time," as embodied in Energy-cycling, begins and ends, as Mass-in-Motion interacts with Mass-in-Motion, within the Gravitational-Field Curvature Frame of Infinite Continuum Time-Space – e.g., the Earth revolving around the Sun and rotating upon its own axis.

According to learned "clock-Time," Revolution takes 365 and one quarter days; Rotation, 24 hours. But even in the absence of "clock anchors-and-indicators," Revolution and Rotation occur in definite Time-periods marked by "Sunrise" and "Sunset," and by duration of Seasonal Cycles.

Yet these "Quantum moments" have uninterrupted stream Continuum or wave-cycling Continuum, e.g., the Sun never stops shining; and the Earth never stops "moving" in Rotation and Revolution, in order that "Sunrise" and "Sunset" take place.

As a line is a succession of dots-points, "Quantum-Moment" and "Continuum-Stream," are contemporaneous, simultaneous, and contiguous: All-in-One! Apparent interruptions and ceaseless Continuum: Are all-in-One! God is an awesome Creator! What a wondrous mystery!

Per the "clock-device mechanism," a thin rotating hand is superimposed over two thicker rotating hands, beginning at the center and nearly the length of the clock's radius, the thinner-rotating-Hand, indicating Seconds that accumulate into Minutes; Next, comes the Minute-rotating-Hand underneath the Seconds-rotating-Hand and superimposed over the Hour-rotating Hand, indicating "the flow of Minutes" that then accumulate into Hours; Finally, comes the Hour-rotating-Hand, the sum of which, covering a span of 24 hours, constituting 1-day.

In accordance with another device called "the Calendar," days accumulate into weeks; weeks accumulate into months; and months accumulate into years etc . . . , following the directional vector-pattern-and-path of Time, which is continuous, contiguous and forward-linear. We remember the Past, but live in the Present, as we project into the Future.

An event has a one-directional path from its "Past," to its "Present," and to its "Future," which, consequently, underlines the fact that Entropy, the Second Law of Thermodynamics, can only be positively measured — "always increasing." Though Conservation and Entropy are occurring simultaneously and contemporaneously in a System that is in dynamic Equilibrium, there is no such thing as "negative Entropy." Hence, Thermodynamics gives rise to cause-and-effects processes and mechanisms requiring that every biological creature, including Human Beings, must have a Genesis or Beginning, a Lifespan or Process-of-Maturity, and then the on-set of Mortality.

Thus, "Thermodynamic Time" or "non-Eternal-Time," proceeds from the Past, to the Present, to then arrive at the Future, to the stages of which, no one can return or travel, except as an acknowledgement of personal Memory or collective Memory. Hence, events can be "re-enacted" but not "re-lived." For, the phenomenon of mental remembrance or Memory of a "Past event" takes place, even as the person is still living in the Present.

Returning to "a Time in the Past" or to a "Past event," is impossible: For reversing the forward-vector of Time is as impossible as it is to "interrupt" or "pause" the March of "Forward-flowing Time."

It is necessary that we have the right-and-correct understanding of Time, while at the same time, being able to differentiate between Eternal-Time and "Thermodynamic-Time." Thermodynamic Time is "embedded" within Eternal Time. Thus, as we biologically exist within the Frame of "Thermodynamic Time," or within the Frame of "Temporal Time," we are also participating in "Eternal Time," which "coincides" or continues to "flow along," with Thermodynamic Time.

Hence, there is no so-called "relic cosmic background radiation" that could indicate or "re-radiate" to us, through visible-Light emissions, how the Universe might have existed "billions of years ago." And when we look at the Heavens, the radiation we observe is being emitted within the Frame of the Flow of Continuum Time, the sequencing units of which, arriving at "the Time-frame of the Present," but from real-time "on-Sources" of Radiation Energy, and not from "Past cosmic structures."

Mass-in-Motion cycling-trajectories can have "convoluted paths" marked by differentiated Motion-patterns, e.g., an acrobatic aircraft; or observations of "process-vectoring" can present "magnitude and direction convolutions" in temporal Time, e.g., meteorological phenomena occurring at the same time, such as an earth-quake accompanied by a tornado and torrential rains; but Time itself, which is not a physical-material Object with such dimensions as can be directly measured, cannot be "curved" or "bent."

Convoluted Energy-processes and particle-trajectories occur in uninterrupted, forward-marching, forward-flowing, Continuum Space-Time — from Past, to Present, to Future! The "curved paths" of particles under observation via electromagnetism-based instruments, technologies, and equipment, does not mean that the Time-period or duration in which they're "moving" as "Thermodynamic Time" can be "bent" or "curved." Objects can follow convoluted trajectories that combine rectilinear, linear, and curvilinear paths and motion-patterns, but they "move" during the flow-of-Time, and cannot "change its course," or "interrupt it."

Thus, traveling towards the Past or into the Past, while "piggybacking" on projected star-Light from far-away Galaxies, is impossible. Because the directional vector of Time is forward-linear and cannot be reversed, nor can the Flow-of-Continuum-Time be, accelerated towards the Future.

Time cannot be "curved" or "stopped;" nor "bent" or "accelerated." But Objects "moving" during the Flow of Continuum Time can "orchestrate" their peculiar Motion-patterns as Time is "flowing-forward" towards the Future.

Even "the Proton-Neutron Strong Force" possesses a tensing-stressing-pressure Force-Energy dynamic "that does not stop compressing" them against each other, either in magnitude or in power. Do not crystals also "vibrate" in oscillating patterns of Motion in the presence of added-Energy-units that disrupt their cycling-state equilibrium?

We acknowledge the occurrence of events, processes, or phenomena in "Temporal Time," or "Thermodynamic Time," but, we have to be alive and in conscious awareness and "presence of Mind" to perform this "mental operation of acknowledgement."

The "big bang explosion assumption" was not witnessed by any physicist or astro-evolutionist, but is nothing but a feat of fantastic conjecture for which there can never be any valid, proven-scientific, reproducible proof.

Therefore, it is an impediment to the advance of mathematical Theoretics, in that, it violates the Continuum Principle, as well as contradicts the Law of Conservation of Energy.

Time Differentials and Kinetic Opposites, e.g. "Sun rise" and "Sun set," Revolution and Rotation, occur due to Curvature stress-pressure-Force motion-Forms, as differentiated by mass-Frame energy-cycling properties, e.g., planets revolve around stars and rotate upon their own axes; Stars emit Magnetic Field-and-Gravity-Field-induced nucleated-Radiation Energy causing rotary Forms of centri-vectored Motion-Force.

Time's "periodicity coordinates," — marked in "units of elapsed duration," as pegged to real Mass-in-Motion co-determinant variables effecting Relativity interactions, — are thus amenable to "dimensionalities of Matter" that "re-translate" or "transubstantiate" Curvature cycling-Motion patterns in the Forms of "Quantum Moments" and "Streams-of-phenomena," the complex transactions of which, climaxing into "the Property of Continuum."

Is not the Electromagnetic Spectrum made up of "quanta of Energy" or of "units of Energy" displaying various range-specific, differentiated wavelengths and frequencies that in "traveling together," climax as "rays?"

The Sun projects "Curvature binding-Energetics" in the Form of a Magnetic Field that gets "re-translated" or "transubstantiated" into "equivalents" of the Gravity-force for centri-vectored Forms of Motion espousing rotary, rectilinear, linear, and curvilinear movement-patterns, as impelled by Solar System ellipsoidal revolutionary plane of tensing-and-stressing-Pressure-Energetics.

Electromagnetically-induced Curvature binding-Energetics engender Magnetic Field Force and Gravity-Field Force "analogs" and "equivalents" that climax into "mechanisms of Motion," or "instruments of Movement," thus, causing mass-Frames to cycle, even as they transform Energy for dynamic, function-specific, System equilibrium.

Energy Transformation takes place in iterative or repeating cycles, due to thermodynamic requirements necessitating that the Input-Process-Output mechanism that animates the Organizing Principle, be fulfilled in "stages-and-steps," "stochastic rates," or "processing periods," unfolding of which, is "encoded-registered" within variegated but unified, sequential, simultaneous, contiguous, contemporaneous, and/or "parallel processing."

Sequential processing, parallel processing, and simultaneous processing are in "entangled Temporality" or "Entangled Simultaneity," as "Quantum Moments" that "orchestrate" an "ensemble of events" constituting "Streams of Phenomena," as are necessary to fulfill The Laws of Thermodynamics; and by that, sustaining The Continuum Principle, within boundaries imposed by the fixed-constant value of the Speed of Light in a vacuum — standing at 186,000 miles/second, or 3.00×10^8 meters/second.

Density-of-Matter determines the momentum-weightiness pressure-Force exerted at-point-of-impact by accelerated-Mass, e.g., Lead is heavier than Aluminum in inertial-Mass, and thus, has greater Weight-and-momentum impact-Force when in-Motion than Aluminum-in-Motion. Likewise, a Feather, being of lesser Mass than a Rock, will have, when in-Motion, a lesser Weight-and-momentum impact-Force than a Rock-in-Motion.

X-rays or Beta particles, though apparently having micro-Mass, do have greater micro-Mass than apparently mass-less Photons; and in addition, X-rays or Beta particles are "excited-Matter," whereas Photons are non-nucleated visible Light wave-particle units-of-Energy.

Particles, in "particle accelerators" or "cyclotrons," when accelerated to even "a fraction of the Speed of Light," will undergo a process whereby, "Transformational Kinematics," will cause them to engage in the Form of "Mass-Energy gains." These "Mass-Energy gains" in the Form of "Nucleated-Momentum-Impact-Force," get "re-translated" into emissions of cosmic radiation, the sum of which, being "the equivalent of their shed mass."

Due to Electromagnetism being "The Primary Universal Curvature-Force" that necessarily engenders "Rotary Forms of Centri-Vectored Motion," as initial Energy-Mass is conserved, "Mass-Energy gains" are then released as "nucleated-cosmic-radiation," in accordance with the "shift-in-speed" relative to micro-Mass momentum-impact-Force, standing opposite to Curvature-force Differentials from which accelerated particles cannot escape. During acceleration of particles in "cyclotrons," Transubstantiation of "Mass-Energy gains" as released cosmic radiation, is comparable to the "phenomenon-of-Entropy" that counters Conservation of Energy in Quantum Mechanics, whereby Electrons "release Energy" as they return to a lower orbital level, after having been "excited to a higher orbital level" away from the Atomic Nucleus.

Things are equivalent "relative" to a Constant with a fixed value, such as the Speed of Light in a vacuum, calculated to be at 3.0×10^8 meters per second, or at 186,000 miles per second. Due to that limiting factor, the greater the momentum of an accelerated particle as it approaches the Speed of Light, the more likely it will emit cosmic rays, the sum of which, being "transubstantiated by-products" of shed-Mass in the Form of "Mass-Energy gains." Thus, no Object can ever have so-called "infinite Mass."

The acceleration of an Object or particle towards obtaining "infinite Mass-Energy gains," in order to arrive at "infinite gravitational cosmic radiation emissions," is not possible — which means that the Sun will eventually stop radiating; which also means, there is a definite limit to "how big a bang" one can get from "a nuclear bomb" within Earth geo-atmospheric constitution! There is a limit to "how big" in Megatons "a nuclear bomb" can be! For, at "infinite Speed" or "Speed-approaching the Speed of Light," all Mass is shed as "Mass-Energy gains" taking the Forms of "emitted cosmic radiation." Thus, eventual "Terminal Entropy," or ultimate "Thermodynamic Death" is certain for "all things that exist," due to the universal limiting factor: the Speed-of-Light in a vacuum!

****

THE "RELATIVELY CONSTANT" UNIVERSE: THE ATOM, THE PLANET, AND THE STAR!

The Universe is "relatively constant" in the sense that Atoms agglomerate or "clump" into molecules to form "greater space-bodies with Mass." The Solar System embodies and encapsulates the macro-constituents of all other "Forms of Molecular Agglomeration."

What are "Galaxies" made of? Atoms-Planets-and-Stars "re-arranged" in "differentiated structural patterns" of Organized Order. Are we not still talking about a different kind or a different Form of Star when we speak of "Quasars," "Pulsars," or "Binary-Stars?" Stars emitting different Units-of-Electromagnetic Radiation Energy, as to differentiated Frequencies and Wavelengths!

In an Object-with-Mass, density-of-Matter indicating the Space-distance between molecules as "degrees-of-compactness," is determined by the ratio of its Mass to its Volume, which explains why the single positively charged Hydrogen proton and the single negatively charged Hydrogen electron do not "short-circuit" each other, even in the absence of a charge-less neutron in its nucleus to serve as "insulation-ground" — Micro-Electromagnetic-Force and Micro-Gravity-Force as co-equal, co-linear, but opposite Forms-of-Force insure that, relative to their nano-Masses, these atomic particles are "kept" at a "respectable distance" from each other. By that, "atomic System dynamic equilibrium" is fulfilled, achieved, attained, and maintained.

The Atom is constituted of the positively charged proton and the charge-less neutron at the nucleus, and the electrons that revolve around the nucleus. Atomic density, relative to each specific Chemical Element, is a property of dynamic cycling equilibrium, as sustained by particulate micro-Curvature distances, electro-motive charges, and gravimetric Motion patterns. The greater the number of Protons and Neutrons in the atomic Nucleus of a particular Element, the greater the number of Electrons at orbital levels nearer to its Nucleus, the greater the number of molecules in a unit of Mass per unit of Volume, and the shorter the distance-Space between molecules of a particular Element forming the Unit of Mass, the greater the Density of that Elementt.

Density-of-Matter is therefore a necessary component of molecular change for electro-motive dynamics of Mass-Formation as Electrons, respectively, keep a certain orbital distance relative to the atomic nucleus: via same-negative-charge repulsion, so that there is no "short-circuiting" between opposite charges that strongly attract.

In the same manner that Rotation and Revolution engender core-centric "Gravity-force equivalents" within the Earth, fast-cycling Electrons revolving around the atomic nucleus give rise to "rotary mass-Field tensing metrics" that compress the strong proton-neutron force, in conjunction with nucleus-centric charge displacement Motion-forces generated by all particulate interactions, especially Electrons with the same negative electric charge that are prevented from "clashing" with one another, even as they maintain their respective distances from the atomic nucleus.

The strong force prevails within the Atom to cause the predominance of pro-centric particulate motions over counter-centric motions generated by the weak force, at the same time that "captured electrons" enjoy great latitude in their ability to bond at outer orbits for diverse chemical reactions leading to molecular change.

Gravity at rest or inertial Gravity is the equivalent of rest-mass weightiness or "rest-Mass Motion Force." When "pressure-Force" is imposed upon an object-at-rest or at inertial-Mass, a certain "Momentum acceleration-Force" is being applied, thus, causing "an indent" in the material whereupon the Object is "resting," e.g., A fork having 4-prongs "resting" on top of a soft-wooden table, upon which, a cook puts a heavy book, will cause the prongs of the fork to make "indents" in the top-surface of the soft-wooden table.

An Object-at-rest still receives the Force-of-Gravity as it resists Motion. Its rate of acceleration, g = 9.8 meters per second per second, relative to its Mass, is still actively "pushing," to exert a pressure-Force as pounds per square inch upon the surface against which the object is resting. It is "this complex ensemble" of traveling-Mass that is accelerated for momentum impact-Force during Space-displacement Motion. At great velocities, a 5-ton vehicle will make "a big dent" when impacting a concrete embankment belonging to a cement-metal bridge; but analogously, at same great velocities near the Speed of Light, cosmic rays will "penetrate" due to their micro-Mass momentum-Force properties. Thus, in the Solar System Frame as in the Atomic Frame, gravitational radiation Curvature motion-Energetics exerts a "pressure-force" that causes tensor-stress Dynamics compelling Objects — planets or particles — to engage in centri-vectored Forms of rotary Motion-force.

As the Earth is revolving and rotating, Gravity-force acting upon objects on its surface and in its atmosphere, displays rotary, rectilinear, linear, and curvilinear patterns of Motion "along the line-of-radius" as potentiated by center-of-mass, center-of-field, and center-of-gravity. Were not the ground or landmass stopping all surface Motion, every object would be driven towards Earth-core center-of-Mass and center-of-Gravity, e.g., theoretically speaking, as Gravity is towards center of the Earth, a hole can be dug in the ground until "dynamic-Mass suspension" is achieved, thus, cause-and-effect mechanisms showing why an object will continue "to sink into a hole" as "pushed-downwards" by the acceleration-rate of Gravity relative to its Mass, until it reaches bottom.

Likewise, a heavy object such as a rock, thrown into the oceans, will continuously sink until it is stopped by the landmass underneath ocean waters, which constitutes oceanic bottom. In Relativity terms, the oceanic bottom operates as the analog or equivalent to Earth-core center-of-Mass and center-of-Gravity — barring "a sink-hole being dug" in the oceanic bottom that is so continuously-deep as to touch Earth-core! Thus, Earth surface and oceanic bottom, serve as "relative substitutes" for Earth center-of-Mass and Earth center-of-Gravity.

Mass-in-motion Differentials and momentum-Force Opposites that allow distinction between Magnetic-Field-force Energetics and Gravity-field-force Dynamics are important for mathematical conceptualization of an all-encompassing "General Theory of Force," because Motion and Force are "in entangled cycling" where Gravity and Field operate.

There is a concerted interest in finding "A Unified Field Theory of Force" that would integrate gravity, electromagnetism, the strong force, and the weak force. Yet, it appears that equivalents of the Field force, the strong force and the weak force, are frames displaying cycling Forms of energy motion-Force corresponding to "analog embodiments' of Electromagnetism, in that, particulate electro-motive charge Space-displacements are present in their operations. However, Gravity is "inert" or "non-electro-charged," for it displays no particulate characteristics, qualities, or attributes that would permit identification or detection of the so-called hypothetical "graviton," a non-existent or yet-to-be-discovered "particle" that is assumed to be a "sub-atomic particle."

An Electromagnetic-Field has an electric Energy component as well as a particulate-characteristic component from which ensues the Property of Magnetism; however, equivalents of the Gravity-field Force appearing to operate as a "Magnetic-Field force analog," do not display particulate characteristics akin to possessing properties-of-Motion that have an "Electromagnetism-based Foundation."

Electromagnetism is a "direct Force," whereas Gravity is an emergent Motion-Force-property derived from Magnetic-field interactions. Velocity, as applied to matter-weightiness already-in-motion, produces accelerated-momentum-mass-Force; accelerated mass-in-motion yields momentum-Mass-in-Motion impact-Force.

Earth traveling through Space possesses momentum Mass-in-Motion force, field tensile boundaries, plane elasticity, torque Energy, working power, tensor-strength and Curvature-stress, the complex dynamics of which, operating to hold the Moon in its orbital motion patterns and artificial satellites in orbit, as the earth "wobbles" and experiences bi-polar oblation in response to solar system ellipsoidal plane tensing-stressing Energetics.

The Earth is comprised of an Atmosphere, of a hydro-sphere and of a landmass whose cycling operations incorporate wind, geological, ecological, meteorological, and soil processes. All conditions of dynamic System equilibrium such as revolution, rotation, radiation processing, 1-G eco-meteorological parameters, etc . . ., being held constant, it is the 93-million mile-distance from the Sun along the Solar System ellipsoid plane of Revolution, in conjunction with the Moon-Mass-induced 23-degree axial tilt that determines "Earth motion-Force cycles," Energetics of which, make possible the "calibration" of daily temperature-and-pressure ranges, as embedded within cyclical seasonal periodicity patterns.

Comprehensive ecological processing of solar gravimetric inputs is dependent upon the "Curvature tensing-stressing-Pressure-Force complex" arising from "embedded entanglements" characteristic of interactions between Electromagnetism and Gravity. Within these "embedded Force-entanglements" emerge "cycles of Thermodynamic-Energy processing" that are transformed into geo-atmospheric events, ecological processes, and hydrosphere properties, the interwoven interplay of which, engineering the maintenance of Earth life-support Systems.

The universal "Energy Transformation Algorithm" is based and anchored upon the dynamics of the Input-Process-Output Mechanism, as "fed" by "cycles of motion-Force

Energetics," relatively producing all the phenomena needed for Earth vibrancy, namely, Rotation recurring in 24-hour periods; and Revolution recurring in an iterative yearly 365 ¼ day period.

The Input-Process-Output Mechanism is an inherent component of all mass-Frames that cycle for dynamic System equilibrium. Because of the properties intrinsic to Matter-Mass, such as the need for reactions and molecular changes that orchestrate functional operations, Thermodynamics requires Motion; Motion requires exertion of Force; and Force requires Electromagnetism and Gravity. Cycling Mass-Frames do overlap, and hence, where they overlap, Cycling Differentials-and-Opposites specific to each Mass-in-Motion cycling Frame, respectively, will engender "relative turbulence" or "relative perturbations" that must be "reconciled" or "re-normalized" via certain "unknown catalytic parameters" usually represented in Mathematics as "alpha-numeric Constants."

The assigned fixed value of a Constant has a "mysterious operation" in that it "re-frames" all Differentials and Opposites in compatible ways that "standardize" a Physics equation, for uniform valid application within all mass-Frames where gravimetric cycling conditions are similar, e.g., Once a particular definition is set for what a Solar System is, then, all validly proven Laws discovered to be applicable to a specific Solar System, will also be applicable to all Solar Systems with similar "gravimetric signatures."

The Universe is a physical entity composed of material Energy-cycling Mass-in-Motion Systems that are ordered and organized according to quasi-iterative or quasi-similar reproducible patterns of Organization, e.g., The simple Fact that where there is Electromagnetism, an object having Mass, such as a Planet, falling within its Field-of-Manifestation will begin to also encounter the Force of Gravity, the sum of which, will cause it to begin "to move" in rotary Mass-centric Motion Forms, e.g., An aircraft with engines-off will follow trajectories along the "line of Radius" leading directly to the Planet's Core-center-of-Gravity.

Absent an external "pushing Force" causing an already-moving Object to "change course," all universal Motion-Forms are rotary-vectored and/or gravicentric/Mass-centric in direction.

Earth Atmosphere is composed of specific layers that engage in eco-Systemic gravitational-Equilibrium activities causing the Planet to "operate" as a "Whole Entity Ensemble."

Each specific atmospheric layer thrives at a certain distance from Earth-Core Center-of-Gravity and Center-of-Mass.

Distance from the Planet's electro-conductive and electro-motive hot iron molten magma Core, in conjunction with respective "processing" of Solar Gravi-Radiation Energy Inputs that determines how atmospheric layers engage in eco-Meteorological activities. For example, according to elemental composition, structure, organization, density, gravity quotient, and interactive properties, the Magnetosphere blocks and deflects Solar Radiation, some of which penetrating the atmosphere, to engineer Aurora Borealis.

The closer to vacuum space is an atmospheric layer, the more electromagnetically active it is; the closer to Earth's core-of-Gravity and core-of-Mass, the more livable the layer for Human and animal life (for Flora and Fauna).

Aurora Borealis is a bi-polarity atmospheric magnetic field activity. The Magnetosphere absorbs solar radiation that gets passed onto the Ionosphere. As these cosmic rays, waves, and particulates are collected, they undergo electro-motive-magnetic "processing," after which they are "re-translated" or "transubstantiated" towards lower atmospheric layers into Forms of conductive ionized cloud activity, — thus, the production of lightning, thunder, and rain for soil fertility, replenishing the oceans with sweet water, and irrigating the landmass.

The physical structures that "embed" Earth geological characteristic, "surface-Space," and topographical features are caused by either "sudden" or "on-going" Catastrophic Phenomena. "Catastrophism" encapsulates the sum of events and processes that periodically "re-arrange" Earth components, primarily its Landmass. Given that Earth Atmosphere, unlike its Landmass and Oceans, has no solid-materials that can be "molded-for-shape," it is dependent upon Landmass and Oceanic-Hydrosphere events for "Motion-activities" that activate meteorological phenomena. Events such as Earthquakes and Volcanic Eruptions from the Landmass, water evaporation from Oceans, and Ocean tides caused by the Moon, etc…, in conjunction with "Solar Inputs Processing," account for triggering atmospheric changes that generate storms Systems. The Earth is a self-contained planetary Sphere, and there is no other place for the Atmosphere, Landmass and Oceans "to go." And, thus, Earth Landmass being the only solid of the three components, all atmospheric (gases) and hydrosphere (ocean waters) activities can only "re-arrange Earth Landmass," and hence, we talk about prairies, valleys, mountains, hills, plains, lakes, rivers, and creeks, etc… High ocean tides will "flood the coastlines," which belong to the Landmass; diluvial rains will cause great accumulation of water in rivers and lakes that will overflow unto the Landmass. Given that Humans inhabit the Earth Landmass-surface, then, cause-and-effect mechanisms triggering natural catastrophes involving the Atmosphere and the Oceans commonly impact urban life, jungles and forests, and agricultural communities.

Even petroleum and natural-gas exploration, as well as metal-and-minerals mining, have to do with the Landmass that is covered beneath Ocean-surface and sea-waters. We utilize the Atmosphere as a natural resource, but mainly, for breathing Oxygen, for "winged-foodstuffs," and for flying aircrafts. The Earth is "a water-Planet" with more than seventy-five percent of its surface consisting of Oceans-and-Seas. Apart from its thermodynamic utility for "cooling-and-warming" the Planet, respectively, for maintenance of Earth life-support Systems, Ocean waters are primarily utilized for private and military navigation and commercial shipping. However, from the Oceans we also obtain sweet-water and salt from desalination as well as sea-foods for fulfilling our nutritional needs.

As the Input-Process-Output Mechanism is operating under solar gravitational radiation Energy Differentials and Motion-force Opposites that affect Earth equilibrium systems, phenomena such as tornadoes, floods, earthquakes, volcanic eruptions, tsunamis, hurricanes, cyclones, etc . . ., will "shake" or "trigger turbulence" within the Earth in-toto Mass-Frame complex.

"Shock" during these events, must "register" somewhere, within the Earth, that is, in its Hydrosphere, Atmosphere and Landmass. And these "shocks" can be "catastrophic" in their effects. The landmass "swings to-and-fro," "suspended" within the Hydrosphere and "cushioned" by the Atmosphere, even as the Moon's Mass proximally imposes "gravicentric pressure" upon the Oceans, thus, causing tidal surges that flood coastal areas.

Solar "curvimetric" or "gravicentric" Field-and-Gravity-induced Motion-Force Energy cycling operations that are engendered from "Catastrophism," — or simply, via the "gravitational processing" of "apparently random" meteorological phenomena whose geo-dynamics "re-shape" Earth topography, — engineer in their wake, "Geo-Hydro-Atmospheric processes" that give rise to "Forms of soil-Mass-movements" that result in shaping-and-molding the Earth Surface, e.g., In such ways was "formed" The Grand Canyon, within the Continental United States of America.

Likewise, it is the constant electro-motive thermal inputs from countless lightning-strikes, in Mass gravicentric Core-vectored actions, that keep Earth Iron Outer-Core "in a Molten Lava condition or state." Lightning-Strikes, like all other natural phenomena or cosmic event, do have their own "strength-of-Force limits:" They do not reach the Inner-Core which remains relatively solid. The Iron Molten Magma Outer-Core's electro-conductive/electro-motive properties, along with their interconnectivity to solar gravitational radiation Energy and Earth atmospheric activities generate Earth's bi-polar or dipole Magnetic Field possessing two oppositely charged ends: The North Pole and the South Pole.

Were it truly factual that the Universe and the Earth "appeared ex nihilo" more than 14 billions years ago, no amount of initial thermal Energy "held" within its Core would have endured into the Present!

The primary role of Lightning-Strikes: Thermal Earth-Outer-Core-Energies are being continuously sustained towards maintaining "its Iron Molten Magma State," only due to electro-conductive, electro-motive, electro-combustive-activities from incessant Lightning-Strikes!

When great-pressure is built-up therein, volcanic eruptions and/or earthquakes may result as a "safety-valve precautionary measure" in order to "vent-off" the excess Core-Energies, so as to restore Planet-Earth dynamic System equilibrium! Upon such life-support processes depends the perpetual and persistent endurance of all Life-Forms on this Planet!

The Earth is a "Whole-Entity Dynamic System" that must seek "Gravicentric Equilibrium," whereby all co-determinant factors of its "triune-component activities" operate interdependently with each other's overlapping processes, e.g., Planet Earth had to espouse a 23-degree axial tilt in order "to accommodate" the Moon's Mass which stands as 1/4th of Earth-Mass; the sum of which, being a "salutary accommodation," given the abundant life-supporting benefits that ensued: Due to its axial tilt, Solar gravitational radiation Energy "only grazes" the Earth at the Poles which allows the formation of icecaps, effects of which, countering the extreme temperature gradients occurring at the Equator where perpendicular solar rays "tend to strike" the Planet "in a more direct-linear fashion."

Lightning-strikes, Earth compactness-pressure-Force, core heat-temperatures, water moisture, geo-hydrosphere Curvature pressure Energetics, Magnetic Field and Gravity tensor-pressure-stress-Motion-Force, and depth of "catastrophic burial" of different categories of sediments, as "interred bio-Mass," or "entombed bio-Mass," interplay under heavily pressurized thermodynamic conditions "to form" fossil-fuels such as coal, natural gas, and petroleum.

Temperature-and-Pressure, being the primary co-determinant instruments of "subterranean-chamber yields," account for molecular formation of metals, rock crystals, minerals, and other known subterranean materials, compounds, and substances, emerging from events and processes, akin to those giving rise to the chemical reactions producing radio-active Elements such as Uranium and Radium, the substrates of which, might also have been originally "formed" during the Planet's initial "Creation-Moment."

Without lightning-strikes engineering temperature Differentials that cause convection processes and Motion-force Opposites that trigger thermo-chemical reactions along with "contiguous-simultaneous proximal-collaboration" from landmass-generated pressure-Forces acting upon "buried materials, substances, and sediments:" Inner-Geo-thermal conditions yielding formative processes and chemical reactions conducive to these transformative "Energy-motion-Force Conversion states," which entails "transubstantiation-of-buried Materials," could not have taken place.

Due to the Law of Energy Transformation, Conservation and the Equivalency Principle are "operationalized" via redundancy of iterative analog structural Forms that are organized to function with ingrained cycling periodicity, from which emerge Entropy-processes, the complex trans-relations of which, permeating Field-and-Gravity-activated Continuum Space-Time Curvature.

The Sun, must, by necessity, contain all Earth-manifested elemental Forms. As nucleated moving-Mass Radiation Energy, the Sun is the thermodynamic archetypal paroxysm of all universal Energy Transformation phenomena!

These Sun-plasmatic Elements are "recapitulated," "re-instated" or "replicated," within Earth gravimetric "non-nucleated" Mass-in-Motion Frame, in analogous, but "congealed Forms."

Earth-activated "Energy-Motion-force states" operating as bounded gravimetric Frames, are only condensed analogs of the Curvature tensor-stress dynamics already operating within the Solar System ellipsoid plane of Revolution, wherein prevails electromagnetically-induced pressure-tensor-stress motion-Force Energetics.

Earth "wobbling," bi-polar oblation or flattening at the Poles, and the perihelion shift of Planet Mercury: All, lie, in "synchronized gravitational Continuum" along with electron trajectories interpreted as "waves of probability," — Planetary Mechanics and Atomic Dynamics "are in gravi-centric synch" — as engendered by Curvature pressure-Force Energetics, respective to both planetary cycling-Mass-in-Motion properties and Quantum mechanical Frame particulate cycling-Mass properties.

"Wobbling" arises from a solar system "gravitational perturbation" just as simultaneously determining momentum and position of a particle appears to be a quantum mechanical "Field-uncertainty." Quasi-similar, but Frame-specific events and phenomena, amenable to similarly prompted cause-and-effect mechanisms, occur within both the Macro-Frame of solar system mechanics and the Micro-Frame of quantum particulate mechanics.

The complex difficulty with discerning the physical laws that cause these Differentials and Opposites at the quantum mechanical level does not vitiate the facticity of iterative Forms of structural organization appearing to be of quasi-similar patterning order.

The solar system is a whole energy system complex embodying mass frames in Curvature cycling Differentials brought about by relative Motion-force interactions between Opposites as characterized by centers-of-mass, centers-of-field, and centers-of-gravity, as they all must interface to sustain universal dynamic System equilibrium.

Universal reality is encapsulated into matter-mass frames operating as "bounded" energy-motion-force states of structured cycling, within the electromagnetically-induced gravitational "theater of Space-Time Continuum Curvature Energetics."

Cycling Differentials and Motion-force Opposites are "re-normalized" or "reconciled" via hidden variables conceptualized as alpha-numeric Constants that work as Time-triggered catalytic mechanisms between "overlapping Frames of functional operations, as "nourished" by the universal Continuum-Force Algorithm animating all mass-Frames for Electromagnetic-and-Gravity Fields-induced Curvature motion.

The Heavens are filled with Galaxies! All Galaxies reflect the ubiquitous proliferation of radiating Stars — Thus, Stars are the original Source and Wellspring of Electromagnetic Field-Force Energetics triggering Continuum Gravitational Curvature Energetics! All Forms of motion-Force in the Universe are amenable to Stars, whose Electromagnetic Properties are "re-translated" or "re-substantiated" in order to generate Magnetic-field induced Gravity-field equivalents!

Space-Time could not have had Continuum until Matter-mass was introduced therein with electromagnetic properties that function as "Energy-cycling" in order to engender gravimetric Forms of rotary motion. Rotary Forms of Motion engender Thermodynamics, "embodying" processes of both Conservation and Entropy in abidance to the Law of Transformation of Energy. Rates-of-Change in "Energy cycling co-determinants" characterize the Law of Energy Transformation as "the unfolding" of Thermodynamics via the "INPUT-PROCESS-OUTPUT mechanism by which is structured the Organizing Principle "ordering" the variegated functions of Mass-in-Motion Frames.

Space had to be empty-void, invisible and yet infinite-indivisible, for uniform expansive Oneness. Matter with cycling energy mass appears to establish "breaks," "divisions," "separations," or "compartments" or "volumes" within Space. Earth-space Volume is contiguous with void-vacuum Space. There is no "solid division" or "solid separation" between Earth Space and void, vacuum Space — except that Earth Space-Volume has a breathable gaseous Atmosphere that is sustained by Magnetic-field Force and Gravity-field Force, whose breadth-of-Strength, is relative to Earth-Core "processing" of Solar Curvature Radiation Energetics.

Earth atmospheric layers are effective in their "anti-Radiation" activities that "shield" the Planet from the more destructive rays and particles "carried" by Nucleated Solar Energy, in that atmospheric "congealed processes" constitute a "natural barrier" against sustained bombardments from nucleated Solar plasma condensate emissions.

Alpha-numeric Constants are always needed in order to "link" or "connect" differentiated but interconnected Frames that overlap in transformative operations for dynamic whole-Energy System equilibrium. Space-Volume is Infinite in the absence of physical Matter and its measurable dimensions that impart "boundaries" to its expanse! Measurable Space-Volume is usually "contained" or "embedded" within the structural limits or boundaries of Mass-dimensions that geometrically Frame its "shape" or "mold." No "measurable Space-Volume" exists without the limited contours of Matter that "contains" it, e.g., Space that is "contained" within a three-dimensional "vessel" such that it is enclosed within the inner-confines of the cardboard box, is "measurable." But because Space-qua-Space has no physical limits or material boundaries, and is hence, Infinite, the dimensional properties that impart Volume to that particular "expanse of Space," belong to the "three-dimensional vessel" within whose inner-confines that "expanse of Space" is contained, e.g., the cardboard box.

Though Space-and-Time belong to a Continuum whose Spectrum is infinite, due to Curvature Energy-motion-force States imposed by Matter-with-mass-in-Motion, Continuum Space-Time appears to have become divisible, finite, visible, detectable, separable, quantifiable and measurable. As these Mass-in-Motion physical dimensionalities ascribe "bounded Volume" to spatial structures "thermodynamically operationalized" within the Framework of Time, Space inherits specific quantities, characteristics, or values that Human Beings can measure, e.g., Planet Earth has a spherical geometrical "shape" that imparts Volume" to the Space "occupied" by its Atmosphere, by its Landmass, and by its Hydrosphere (Oceans, rivers, lakes, subterranean wellsprings, etc…)

Thermodynamics — presenting a Beginning, a Span, and an End — necessitates quantifiable operations due to the boundaries and limits imposed by the fixed-constant value of the Speed of Light in a Vacuum — 186,000 miles per second; or, 3×10^8 meters per second.

Differentiated cycling Dynamics, — e.g., Planet Earth is in a state of "congealed Energy;" whereas the Sun is in a state of "excited Matter," — and Opposite Motion-force Energetics, — e.g., Rotation of Earth Mass-in-Motion is anticlockwise or counterclockwise; whereas Revolution of Earth Mass-in-Motion is clockwise — are thriving within the span of measured-Time, to engender "stochastic Input-Process-Output patterns" as "organizationally structured" by "the modus operandi" of the particular "Mass-in-Motion Frame," due to Accelerated Energy-Motion-Momentum-Force that "pushes them" under "the urgent prompts" of Curvature pressure-tensor-stress Energetics requiring "super-position of quantized processes" over Continuum Space-Time, e.g., The Orbital Velocity of the Earth stands at 29.8 kilometers per second, whereas Jupiter's Orbital Velocity proceeds at 13.1 kilometers per second; The rate of acceleration of an Object by Gravity on the Earth stands at 9.8 meters per second squared (or 9.8 m/s^2) whereas on Jupiter, the rate of acceleration of an Object by Gravity is 23.1 meters per second squared (or 23.1 m/s^2); On the Earth, the length of the day is 24 hours whereas on Jupiter it is only 9.9 hours.

Continuum, therefore, is constituted of "quantized-stochastic," but not random or accidental, patterns of Energy Conservation and Entropy Processing, because all events, processes, and phenomena occur and proceed, develop and unfold, only according to specifically proven-scientific physical Laws that together operate-to-keep the Universe in "perfect working Order."

These "quantized-stochastic patterns" engender an "encoding dynamic" that yields a Mega-Force Theory, the sum of which, necessitating integration of Motion, reconciliation of Differentials and Opposites, and renormalization of Forces, via superposition of alpha-numeric

Constants that "uni-Temporize" them into "smooth-waves" of Relativity-interactions within their common Framework of Continuum Space-Time.

Matter-Mass-Units in Field-and-Gravity-induced Curvature pressure-Motion-Force, gives physical "Energy-cycling dimensionality" to Continuum Space-Time Energetics.

As Mass-in-Motion Frames cycle under bounded equilibrium conditions embodied in specific rates of Energy Conservation and Entropy rates-of-processing, gravimetric Differentials and Motion-force Opposites arise to "cause-and-effect-establish" the "Continuum Frame," which then necessitates alpha-numeric Constants as "substitutes" for hidden "mechanisms of normalization." The strong force prevails as energy conservation engenders pro-centric forms of cycling motion wherein both Conservation and Entropy "entertain" thermodynamic processes of Entropy. Resources are generated and utilized, by-products and "waste" are "recycled as New Inputs for "the next Processing Cycle" yielding newly produced Outputs. The Universe and Nature are not akin to an experimental laboratory with "manipulated conditions." Within the Universe and in eco-Nature, "nothing is wasted" because every event, process, or phenomenon takes place as prescribed, respectively, by pertinent, specific, "overlapping physical Laws" effecting the unfolding execution and developmental implementation of their modus operandi.

A Constant is always needed for "Continuum-Oneness," due to Gravitational-and-Field Curvature Differentials and Motion-Force Opposites that must "connect" or "interface," for "equivalents" of the Gravity-Force, so as to integrate overlapping "Mass-Energy-cycling Units" into a "relatively unified sphere" of interdependent-and-interactive manifestations or a "relatively unified theater" of operational influence.

Alpha-numeric Constants restore coherence and balance where needed, and equilibrium and consistency where necessary, to the scientific paradigm that is internally designed to "embody" cosmic operations of Unified Universal Continuum.

Constituent components of numerous Differentiated Mass-Frames must "overlap" within Relativity-conditions of operation, as Continuum Space-Time Curvature functions to elicit Motion-force Opposites that correlate with actual planetary movements, e.g., Rotation and Revolution, as well as to synchronize apparent gravimetric "anomalies" with regular Solar System functioning, e.g., Perihelion Shift of Planet Mercury; — In sum, all Continuum Space-Time inter-exchanges, interactions, and relationships must be accounted for, within an equation of proportional equivalency as potentiated by "conversion factors" that can be "operationalized," executed or implemented towards useful technological applications, e.g., Creating or inventing a new type of rocket propulsion system designed for launching spacecrafts into Earth orbit and beyond.

When a System is already operating in dynamic equilibrium, a change in values, or a modification in Motion-force vectoring, as applied to one "conditional parameter" or "conditional co-determinant," constitutes "disruption of equilibrium," or "suspension of equilibrium," hence, necessitating restoration.

Dynamic equilibrium must be restored by re-synchronizing "operational parameters values" attached to controlling variables, in order to accomplish optimum "re-arrangement," or "restructuring" of operations towards "Entropy-reducing re-formalization."

Conservation operations within a System already in dynamic equilibrium are meant to thwart "the advance of Entropy," e.g., Even as a ten-year old Human Being is "aging" while growing and maturing towards twenty-five years of age, mortality is "kept in check" via bio-metabolic and physio-hormonal processes that thwart, inhibit, or delay the onset of terminal death, which is expected to occur at seventy-five years of age or older.

Thus, Energy Conservation must override or inhibit "the promptings" of terminal entropy in order that Continuum can proceed via quasi-similar iterative periodic cycling. For example, the Earth erupts into volcanic lava flows when molten outer-Core thermo-pressures are so extreme under intense electro-motive lightning stimulation, as to require "safety-valve-like release." Or, under such extreme conditions, tectonic plates "move," and by that, make "vibrations-and-shocks" travel within a correspondingly attainable "swath of the mantle" of the Earth, to arrive through the crust, so as to cause us to experience an "earthquake."

In the same vein, as the Sun is "an imploded nuclear fission" as well as "an explosive nuclear fusion," these "convection Differentials" generate Opposite Inner-Forces — "fused explosion" and "fission-ed implosion" — that cause it to display "Sun spots," emit solar flares, propel solar winds, throw plasma rains, and expel coronal mass ejections as well as other "electromagnetically-toned" radioactive materials.

These "convulsed solar emissions" take place when inner-convection motion-Pressures from reactant-chains of nucleated Forces "trigger safety-valve mechanisms to engage," such as imposition of greater Magnetic Field Strength, which in turn operate to thwart "collapsible-inwards implosion" and/or "explosive disintegration" of potentially "de-pressurized core-densities."

The Sun possesses many nuclear plasma condensate fuel cycles that sustain System dynamic equilibrium, even in the presence of inner-core perturbations and "convection-Zone" upheavals, e.g. Every eleven years, there are solar winds and solar flares due to magnetic polarity change or displacement; Every twenty-two years, there is a "field polarity change" accompanied by radio-active discharges that may cause magnetic storms and meteorological turbulence.

Continuum Field-Momentum-Motion-Force that triggers rotary Forms of motion by Objects within its "range-of-Force-strength influence" is a function of Energy as Radiation, e.g., As in the Star, Energy-as-radiation is a function of "excited" Electromagnetic-Mass "trapped" within "implosive nucleated reactions" by "plasma convection."

Radiation, Mass, and Motion, as confined within solar gravitational pressure-force stress-tensing Dynamics, form the basic foundation of Energy-force-motion-Frames, such as Planets, that "process" "Curvature Energetics" as "binding-energies" marshaled for operating in thermodynamic cycling equilibrium.

Radiation-Mass pertaining to an Object already in Motion, such as the Electromagnetic Spectrum Energy Units, is a function of cumulative Matter-particles having accelerated in "Electromagnetic weightiness" yielding Force — (either as "congealed state," e.g., the Photon of Visible Light; or as "excited Mass," e.g., Radioactive emissions such as X-rays) — due to a condition of being "caught" or "trapped" in momentum-accelerated Continuum, geared-for Space-Time Motion.

Because when an Object "falls" within a Magnetic Field it "automatically" or "naturally" begins to rotate/revolve, — e.g., An electric fan's blade-stem-assembly; or a Planet, — then, great difficulty arises in ferreting out, the cause-and-effect mechanisms accountable for that phenomenon — because "that's just the way it is."

Projected Radiation-induced Solar Curvature pressure-force Energetics, "re-translate" the electromagnetic properties of Matter into Magnetic field-Force and Gravity-field Force "equivalents" that engender centri-vectored Forms of rotary, curvilinear, linear, and rectilinear patterns of Motion. Atomic nucleus-centric particulate Motion, geo-centric Moon orbital Motion, planetary helio-centric revolutionary Motion, respectively "embody" or "in-carry" gravimetric Differentials and Force-Opposites that "overlap" in uniting all mass-Frames, as they cycle in counter-vectoring trajectories and paths to sustain Energy Conservation and Entropy Processes aimed at "replenishment of Inputs" geared for maintaining Continuum Dynamic System Equilibrium.

An integrated conceptual-theoretical reference frame or scientific paradigm that aspires to unify all Force analogs-and-equivalents into a "Field frame" — such as Electromagnetic Field-force and Gravity-field Force — ought to consider differentiated values for thermodynamic Inputs respective to each Mass-in-Motion-Unit, as well as Opposite motion-Force Forms engineered by "equivalents of the Gravity-Field" that do not display particulate charge-displacement-motion properties akin to Magnetic-Field exertion.

Pertinent to the fact that all universal phenomena answer to very specific scientific physical Laws, to say that "Things are in a so-called 'free fall'" is misplaced opinion, given that they do not proceed "at random." Randomization and Probability are human inventions designed to mask either inability to fully comprehend a phenomenon, or inability to quantifiably measure such phenomenon due to its "unknown Source" or "unfathomable Origin."

""Sources unknown!" or "causal mechanisms unknown," appears to govern the use of such mathematical expressions as "Infinity," "probability," "randomness," or the "creeping" of common parlance terms such as "chance," and/or "accident" into Mathematics and Physics.

Given that the Gravity-field-motion-force is an emergent property of the Electromagnetic Field-motion-force, then "the strong nuclear force" belonging to Electromagnetism impels the Universe to Energy Conservation, even as Entropy Processes are "on-going" as well, e.g., The Sun "keeps" or "maintains" a greater amount of Energy than that being "carried-out" by its regular or infrequent "emissions," such as on-going radiation, solar flares, solar winds, coronal mass emissions, and radioactive decay, etc…

Energy Conservation is primary because "the strong nuclear force" prevails over "the weak nuclear force," as centripetal or centri-vectored motion-Forms predominate, hence, "giving rise to Gravity" or "its equivalents" in the forms of co-equal and co-linear Forces, but the sum of which, being Opposite to Electromagnetic Force-Forms, e.g., the centrifugal or counter-centric Force.

All Human Movements on the surface of Planet Earth participate in these two primary Motion-Forms: Centripetal (Mass-centric Force) and Centrifugal (counter-centric Force), e.g., Jumping from a mountain-top with a parachute towards Earth Surface or Earth-Center-of-Mass (centripetal); Pole-vaulting athletic aerial jumps away from Earth-center-of-Mass (centrifugal); or an aircraft being piloted for aerial acrobatics, that, by the use of a motorized engine (propeller-plane or jet-powered aircraft), can perform movements towards Earth-center-of-Mass (Mass-centric) or away from Earth-center-of-Mass (Mass-counter-centric) — thus, engaging in both centripetal and centrifugal Motion Forms, and/or in composite-combinations thereof.

"Relative Forces" that are active between mass-Frames in-motion, and between-within the Spaces they occupy as — these Mass-in-Motion-Frames "travel" by-way of gravitational propulsion power for Continuum Space-Time energy cycling, — are not uniform in patterning structure, operational vector, causal constitution, and original composition. Forces can be quasi-similar, co-equal, co-linear, but also opposite to each other, as performed by differentiated Mass-in-Motion Frames purposed for variegated operational functioning.

Mass "bends" or "curves" the Space around it or within its vicinity, but as "contained" within the physical boundaries of its "interactive Field of influence." This Relativity Field of Interactive Manifestation is limited by intersecting or overlapping variables, parameters, and co-determinants that are embedded within the Energy-motion-Force Form-Complex, as applicable to each Mass-Frame, the sum of which, engendering gravimetric Differentials and Field-motion Opposites, consonant with "their sensitive co-dependence" upon the constant standard-fixed value of the Speed of Light.

Thus, possibilities for enumerating, defining, and framing the nature of physical Laws are not limitless and infinite.

Continuum Space-Time is infinite and boundless. However, "gravimetric Energetics" ensuing from its interactive Magnetic-Field and Gravity-Field "equivalents" is finite in magnitude, projection, and strength because they are properties of cycling electromagnetic mass-Frames, motion-Forms of which, necessitating rotary operations or movements consistent with Curvature Motion-force.

The Laws of Thermodynamics cycle in quantized, measurable, relative proportions-and-ratios that are predetermined by "the values-limits" circumscribed by "Mass-frame boundaries," — e.g., Each Planet's "structural arrangement" has its own "gravimetric Signature:" Gaseous Jupiter has greater Gravity-force strength than "terrestrial Earth," — as imposed by execution of the Input-Process-Output Mechanism that is "embedded" within their respective operations by implementation of "The Organizing Principle."

A line is a succession of points-dots. A particle's dimensions consist of "point successions." There are "elemental particles" for Matter because Matter possesses objectively measurable physical dimensions, e.g., the Atom's proton, neutron, and electron that are differentiated in micro-Mass, size, and electric charge, and their respective Energy-forms that have wavelengths and frequencies.

There are "particles" for radiation because Energy, being a property of electromagnetic Mass-in-motion, also possesses materially measurable physical dimensions, e.g., cosmic rays such as X-rays, Gamma rays, Alpha rays and Beta particles; and their wavelength in centimeters, frequency in cycles per second, and energy units in electron volts.

Energy is an intrinsic property of electromagnetism. Electromagnetism is an intrinsic property of particulate charges in Motion or "bodily-displacement in Continuum Space-Time."

Energy, as "congealed" or "excited," is a measurable property of Mass-in-motion due to particulate "micro-electro-motive dynamics" constituting Matter at its most fundamental or basic level, as it continuously partake or participate in "Universal Energetics." Particulate-charge micro-Motions engender "macro-resonance" as "Molecular Space-Time displacement-Forms," e.g., electrons revolve around the atomic nucleus; planets revolve around the Sun. "Energy is never created nor destroyed but always transformed:" Redundancy of symmetry in iterative-Forms as organizational patterns that replicate the same structural Forms!

How do "Continuum Relativity Interactions" between Space-Time, Matter-Mass, Electromagnetism-and-Motion, and Gravity engender Curvature pressure-force-tensing-stress Energetics?

Mass, Radiation Energy, and Motion are "the fundamental States" espoused by "re-translated Mass-Electromagnetism" — The Universe = Matter (Electromagnetism) + Continuum Space-Time.

Within "those Mass-Energy-in-Motion States" as "set" by Electromagnetism and Gravity, are "embedded" all the "physical dimensionalities" of Matter, — quantifiably measurable parameters — such as those that can be ascribed to or that pertain to "parametric in-Relationship-values," displaying variable "ranges-of-effective-projected-exertion," — which are manifested as proportions, ratios, and/or exponent conversion factors that establish relative equivalencies between Motion-Force co-determinants, the complex resolution of which, operating to endow cycling-Frames in Continuum Space-Time Curvature with dynamic System equilibrium.

Continuum Space-Time Curvature Energetics — comprising every "powered Motion-Force function" or "Fields-induced operation" pertaining to a specific Mass-in-Motion Cycling-Frame — is the perennial Source and wellspring of all Forms of Motion-Force that can be generated from Magnetic-fields and Gravity-fields.

"Continuum Curvature" obtains its "Force equivalents" from Electromagnetism and Gravity, as Force-fields engendered by the complex interplay of opposite particulate charges, hence, summoning "the strong force" and "the weak force," which constantly remain in "Continuum entangled interactive exertions."

In other words, only two primary Forces "move" the Universe: Electromagnetism and its "emergent-co-laborer," Gravity!

Thus, functioning Cycling-Frames constituted of different Forms of the "Mass-Energy-Motion-Complex," embody "re-translated Field-forces" or "transubstantiated electromagnetism," from which derive, "the strong force" and "the weak force," all of which climaxing into Curvature Motion-pressure-tensor-stress Force, taking the Forms of Magnetic-Field and Gravity-Field "Force-equivalents."

The universal Input-Process-Output Mechanism that activates thermodynamic processes of Conservation and Entropy leads to the formation of "steady-state Star-masses" — thriving in dynamic System equilibrium — that possess specific radio-metric Signatures and gravimetric Plasma Characteristics that differentiate them from other Space-Mass-Units, such as Planets.

For example, the Sun in our Solar System has been "shining-and-radiating" for thousands of years as a reliable-predictable Space-Mass-Unit yielding the Electromagnetic Spectrum — thus, "Core-centric nuclear micro-turbulence" engendered by solar nucleated Field convection plasma dynamics is "so distilled and homogenized" throughout the Solar-System-Mass for Energy Transformation, that it climaxes into a "Steady-State Mass-Energy-Unit" continuously "re-calibrating its parametric values" due to "the variable range" of its Magnetic Field Strength, in order to attain, accomplish, achieve, fulfill and maintain dynamic System equilibrium.

It is in the same manner that the Human Body is said to a "homeostatic system" with a relatively constant optimum body-temperature standing at 98.6 degrees Fahrenheit (macro-Stability at dynamic System equilibrium) even as blood pressure varies within a life-affirming healthy range in diastolic and systolic values, and even as heart-rate or pulse, possesses its own "variable range" of "beats-per-minute."

In the same vein, there appear to occur, various "values re-calibrations" at the Quantum Mechanical Frame where negatively charged Electron particles are "whirling" around the nucleus or center of the Atom that is constituted of the positively charged Proton particles and of the "un-charged" Neutron particles.

Yet, the Atom always has "macro-Stability" or is always standing at dynamic System equilibrium, allowing molecular change and chemical reactions leading to greater agglomerations of Matter, such as a Planet, though things appear to be "uncertain," "turbulent" or "probabilistic" from a "micro-particulate-perspective."

Phenomena, events, and processes at the atomic level are not to be categorized as "micro-turbulence" with "quantum probability," for they take place according to very specific physical Laws that impinge upon the minutely infinitesimal Mass-units that are embodied by each atomic particle.

Though these physical Laws still elude our comprehension for discovery, according to their principles, particles adhere to the "range-determined" trajectories and paths, movements, displacements, and Motions imposed upon their micro-Masses, by the "proper-Strength" that is "fit for the Atomic-Frame," corresponding to Magnetic-Field-Strength and Gravity-Field-Strength, causing them to engage in such "cycling patterns."

Steady-state whole-Energy systems operate with all conditional parametric Values synchronized and integrated, for dynamic equilibrium ranges-of-efficiency, except when equilibrium is disrupted and restoration activities are undertaken for Qualitative Conservation, thus, overriding "the prompts of Entropy." For example, revolution and rotation are Continuum gravimetric signatures of Planets, which are "congealed electromagnetic Matter-Mass" or "non-nucleated Mass-Units," in conjunction with solar gravitational radiation field-Curvature tensor-stress-pressure-Force conditions that cause "wobbling" or "bi-polar/dipole oblation," or even a mercurial perihelion shift.

Magnetic polarity change is a radio-metric signature of Star-types, like the Sun, as solar plasma convection "Transform-dynamics" climax into gravi-field Differentials and Motion-force

Opposites that necessitate re-calibration of equilibrium parameters Values, via oppositely vectored Field and electro-motive Force-conductivity.

There is no accelerated Motion without Curvature-pressure-Force-momentum effecting field-binding energies and gravity-tensing force-equivalents. There is synchronized symmetry in universal patterns of "energy-force-motion Frame embodiments" harmonizing cycling Differentials and force-Opposites for solar system ellipsoid plane elasticity, tensile strength, and equilibrium stressing dynamics, climaxing into each Planet having its own place, location, and position from the Sun.

The Earth is "in third place" from the Sun. The perihelion shift of planet Mercury is due to a change in the rate of momentum-Curvature pressure Force, as it gets closer to the Sun, causing "Mass-acceleration" as "computed-calibrated" by Relativity interactions between centers-of-Mass, centers-of-Gravity and centers-of-Field from projections of Continuum solar gravitational radiation Energetics, cycling-Differentials and force-Opposites of which, espousing quantized "stochastic rates" of emission.

Mercury reacts to solar-Differentials in interaction with neighboring planetary cycling-Frames that are also receiving tensile-stress and tensor-pressure-Force from ellipsoidal plane Inputs. Mercury undergoes "a global change" respective to the values configuring its gravimetric conditions, thus, necessitating a corresponding "re-calibration of Cycling-rates" designed for restoring dynamic System equilibrium.

"Planetary displacement" presents a range-of-Motion yielding "momentum-acceleration-Force" within which the Planets must revolve and rotate; and Mercury's perihelion shift is a disruption in the range of its cycling-Values that "clothe" its dynamic equilibrium repertoire — The shift, is Mercury's "response" to close-proximity Solar Inputs causing changes in its "cycling-Values."

Due to Mass, Volume, Motion, and Pressure Differentials between its constituent components — landmass, oceans and atmosphere — Earth Rotation and Revolution motions also "carry" residual accelerated momentum-Force that causes it to "wobble" as its Poles are "slightly flattened." "Carried residual accelerated Momentum-force" is also accountable for Landmass, Wind, and Oceans "movements," such as surf-and-tide, relative to all solar Inputs and Earth-internal ecological processes yielding "gravimetric turbulence."

The Moon geo-synchronously revolves around the Earth climaxing into a range-of-Motion that allows natural recalibration of its orbital path in "degrees of variability," e.g., as a few millimeters away from its regular trajectory, while maintaining its periodic phases which, with other factors, continue to cause tidal surges in Earth oceans.

In short, though there are "degrees-of-precision" or "Constant Curvature-clearance" in Moon orbital ranges-of-Motion so that it maintains its distance from the Earth relatively constant, still, there is also "flexibility-and-play" or "Curvature-tensile-clearance" in Earth projected magnetic-Field Strength, allowing for "gravitational free-fall," — "free-fall:" "Movements-as-is" by all bodies in Space, e.g., the Moon, without Human intervention or external Force. "Curvature-tensility" encapsulates Motion-range fluctuations, permitting the Moon to oscillate back-and-forth, during which, it appears to "move-away slightly" or "escape" from "Earth Gravity-pull." This "Moon-escape displacement-Motion" is estimated at 3.8 cm/year, the sum of which, being apparently within its normal "gravitational Curvature-range-of-Motion," due to Continuum-Motion-Force from solar Inputs as affected by "stochastic plasma convection perturbations" necessitating "re-normalization" or "corrective re-calibration."

In Nature and the Universe, because Energy Transformation must cycle in "stochastic rate-patterns of consumption" that allow for both Energy Conservation and Entropy processes, cycling-Mass-in-Motion Units alternate between Conservation and Entropy processes, while simultaneously undergoing "resource Replenishment." This "unfolding lapsed-development of Continuum processes" in gravitational Space-Time Curvature requires that phenomena and events operate in "stages," "steps," or "periods-of-Time," while parameters and variable operationally function "within allowable ranges-of-Values" as sustained by inherent physical Laws that undergird "on-going Continuum," e.g., While farmers look towards harvesting their crops at Summer's ending, Earth Rotation on its tilted 23-degree axis takes place every day as Solar Inputs are "renewed" and ecological processes "revitalized" for maintaining Earth life-support Systems in dynamic System equilibrium.

Earth life-support systems are thus protected from "radioactive-Electromagnetic overload," as all living organisms enjoy steady processing of gravitational Inputs within periodic recurrence of measured allotments of Energy in proportions and ratios that allow Life and Earth ecology to continue to prosper.

Every functionally operating thermodynamic Mass-in-Motion Unit or "Moving Body-in-Continuum Space-Time" must subscribe to the Law of Transformation of Energy: Its organizing mechanism is embodied in the inescapable Universal Thermodynamic Principle: Input-Process-Output.

Conservation and Entropy requirements — for "conversion of Solar Inputs" into ecological resources that must be "consumed" to sustain dynamic System equilibrium, — compel natural processes to cycle, as "corrective measures" that thwart "terminal exhaustion" of "ecological assets" are activated. These "thermodynamic safeguards," as induced" from "initial Genesis programming," or by "an encoded mega-universal Algorithm," allow for "recalibration" of Continuum gravitational Curvature-Field-values that obtain their sustainability from solar radiation Energy Inputs.

Thus, thermodynamic cycling Mass-in-Motion operations constitute "a cybernetic System process" pursuing dynamic System equilibrium that allows for "responsive recalibration of parametric Values" by each inter-relating Space-Body that is in Motion within the same Solar Ellipsoid Plane as necessitated by oscillations, perturbations, fluctuations, vibrations, undulations, and variations that inhere in the complex Framework of planetary "Inputs-digestion," "homogenized diffusion," or "global distillation" of solar gravitational field-Curvature-force Inputs, respectively.

Earth Core-centric motion-Force mimicking or emulating "strong-force-Continuum" is also in consonance with Energy Conservation for ecological functionality and cycling equilibrium.

Such is the way in which "Continuum Thermodynamics" is sustained: Via "cycles of "ratios-and-proportionalities," such that "Differentials-processing" reconciles and renormalizes all Motion-Force-Opposites, consistent with "rates-of-fulfillment" designed "to trigger" dynamic "cybernetic Values-re-calibration" of Continuum Energy-Transformation-Inputs, for functional range-specific operations singularizing each mass-Frame's "rates-of-Consumption," e.g., Due to "Wobbling," periodic "bulging at the Equator," and "slight dipole oblation," Solar gravitational Field-radiation Energy Inputs do not "strike" Earth volume and surface, and thus, nor trees, plants, and shrubs, at the same specific angles, the sum of which, every day displaying "slight angular variations" in "degrees-of-deviation;" The angle(s) at which flora is "radiated" — pertinent to regional geographical position, temperature zone location on the Planet, and idiosyncrasies of each particular Seasonal Cycle, — play a corresponding role in establishing soil fertility, in determining atmospheric and soil moisture quotient, in forecasting plant growth gradient, and in fostering plant Species diversification, e.g., Fauna and flora in the Amazon Forests show extreme differentiated diversity as compared to their counterparts in Northern America.

Natural processes and Solar System events and phenomena yielding dynamic System equilibrium, cause mass-Frames to operate-and-function as Steady-State-Whole-Energy Systems, because their cycling patterns embody "symmetrical equivalency" as they converge into interdependent-and-overlapping iterative quasi-similar Forms, the complex sum-and-interactions of which, "thriving together" in pursuit of coherence, unification, and integration, within gravitational Continuum Space-Time Curvature, e.g., All cycling Mass-frames display pro-centric rotary motion-Forms sustaining Energy Conservation in propagation of "strong force Continuum," even as Entropy processes ("the weak force") are on-going in "recycling Outputs" as "new Inputs."

Symmetrical synchronization of Conservation and Entropy "processing-patterns" in differentiated cycling-Frames for whole-System dynamic Equilibrium is a kind of "universal impetus Algorithm" imprinted into "things-that-Move" as "gravitational structures" by Almighty God, our Creator: In that they seek to preserve the originally created "Force-programming" that initially "impelled-moved" their interconnections, transactions, interactions, and relationships during Energy Transformation, into convergent compositions and structures, divergent organizational-Motion-patterns, and quasi-similar functional operations.

The Organizing Principle engenders a multiplicity of "corrective mechanisms of Sustainability" that impede-and-thwart "useless waste," as well as decrease the rate-of-Consumption of gravimetric Inputs, in order to prevent "terminal resource(s) exhaustion," — "Nothing is wasted in Ecological Nature." — These "Continuum-preserving Safeguards" attesting to redundant patterns of operation in iterative fulfillment of the Input-Process-Output Mechanism prevail respective to each Cycling Mass-in-Motion Frame, even as "reconciliation" of overlapping Energy cycling Differentials and Motion-force Opposites functionally particularize their properties into specific "range-bounded processes."

****

MOTION-FORCE: THE CONTROLLING PARAMETER IN CURVATURE ENERGETICS

"In the beginning God created the heavens and the earth. And the earth was formless and void, and darkness was over the surface of the deep; and the Spirit of God was moving over the surface of the waters. (Genesis 1:1-2, NASB, Holy Bible).

God's Spirit "was moving over the surface of the waters" during the process of Creation. Life is in the Spirit. Creativity is from the Spirit. Motion-energy-force is engineered by the Spirit. Light is an endowment from the Spirit. Objective materiality is a dynamic transformation by the Spirit. Spiritual transformation includes the renewal of the Mind. Creative functioning for productive living means to engage in Motion. Motion is consistent with the electromagnetic substrate of Matter. The unfolding of thought processes constitutes the analog of "Mind movement."

The Universe consists of overlapping cycling-Frames, replete with motion-Forms that are structured as "Transform-Differentials" due to the Equivalency Principle, hence, necessitating alpha-numeric Constants that "renormalize" or "reconcile" the apparent "stochastic nature" of thermodynamic Curvature pressure force binding energies, for "homogenization" into Continuum streams or waves, or quanta, of transformed or converted Energy.

Continuum is made up of iterative cycling periods by quasi-similar Frames that embody the mass-in-Motion Energy-force complex. The Solar Nuclear Plasma Complex projects gravitational tensing-stressing pressure-force Energies that "re-translate" the electromagnetic properties of Matter into "theaters of force interactions" comprising Field and Gravity and Curvature "equivalents" constituting Solar Gravitational Energetics that causes mass-Frames to cycle in patterns of centri-vectored rotary Forms of Motion.

Every thing in the Universe and Nature has "Motion." Every thing that exists must "move" in order to properly or healthily function. The Earth revolves and rotates; a ripe apple will fall on the ground to go into seed for new tree growth; the Human heart pumps oxygenated blood even as we sleep; the lungs breathe air as Oxygen is inhaled and Carbon Dioxide is exhaled; Human digestive system has to do with "muscle movements" called "contractions;" Nitrogen-fixing bacteria "move" the soil in order to engage or trigger "fertility processes."

In the cosmological Frame, gravitational Motion common to Planets takes the rotary Form — (things move in a circular pattern, such as via Revolution and Rotation,) and the radial Form — (in a spherical mass-Frame like the Earth, things move "along the line-of-radius" towards its center-of-Mass and center-of-Gravity, such as an apple falling from its tree-branch; or a vehicle traveling on Earth surface apparently in a linear, rectilinear, or curvilinear pattern, but which is really "carving an arc" along Earth-line-of-Circum-sphere, and, but for the surface, would be "drawing a radial trajectory" towards Earth-center-of-Mass).

Rotary, arc-ward, and radial Forms of Motion are fundamentally centri-vectored in a planetary-circum-sphere. As a Unit of Mass-in-Motion Frame thermodynamically cycling for yielding Energy-force from rotary Motion-Forms such as Rotation and Revolution, fulfillment of Continuum for the Planetary-Frame, is primary, but within gravitational Space-Time Curvature, even as the planetary spherical Mass must pursue maintenance of dynamic System equilibrium.

Particles and Planets are also "stirred" by "composite-recombinant Forces" that are "in interactive entanglement" with rectilinear and curvilinear patterns of Motion, in compensation for "resistance" to pro-centric and counter-centric Forms of force. While the strong force sustains Conservation of momentum, centripetal forces sustain Energy Conservation. Conservation is inner-Spherical and pro-centric, while Entropy is exo-spherical and-counter-centric. Thus, in cycling "Mass-in-Motion-Frames with Momentum Force-energy," rotary, arc-ward, and radial Forms of Motion are consistent with "the strong force" aligning Conservation of Energy with centripetal Motion patterns. Conservation tends to coalesce or concentrate while Entropy tends to disperse or disintegrate.

The Moon follows a trajectory around Earth plane of rotation. Likewise, artificial satellites circle the Earth in orbital patterns that align with rotary Motion. Motion upon Earth surface takes apparently curvilinear, linear, and rectilinear Forms, dependent upon "angular positional-location of obstacles" standing in the way of centri-vectored Motion along its line-of-radius towards its center-of-Mass and center-of-Gravity.

Consequently, all Forms of Earth-surface Motion are "interrupted," "arrested," or "negotiated" Forms of centri-vectored Motion along the Planet's line-of-radius towards its core-center-of-Gravity and core-center-of-Mass.

From a cosmographic standpoint, Curvature pressure force-Motion is "re-translated" into the helio-centric rotary patterns, along the ellipsoid plane of Revolution. And from within-a-Planet as a reference-point or standpoint, Curvature pressure force-Motion is "re-transmuted" into geo-centric radial Forms of Motion, "clothed" in rotary, linear, curvilinear and rectilinear patterns that seek to be aligned with Earth "center-of-Mass equilibrium" or Earth "center of Gravity equilibrium."

While Revolution and Rotation are "Earth-global" or "exo-terrestrial" rotary Motion-Forms activated "externally" by solar magnetic-Field Curvature Energetics, internal geo-centric Forms-of-Motion possess Earth-radius-oriented, multi-vectored-trajectory patterns that are "locally induced" by "equivalents of the Gravity-force."

Planetary-surface equivalents of the Gravity-force, — e.g., A vehicle traveling upon the surface of the Earth subject to an acceleration rate of $g = 9.8$ meters/second/second, — are emergent Force-properties ensuing from Earth-centric Magnetic-field and Gravity-field-induced Curvature tensing-stressing pressure-Force Energetics. These Force-Energies are "re-translated" as "within-Earth" rotary Forms, given that they also follow "the circle of the Earth," in paths or trajectories "ultimately directed" towards Earth center-of-Mass.

These "equivalents of the Gravity-force," as imposed upon Mass-bodies that are either at rest or in Motion, impel those Mass-bodies to travel along Earth line-of-radius — akin to centri-vectored radial-arranged equidistant pokes of a bicycle wheel whereby all Gravity-force equivalents impinging upon its rim via the pokes "ultimately seek" a directional destination that is aimed towards the wheel's center-of-Mass.

In the universal Continuum Space-Time Curvature Frame, all Motion-Forms are necessarily "pro-centric" or "centri-vectored," due to "their primary engine of propulsion:" Electromagnetic Field Motion-Force, the complex sum of which, being anchored in Relativity-interactions configured by a constellation of transactions between bi-polar/dipole electro-motive charges causing "Space-displacement trajectories" that must "negotiate the Momentum-Mass cybernetics" engendered by Gravity-field-induced Motion-force Opposites.

All centri-vectored or pro-centric Forms of Curvature pressure-tensor-stress Motion-Force impacting a planetary Space-body, — e.g., rotary, rectilinear, curvilinear, radial, linear, and composite-combinations thereof, — are engendered by Magnetic-field Force and Gravity-field force "equivalents." These "force equivalents" operate to "re-translate" solar projected gravimetric Curvature binding-Energetics, the exerted Force-dynamics of which, channeling Motion-force into Frame-specific thermodynamic conditions of Energy Transformation for "cycling processes" that seek to fulfill dynamic System equilibrium, e.g., Though the Earth "wobbles" and is "slightly flattened at the Poles" while it "bulges at the Equator" during Moon-gravity-induced Ocean tides, as a Mass-in-Motion System, it must "continuously re-calibrate its gravimetric parameters" in order to regain stability for returning to dynamic System equilibrium — "after the dust settles," so to speak, dynamic System equilibrium must be restored, in the same manner that the Human Body, when convulsed by high fever from infectious disease, must go through various "bouts of immune system responses," — relatively framed as "a feedback system" — as engendered by the "Bio-Organic Cybernetic Mechanisms of Physiological Self-Defense," after which, with the help of medicinal pharmaceuticals, such as antibiotics, homeostatic equilibrium is restored to 98.6 degrees Fahrenheit within healthy blood-pressure range.

Curvature tensing-and-stressing pressure-Force gives rise to "equivalent Forms" of motion-Force that emerge from exertions of Magnetic-field force and Gravity-field force"— or as "interactive spheres of Field-influence."

It is crucial and necessary that rotary Forms of planetary Motion-force be conceptualized and operationalized, as "counter-vectored," or "Oppositely entangled Field-expressions" that have their "Sources-of-Propulsion" dynamically anchored-structured by "Mass-in-Motion Specific Ranges-of-Motion-Force Strength," complex transactions of which, being peculiar to Magnetic-fields and Gravity-fields that are incessantly "in entangled inter-relations," e.g., An Electro-Magnetic-field must induce rotary Motion-Forms from which emerge Gravity-force equivalents, allowing or effecting variegated rotary pro-centric Motion-forms along Earth-line-of-Radius. This "Field-sourced Rule for Emergent Gravity-Force(s)" applies to all Objects, from an aircraft to a falling apple, from a rock rolling down a mountain slope to a semi-truck traveling at high speed on a highway, regardless of the "Forms of Traction-Force Engine" accountable for initially propelling them into Motion, e.g., Human muscles or an electric motor.

All traveling, moving, or falling Objects must seek Earth-center-of-Mass and center-of-Gravity along its line-of-Radius, whether a jet-powered aircraft or an apple unhinged from its tree-branch!

As micro-Structures, — e.g., Atomic reference-Frame inherent in all types of Matter, whether as solid, liquid, or gas as assembled into greater quantities of molecules giving rise to Earth Landmass, Oceans, and Atmosphere, respectively, and then to the Stars; or the biological Cell intrinsic to all living things on Planet Earth agglomerating from molecular change to form bacteria and viruses as well as various organs such as Lungs, Kidneys, and Cerebral Lobes belonging to Mammals — give rise to macro-Frames of "increasing levels-degrees of

Complexity," every thing that exists must engage in "locally-anchored self-reflexive Motion" that interfaces with universally unified "Cosmic-global-Motion." For there is "Continuum Affinity" between the two Forms of Motion, e.g., "The strong force," as exerted upon the atomic nucleus, "compresses the Proton-Neutron Complex" into "simulated inertia," while "the weak force" allows for Electron orbital Motion around the atomic nucleus.

In the same manner, as atomic-Frame Motion animates molecular change to form greater Mass-bodies, Planets also display rotary Motion-Forms, such as Revolution and Rotation that animate the Solar-System reference-Frame. "The weak force" also allows for "isotopic decay," and radio-emissions. Planets engage in revolution and rotation even as the Moon revolves geo-synchronically around the Earth. The Solar System "moves" within the Milky Way Galaxy, which itself, "revolves" around the Andromeda Galaxy. In other words, it can be said that "Everything is "moving around" Everything!"

In short, the Atomic-frame micro-Structure displays rotary-Motion, and thus "move" its greater molecular arrangements, from Planets to Stars, regardless of "levels or depths of complexity," — Affirming the Principle of Iterative Symmetry in Structural Patterning modeled within Mass-in-Motion Frames that "fill" the cosmic expanse of Continuum Universal Space-Time Curvature!

In addition, where Mass-bodies-in-Motion Frames intersect to yield overlapping "regions of turbulence," or overlapping "Fields of Permutations," there must appear "cycling Differentials" that give rise to motion-Force Opposites yielding perturbations such as the perihelion shift of Mercury.

Curvature pressure-Force Motion Energetics permeates cycling Mass-Frames for "Relativity interactions" and "Dynamic Transactions" — as engineered by The Organizing Principle's "Relativity-Engine:" The INPUT-PROCESS-OUTPUT MECHANISM.

These interactions and transactions, e.g., between Revolution and Rotation whose "lines-of-trajectories" continually "interface-intersect-merge-cross" even as the Earth is "traveling through Space" — cause "overlapping regions of turbulence" wherein occur periodically cycling manifestations of natural phenomena and ecological events, such as Tornadoes, Cyclones, Hurricanes, and other Geo-atmospheric Hydro-meteorological events, e.g., "Overlapping-Intersecting regions of Turbulence" on the Earth, — as engineered by "rates-of-change" in thermodynamic Variables that "transmute-transubstantiate" into "Cycling Differentials" due to Motion-force Opposites engendered by Revolution (clockwise) and Rotation (counter/anti-clockwise), additionally "conjoined" with Solar gravitational radiation Energy Inputs, — cause to arise, the North-to-South Jet Stream from the Arctic and the South-to-North El Nino Jet Stream from the Antarctic, as well as other phenomenal atmospheric "Storm-vortices," the complex interactions-processes of which, effecting seasonal and other recurring meteorological phenomena.

"Regional Turbulence" "or Meteorological Violence" will arise where clockwise Revolution "intersects-overlaps-merges-interfaces" with anti/counter-clockwise Rotation, in conjunction with all other parameters, co-determinants, and variables accountable for Earth cyclical seasonal ecological processes, such as Solar Inputs and Radiation-Energy "angles-of-penetration" — constituting the primary "Engines of Motion-force" engineering Earth meteorological-geo-atmospheric phenomena and events, to which are inextricably "conjoined," or 'factored," in "cycling entanglements," both Energy-Forms Differentials and Curvature-force-Opposites that effect Relativity-interactions yielding "cause-and-effect mechanisms" driving active operations of Pressure, Temperature, Motion, and Force — constituting the four primary components activating operational processes inherent in a Chemical Reaction Chamber.

"Energy is never created nor destroyed but always transformed." Thus, we are only discovering, through Transformation of Energy, all the variegated, yet limited Forms, Energy can "take," via numerous "transmutations," "permutations," and "transubstantiations," as "Entropy processes are trailing not-to-far behind," e.g., Coal liquefaction, Coal gasification; transformations proceeding from Petroleum to Plastic, or to other petro-chemical by-products; — Styrofoam is a "dead-end by-product."

For molecular change consolidation that yields specific compounds and substrates, chemical reaction chambers mimicking "containment-by-Magnetic-field," are necessary to "confine entangled perturbations" yielding Motion and Force, within specific pressure-quotients and temperature-gradients that quantify "cybernetic controls" intrinsically processed and inherently utilized by particular chemical reactions, as attuned to "entangled dependence upon initial conditions of Creation that originally formed the Elements," precipitating or coalescing, bonding or dissociating, depending upon their "reaction-affinity" for each other; hence, the concepts of "specific pressure" and "specific temperature," in order that Atoms chemically react for "new" elemental and compound formations.

The chemical reaction chamber's walls impart "confinement" as well as Motion and Force to its internal Volume during "congealed" or "non-nucleated processes;" which is analogous to the necessity for the presence of a strong Magnetic-field Force that "contains" or "confines" convection of "excited Solar plasma dynamics," so as to restrict Energy dispersal common to greater-Mass formations in which inheres "internal combustive turbulence," or "internal nucleated plasma convection violence," as well as to "preserve" specific Temperature and specific Pressure gradients attuned to "elemental chemical reaction-affinity," e.g., The hard-casings of a grenade engineered for more powerful explosive-force; the "hardened bunkers" built to "contain" or "confine" nucleated radioactive emissions from a nuclear power plant.

In Quantum Mechanics, electrons move around or revolve around the atomic nucleus. In molecular change, electrons engage in bonding at outer energy Orbitals, in order to form greater Mass consolidations or agglomerations.

In Relativity dynamics, planets don't merely "attract each other," but they revolve around the Sun as they "curve" or "bend" each other's Space, thus, affecting or impacting the trajectory or path of an Object "falling" within their "exerted-range-of-Force-Influence," e.g., Planet Earth has only one Moon or natural satellite, but The Star in our Solar System "holds" nine Planets.

As Planets "re-calibrate" or "re-negotiate" gravimetric parameters, in response to, or "to navigate through" the complex interplay of Relative interactions between centers-of-Mass, centers-of-Field, centers-of-Gravity, Energy-Forms Differentials, and Cycling Motion-Force Opposites where "overlapping-intersecting regions of turbulent perturbations" arise, the complex transactions between these Gravimetric Parameters get "re-translated" into "compensatory patterns of thermodynamic cycling," akin to the perihelion shift of Planet Mercury, in order to maintain-restore dynamic System equilibrium.

Within the solar nucleated inferno's convective Magnetosphere, Curvature pressure-force Energetics engineered by its extremely potent Gravity-and-Magnetic-field Strength, is core-centric, even as "fusion implodes" nucleated Mass for contractions; and "fission explodes" to "push-and-disperse" radiation Energy into dilations conducive to "convection dynamics."

Convection processes taking the Forms of pro-centric and counter-centric Force exertions, cause transubstantiated plasma particulates-and-molecules and their combined-composite transmutations, to "change Energy levels," "directions," intensities, Motion, force, pressure and temperature, etc…, thus, engendering propulsive plasma flux towards "the radiative zone" for corona-directed ejected emissions.

 This "compression-dilation dynamic" climaxes into counter-vectored attractive-repulsive force Energetics that causes heavier elements to fission, hence, distending the convection zone, after which these oscillations stabilize into prevalence of compressive forces.

Heavier elements undergo fusion-and-fission at different cycling rates, pressures, and temperatures, as lighter elements like Hydrogen fuse with more consistency and regularity.

Elements must reach "fusion and fission equilibrium" so that nuclear chain reaction processes transmute into particulate charge entanglements and molecular change transubstantiations that sustain Whole Energy System Cycling Coherence and Force Integrity for "fuel conservation."

The strong Magnetic Field that engenders core-centric motion-force predominance ensures "strong force Continuum" for Conservation of Energy, even as "the weak force" engenders periodic emissions such as solar flares, solar winds, coronal mass ejections, and the regular Electromagnetic Spectrum rays, particles, waves, and photons.

Core-centric fusion and corona-vectored fission simultaneously cycle into convection dynamics thriving under heavily pressurized conditions to create omni-directional particulate Motion patterns in "compressed cycling Time-Frames," climaxing into plasma condensate radiation emissions: such as X-rays, cosmic particles, waves, and visible Light.

Mass Energy gains, due to exponentially accelerated momentum kinematics, cause particles and plasma condensate molecules to "fast-cycle" as they undergo "entangled Magnetic-field pressurized collisions," through changes in pressure, intensity, velocity, and temperature, so that they must "shed their Mass Energy-gains," as gravimetric radiation emissions, the closer they get to approximating "Speed of Light boundaries."

Not even inner-core solar particles and plasma condensate molecules can travel faster than the Speed of Light — hence, "this limiting Factor," ensuring a steady, relatively stable, "Thermodynamic Platform" for A Continuum-Stream System, of Universal Energy!

 Particles in a cyclotron, accelerated to a fraction of the speed of Light "proceed" in like manner: they "re-translate" their mass energy gains into "shedding of mass" by radiating cosmic Energy.

There is Continuum in the Space-Time Complex of Motion-Force Curvature Energetics, due to Frames "confined" by specific Mass-Energy-Force quantities, amounts, ratios, and equivalencies, complex Relative interactions of which, must engage in "entangled transactions" conducive to "equilibrium-seeking thermodynamic routines," — the imposition of Curvature pressure-force motion-Forms climax into "stochastic patterns of Energy cycling," — "rates-of-Energy-quanta" changed-transformed into "Streams-of-Radiation-Waves" — which engenders gravimetric energy-Cycling Differentials and Motion-force-Opposites, through which, they have to interface.

Solar Mass is "confined Mass!" Mass quantities engaged in fulfilling thermodynamic "Input-Process-Output law requirements," within the limits of Temporal-measurable Time for necessary resource production-utilization-and-replenishment, must also cycle, even as they transform Energy; and Energy cycling has an electromagnetic foundation that necessitates engagement in Motion — for those reasons anchored in "cause-and-effect Relationship Routines," one Earth-day or rotational-period, is, predictably or relatively 24 hours, within a $365^{1/4}$ yearly Revolution-period, — in the same manner that, a 15-degree-Arc of Earth circum-spherical Mass-Volume, is equal to, 1-hour of Revolution-Rotational Time-span. (360-degree-circum-spherical Arc / 24-hour thermodynamic Time-cycling period = 15-degree-Arc-of-Rotational Travel within 1-hour of Rotary-Traveling Motion).

Time, therefore, embraces Motion, from its reference point, as "a Form of travel" from one state to another; as "a Form of movement" from one point to another; as "a Form of elapsed unfolding" from one stage to another; as "a Form of transubstantiation" from one Form to another; as "a Form of transposition" from one distance to another; as "a Form of development" from one condition to another.

For example, in 24 hours, the Earth "travels" upon its own axis, or around its own circumference, in a circum-sphere pattern having 360-degrees/arc — the Solar System rather has an ellipsoid Plane of Revolution. But, for purposes of illustration, a sphere has 360-degrees arc/circumference, which can also be represented in terms of kilometers or miles around Earth circumference. Thus, in one hour, or 360 divided by 24, the Earth "travels" 15 degrees arc-hour distance on its circumference, each degree-arc having its own Dynamic Ecological Conditions of Equilibrium within that "swath of Earth Mass-Volume!"

"Electron orbital travel" around the atomic nucleus occurs in "compressed Time dynamics" that causes tensing-and-stressing of Curvature pressure-force Motion to embody trajectory patterns that appear as "quantum events" interpreted as "waves of probability."

But these expressions that encompass uncertainty or doubt only mask failures of Human capacity to comprehend, understand, or discover the factual physical Laws that activate the cause-and-effect mechanisms-of-interactions that are accountable for "Electron travel-routines."

Electron velocity is phenomenal as compared to the slow 24-hour velocity of Earth Rotation and to the $365^{1/4}$-day velocity of its Revolution around the Sun's ellipsoidal circumferential plane.

The smaller are steady-state entities, the faster they appear to move — the Earth is an extremely massive Object as opposed to an Atom which is the smallest indivisible quantity of Matter.

Planetary rotational Motion is the analog to "particle spin," whereas planetary Revolution is the analog to electron revolution around the atomic nucleus.

We witness "redundancy of Forms," or "Iterative Symmetry of Structural Patterning" as expressed in specific bounded-states, where Motion itself is uniformly practiced by all mass-Frames that interact within the range of their respectively "confined," or "contained," thermodynamic parameters that comprise: Field, Gravity, Mass, Motion, Force, Pressure, Temperature, and Curvature.

Curvature tensing-pressure Force, field stress-dynamics, and equivalents of the gravity force are "reconciled" through patterns of Energy cycling Motion-Forms that engender overlapping pressure and temperature Differentials and Motion-force Opposites needing "re-normalization" via alpha-numeric Constants that re-establish relative, overlapping, interactive, "operational equivalency."

From the quantum to the molecular, from the planetary to the radio-gravitational, from the "congealed or ground State" to the "excited or nucleated State," from the "combustion Frame" to the plasma-Frame: Curvature Binding Energetics operate to unify all Mass-in-Motion Frames that partake in "re-translating" the Electromagnetic Properties of Matter as the initial source-wellspring of Gravitational Energy Dynamics.

Living means Motion; function means Motion; operation means Motion. Bio-motion lies in a Continuum with eco-Motion. Our heart "moves" at approximately 80 beats per minute as the Earth "moves" within a 23-degree plane of axial inclination. Even our DNA-RNA Complex engages in Motion for "Continuum Replication" even as it is "angled" at 3.4-Angstrom units.

The Solar System is in constant Motion; the Earth is in constant motion; the Human body is in constant Motion. Even as we sleep, the Human body internally undergoes continuous Motion — blood makes its round through the circulatory system; the heart is actively pumping oxygenated blood unto arteries, vessels, veins, and capillaries; the kidneys are engaged in filtering all fluids; the lungs breathe the air for Oxygen replenishment as we exhale Carbon Dioxide; and the brain keeps the body globally functioning via its "command-and-control-Center" hormonal mechanisms and neuro-electrolytic instructions.

A General Theory of Force would have to account for the fact that Motion is present in all mass-Energy frames as patterned by thermodynamic cycling Differentials and Opposites that engender relative Continuum System equilibrium.

The Solar System is the quasi-equivalent macro-analog structure to the micro-atomic structure — both electrons and planets revolve, but in different Forms, via different mechanisms, yet, as animated and activated by quasi-similar expressions of the "self-revealed Forces," namely: Electromagnetism and Gravity.

"Congealed atoms" do not radiate; but "excited or nucleated" isotopes radiate. Earth-bound, gravity-anchored, and "congealed fuels" undergo petroleum-gasoline combustion, whereas, Sun-bound Hydrogen, as "excited" radio-gravitational nuclear fuel undergo fusion and fission for molecular plasma condensate ionized transformation.

In the total inclusive Universe, "combustion-radiation," e.g., infrared, and "nucleation radiation," e.g., X-rays, are on a Continuum-Spectrum, as the Laws of Thermodynamics find their respective expressions within each cycling Mass-energy-in-Motion reference-Frame.

Symmetrical Iterative Redundancy of Structurally Arranged Forms insures relative quasi-similar "equivalency of functions," as "conversion factors" re-translate Cycling-Differentials and Force-Opposites in operational terms and conditions that simultaneously sustain Conservation and Entropy, even as "hidden variables," discovered for computable utilization as alpha-numeric Constants, "re-normalize" stochastic thermodynamic rates-of-change into Continuum System equilibrium.

For example, on Earth, things burn due to Oxygen, as tempered by the presence of Nitrogen and other "trace-Elements" in the Atmosphere such as Neon and Argon.

In the Sun, the nucleated Hydrogen cycle is tempered via other fuel cycles such as of Oxygen, Nitrogen, Iron, and Carbon that stabilize convection dynamics and compressed-pressurized Energetics, for controlled heat and overall energy retention, as well as for measured rates of radiation emissions.

Hydrogen, being the lightest element without a Neutron in its nucleus, possesses an affinity for nucleated processes that cause it to cycle from the Hydrogen positron to Deuterium, then to Tritium, and then to Helium.

Solar Carbon cycles, Oxygen cycles, Nitrogen Cycles, as well as cycles from heavier elements, such as Iron, combine in sustaining Energy Conservation so as to thwart premature exhaustion of the Sun's Hydrogen fuel supplies.

Extreme internal Curvature pressure force Energetics engendered by the Sun's massive magnetic field, potentiate and simultaneously confine fusion and fission, contraction and dilation, tensing and stressing, explosion and compression, via its electro-motive dynamo and magnetic currents that inhere in complex particulate trajectories embodied in core-centric Forms of condensate plasma Motion.

In all these field-induced, Curvature tensing-and-stressing pressure force Frames, Motion is uniformly ever-present due to "real-Time cycling processes" consonant with "overlapping-intersecting turbulence" engendered by force-Opposites, Fields-confluence. "Gravity-trans-flux," and Conservation-and-Entropy Differentials.

Mass-Frames are structured within distinctive, yet quasi-similar or iterative patterns of functional operation, consistent with the specific ways in which the Input-Process-Output Mechanism finds thermodynamic Cycling fulfillment.

Gravitational radiation induces projected Curvature pressure force Motion Energetics that engenders Magnetic field-force binding-tensing-stressing rotary-Momentum, giving rise to equivalents of the Gravity force for Motion along "lines-of-attraction" (centripetal) and/or "lines-of-repulsion" (centrifugal).

Continuum Field and Gravity Curvature Motion-Forces accompany each other, as manifested within each Frame in differentiated energy cycling rates, motion patterns, and Forms of "resultant Momentum."

Each Frame finds "embodiment" into a "Mass-energy-motion-force Complex" within Continuum Space-Time via gravimetric Differentials and Curvature-induced force-Opposites that effect, on Earth, for example: Geo-hydro-atmospheric changes conducive to "re-calibration" of ecological parameters; to magnetic field strength variations; to gas molecular ionization rates for cloud formation; to cycles of sweet-water evaporation from Oceans for atmospheric condensation; to lightning generation and to rain precipitation, etc. . .

These cycling changes, in rates that are "naturally computed" by the Planet's "auto-Algorithm" are aimed at restoring dynamic System equilibrium, after which they get "re-translated" into Relativity proportions caused by re-calibration of co-determinant Variables at "points-of-intersection" between centers-of-Mass, centers-of-Gravity and centers-of-Field, due to solar gravitational inputs being "processed" and inter-planetary perturbations "registered" within the Planet's circum-Sphere.

The Moon in turn responds by varying its orbital trajectory: As its distance from the Earth decreases, the complex interactions between these gravimetric Centers engender ocean tide surges.

Motion is also active during molecular change as caused by particulate charge displacements-in-Space when Electrons bond at outer Orbitals as they are reconfiguring structural arrangements between reacting elements. Molecules form-and-bond at specific angles, just as the Earth "moves" within a 23-degree axial-tilt Curvature pressure-Force. DNA molecules "move," as RNA-DNA replications, having a 3.4 angstrom-angle of Curvature pressure-Force.

Given that all Forms of Motion-force have their wellspring or Source rooted in The Electromagnetic Properties of Matter, then, Electron Motions during molecular change, are graphically patterned into a "Curvature matrix of multi-dimensional vectors," from which derive "computational formulations" that are quasi-iterative of "particulate Motion-dynamics" peculiar to Quantum Mechanics.

However, due to pressure-and-temperature gradients and intensities varying within reaction-specific ranges-of-cycling-Motion, respective to each "ensemble of reactant Elements," during elemental, compound, or substrate formation, thermodynamic Differentials and force-Opposites distinguish molecular change from "particulate Curvature routines" prevailing within the micro-atomic reference-Frame where "particulate Motion-routines" are interpreted, from the paradigmatic standpoint of "random-chance-Evolutionism," to be "moving" as "waves of probability."

Respective to Mathematical Theoretics qualifying Continuum approaches to conceptualized understanding of universal phenomena — for example, in the tradition of equations like: $F_g = G\, m_1 m_2 / r^2$; $f = ma$; and/or $E = mc^2$ that incontrovertibly prove that Relativity-interactions are usually "framed as working-Equivalencies," — a General Theory of Motion-Force would engage theoretical Physicists into "an Equivalency formulation" that accounts for:

(1) Curvature pressure-tensor-stress-Motion-Force Energetics as "embodied" in:

   a.   Electromagnetic Field-induced rotary Motion-force Dynamics;

   b.   Gravity-field Omni-vectored Motion-Force Dynamics;

(2)   And Electro-magnetic charge Space-motive-displacement-Force, amenable to account for "the strong nuclear interaction" and "the weak nuclear interaction," as "re-translated Forms," or "transubstantiation Instances" of The Electromagnetic Properties of Matter, the Complex Expressions of which, being made manifest from the Sun's Field-gravitational projections of: Radiation Energy, Electromagnetic Spectral Emissions, Solar Flares, Solar Winds, and Coronal Mass Ejections.

Matter-embodiments manifested as Units of the Mass-energy-motion-force-Complex, are framed for thermodynamic Energy Transformation via cycling Differentials that need "reconciling" due to exertion of "Gravity force equivalents" taking many distinctive Motion-patterns and Vectoring-Forms.

"Mathematical reconciliation" of Differentials and Opposites embedded in cycling Mass-in-Motion Frames, respectively, via alpha-numeric Constants in conjunction with "conversion factors" that re-establish Equivalency, would "re-normalize" their Relativity-interactions, for transactional-inter-exchanges conducive to "Whole Universe Energy System Unification."

Within its respective range of verifiable applicability, Newton's discoveries are as valid and scientific as Einstein's discoveries. They can only complement each other as they "intersect" within overlapping regions of common interface, with natural and cosmic phenomena. Every proven-valid Law of Physics must accord and agree with every other proven-valid Law of Physics, each, within its verifiable range-of-applicability, respectively.

As Newtonian Mechanics explains how the Earth is kept in steady orbit around the Sun and in steady rotational Motion upon its own axis, and how the Moon remains in steady orbiting trajectory around the circle of the Earth, Special Relativity explains the relationship between Matter and Energy — Energy is "excited Matter," and Matter is "congealed Energy." And General Relativity extended the Law of Gravity and the Laws of Motion via the proposition that mass does not merely "attract" mass, but also "curves" or "bends" the Space within which it moves. Newtonian mechanics explains Gravity as an attractive Force. Special Relativity and General Relativity explain: (1) how Energy and Matter are transformable into each other; (2) how centers-of-Mass and centers-of-Gravity "curve" or "bend" each other's space; and (3) how another applied Force can become "an equivalent substitute" for the Force of Gravity.

However, neither Newton nor Einstein arrived at explaining the Relativity-interactions between the Electromagnetic Field and the Gravity Field.

General Relativity does not explain the "Force-mechanisms" by which "tensor-stress-Pressure Metrics" are expressed when conjoined with Electromagnetism and Gravity, via which Space-Time is "curved," as it maintains Continuum Motion.

It would be logical to consider the proportional equivalence of "gravitational Energy as Motion-Force," — $c^2 = E/m$ — in that "the mc^2 as accelerated exponential Momentum-Mass-in-Motion" performs a "dynamic function of transubstantiation" in being also "the Complex Operant" accountable for "the Force" that engenders Curvature, Magnetic Field, and Gravity-Field "force equivalents" that emerge from the Sun's Radiation-Mass tensing-stressing Momentum pressure-force Energetics, which is, iteratively quasi-similar to particulate charge displacement Motion-in-Space Forces, inherent in the Quantum Mechanics Frame that then "transform Electromagnetism-and-Gravity" into a Motion-Force-Energy Complex that also acts as a Field(s).

Electromagnetic Field equations cannot be "reformulated" as the Force of Gravity, but both can be "merged-combined" into a "composite Force" that yields "an interactive Field of relative Curvature," but having, respectively, differentiated functional "processes of operation."

 For the Gravity-Field does not display any verifiable or machine-detectable electromagnetic or particulate charge Space-Motion displacement characteristics — The Gravity-Field is not "electro-motive."

Rather, the Gravity-Field "operates" as an emergent Motion-Force that "acts in consort with the Electromagnetic Field," but, which thereby, "allows" many Motion-Forms that can oftentimes "counter-move Objects" in ways not allowable by the Electromagnetic Field alone.

 These resultant Gravity-Field-induced "counter-Magnetic-Field Motion-Forms," are variegated and differentiated, e.g., curvilinear, linear, rectilinear, from the "uni-vectored rotary Motion-Forms" that are commonly elicited by the Electromagnetic Field, i.e., Rotation and Revolution.

It is not that "the graviton particle" needs to be discovered or detected. Rather: That Gravity does not exist "in particulate Space-displacement Motion-Form," is a foregone conclusion. In other words, there is no such thing as "a graviton particle."

The Force implied by General Relativity has not been detailed in interconnections with Special Relativity dynamics. As encapsulated in $E = mc^2$ when "departing" from the Equivalencies that the Equation reveals, a General Theory of Motion-Force would unite all Continuum Space-Time Frames that are made manifest as "embodiments" of Mass-Energy-Force-in-Motion. "Via mathematical formulation of those Relationships, a "reconciliation" of cycling Differentials and Motion-force Opposites that distinguish operations of the Input-Process-Output Mechanism from Frame to Frame, can be performed by interposing alpha-numeric Constants for "re-normalization."

How would Special Relativity and General Relativity "interface" with Electromagnetism and Gravity, in formalizing a General Theory of Motion-Force? This "interface" can be "measurably quantifiable," by "merging" Curvature and Continuum via mathematical codification of Curvature as a common parameter to both Electromagnetism and Gravity as "co-inducers of Motion-Force" within the Frame of Continuum Space-Time.

Curvature, is the common parameter to both Electromagnetism-and-Gravity, complex interactions of which, being "co-inducers of Motion-Force," which establishes "commonality of place" or "commonality of belonging" to their overlapping-intersecting operations within Continuum Space-Time.

Special Relativity and General Relativity are only expressions of the ways in which Curvature makes manifest the rich diversity of Motion-Forms that both Electromagnetism and Gravity make available as "conjoined Continuum Fields."

Mass, Curvature Pressure-Force Energetics, and Motion are differentiated embodiments of Energy-cycling via "re-translations" of the Electromagnetic Properties of Matter. Curvature is "the background transmission" that "torques" Electromagnetism and Gravity into "Momentum-Motion-Force."

What is missing is an equation explaining solar gravitational Energetics as the "Comprehensive-Globalytic Emissions Complex" that effects Curvature Motion-pressure Force, Magnetic Field-force, and Gravity force equivalents, with "the strong force" and "the weak force" as the intrinsic substructure of molecular change for formation of greater Bodies-with-Mass, such as, a "Planet."

But, what is a "Planet?" We call a greater agglomeration of Atoms specifically arranged into a peculiar spherical molecular structure that together Form a non-nucleated ensemble of Matter-Mass in "congealed form: A "Planet!"

What is a Star but a greater aggregate ensemble of Atoms configured as a structure of particularly arranged molecules that "bond together," as a nucleated plasma Matter-Mass-Unit, or that "bond together" in "excited Form?"

Are not Galaxies greater recombinant configurations or greater combined re-arrangements of the same Matter-Mass comprised of Atoms, Planets, and Stars?

Given that:

(1) It is due to Energy Conservation that iterative quasi-similarly arranged Forms are structurally repeated in analogous patterns of organization by Mass-Frames in Thermodynamic-cycling, e.g., the Atomic Frame of Quantum Mechanics is "quasi-similarly arranged" as the Planetary Frame of Solar System Energetics; and given that:

(2) The Atom is the fundamental indivisible constituent of Matter displaying Electromagnetic Properties of Matter-Mass-in-Motion as "a Field-Force;"

Then, conclusions are that, at the Macro-referential Frame, greater manifestations of Force are only "replicate transmutations" or "re-transubstantiations" of these micro-structural "spheres of interactive Field-Force influence."

"The strong force" that holds Protons and Neutrons together is the prevailing Force at the center of the Atomic Nucleus accountable for Energy Conservation, whereby "the weak force" is accountable for "dissociative events" or "disintegrative processes" that result in radiation decay or Energy emissions. Thus, due to "the strong force," intense heat-Energy is needed to "separate" a Neutron from a Proton, — that is, to "split/divide the Atom's Nucleus" — which results in a "nuclear explosion" called "Fission," e.g., "Atomic bomb."

Conversely, the reverse chain-reaction also yielding a "nuclear explosion," is called "Fusion," whereby two positively charged Nuclei that must "repel" each other — like-charges "repel" each other; opposite charges "attract" each other — are "forced to merge together" or "compelled to fuse together," via a catalytic supply of intense heat-Energy, e.g., "H-bomb."

Given that revolving Electrons "carry" a negative Electric charge as opposed to the positively charged Proton at the center of the Atomic Nucleus, where the Proton is conjoined with the "charge-less" Neutron, therewith generating a Magnetic Field: Then, resultant predetermined yields are, that Pro-centric Rotary Motion-Forms predominate in the Universe over other Forms of Motion, such as, rectilinear, linear, and curvilinear Motion-trajectories, e.g., While the Sun's massive Electromagnetic Field "triggers" clockwise planetary Revolution, then, simultaneously and contiguously, Earth own Magnetic Field causes its spherical Mass-in-Motion to rotate upon its own 23-degree tilted-axis in an anti-clockwise or counter-clockwise vectoring direction.

In the Universe, rotary Forms of Motion, i.e., Revolution and Rotation, are primary, due to Electromagnetism from which emerges "the strong force," and hence, "inducing" Conservation of Energy.

Gravity, — emerging from Rotary Motion-Forms caused by Electromagnetism — as an "indirect Field-Force" that also arises from Mass-in-Motion, "triggers" other Forms of Motion-force vectors — as above-listed such as linear, rectilinear, and curvilinear — and hence, "inducing" development of Entropy processes, due to simultaneous and contiguous oppositely vectored Motion-forces of Electromagnetism and Gravity-force equivalents that are "triggered" as clockwise Revolution and counter-clockwise Rotation.

As the Earth "travels through" void-vacuum-Space via Rotation and Revolution, Earth spheroid-Volume Mass-in-Motion — as a relatively "solid ensemble" constituted of Atmosphere, Landmass, and Ocean-Waters "imbued" with "gravimetric tensility/pliability/flexibility" — "is colliding" with the Sun's particles emissions, rays, winds, flares, and ejections.

Consequently, "wobbling effects," "dipole oblation effects," and "bulging-effects" at Earth Equator, are metaphorically and analogously, relatively comparable to "friction/collision effects" such as those resulting from two Objects with Mass-in-Motion "rubbing against each other" in co-equal, co-linear, but oppositely vectored Motion-Forces, e.g., two pieces of flint-rock "rubbing against each other" in opposite directions, will "trigger" emergence of sparks from which firewood can ignite.

Electromagnetism is "what makes everything move" in the Universe within Continuum Space-Time Curvature as the Sun generates "Gravimetric Energetics:" Gravity affords us "freedom of Movement" within Earth spheroid-Volume Mass!

Electrons "have no choice," but must revolve around the Atomic Nucleus while "keeping a respectable distance" there-from, within their "allowable Orbitals" — possessing both "primary" and "secondary Orbits" — due to the scientific fact that the universal primary-major "Curvature Force" that "moves everything" comes from the Micro-Electromagnetic Field

Motion-Force "initiated" by confluent but oppositely charged atomic Particles, — from which emerges the Atomic Frame's micro-Gravity-Motion-Force.

In the Solar-Planetary System comprising spheroid Mass-Bodies in-Motion, the Electromagnetic Field is always accompanied by a Gravity-Field, the complex interactions of which, generating "Curvature Energetics" that gives rise to rotary Motion-Forms such as Rotation and Revolution.

And Energy Conservation overcomes Entropy processes — "temporarily," that is, until the onset of "Terminal Entropy," e.g., biological death — respective to phenomena induced by the Input-Process-Output Mechanism that are peculiar to the specific Mass-in-Motion-Frame upon whose particularly structured organizational arrangement, its applicable Principles, Operations, and Processes are imposed, e.g., The Sun "is aging-and-dying" even as its powerful plasma nuclear fuel chain-reactions are projecting all the "radio-emissive spectral effects" that uphold Earth life-support Systems.

In the same manner that Electromagnetism "induces" prevalence of "the strong force" in Energy Conservation, Gravity "works" with "the weak force" in Entropy processes that lead to radioactive decay, on the one hand, and molecular formation, on the other.

At the Atomic Frame of Quantum Mechanics, while the "micro-Electromagnetic-field effect" accounts for Electron-revolution around the Atomic Nucleus, it is the "micro-Gravity-field effect" that "allows" Electrons at the outer Orbitals to "react" for formation of molecules conducive to greater aggregates such as chemical compounds and substances.

From the micro-atomic frame to the macro-planetary structure, all quantifiable categories of Matter-Mass, such as: Force, Motion, and Energy that cycle in iterative quasi-analogous or quasi-similar, but not in uniformly identical Forms, the sum of which, specializing in fulfilling the Input-Process-Output Mechanism, respective to each Mass-in-Motion Frame's functional requirements for dynamic System equilibrium.

Thus, Gravity, Curvature Force, and Field force and their "composite expressions" are macro-analogs or "quasi-replicants" of "the strong force" and "the weak force," as embodied in Mass-in-Motion-Energy-Frames where "spheroid theaters of Force influence" or "areas of manifestation," cycle in "re-translating the Electromagnetic Properties of Matter.

The Atom constitutes the patterning model for all equivalent replications of quantifiable categories that "re-translate" its Electromagnetic Properties. Within its micro-Frame properties are encapsulated: Curvature pressure force Energetics, Field-Force Dynamics, Gravity Force Mechanics, and Electromagnetic Field, as "embodied" or 'packaged" within the Frame of electro-motive, oppositely charged particulate displacement-Motion, "spheres of influence."

"The strong force" causing nucleus-centric Electron revolution is "recapitulated" in the macro-Frame of helio-centric planetary Revolution, as the Sun's powerful Magnetic Field "induces" Rotary-motion Effects.

Energy Conservation preserving atomic Frame integrity finds replication in Planet Earth functioning as a whole Energy system in dynamic ecological equilibrium, while gravitational solar Energetics is held "relatively constant."

Likewise, atomic Frame integrity preservation accounts for dynamic functioning equilibrium in Earth atmospheric gas-O-sphere, landmass terra-sphere, and oceanic hydro-sphere, complex inter-operations of which, geared towards "relative planetary stability."

Solar system ellipsoid plane tensing-and-stressing Energetics, gravi-magnetic Spectral radiation, and Curvature motion-Forces as projected by electro-magnetic radio-activity cause the emergence of Field-induced Revolution and Rotation, Moon orbital motion, Electron motion around the Atomic Nucleus, molecular change Motion-forms, "the strong force," "the weak force," Earth Human-initiated Earth-surface Motion-patterns, and Earth atmospheric motion Forms.

Gravimetric Differentials and Motion-force Opposites pertaining to overlapping-intersecting steady-state Systems with "relatively common regions of Cycling-activity" seeking dynamic Equilibrium via rates-of-change in Gravi-centric Cycling and Field-centric Cycling, engender the cause-and-effect relationships "triggering" the mathematical necessity for alpha-numeric Constants, whose "interceding multi-variegated operations," will "link-re-normalize/reconcile," all overlapping "spheres of influence," so as to "replicate" true-to-Reality universal Phenomena, Events and Processes, via proven-scientific "formulaic modeling," as encompassed and circumscribed within a Mega-Theoretical-Frame of Relativity-controlled Continuum-Motion-Force interactions.

From Quantum Mechanics to the Sun-Planets Frame; from "congealed" molecular dynamics to the solar-System nucleated plasma-convection Frame; from the motion-Force Energetics of Electromagnetic-and-Gravity Fields to the solar system ellipsoid Revolution-plane: Motion is ubiquitously present, but made-manifest in differentiated Energy-cycling structural "patterns-of-Organization" due to Mass-in-Motion-Frame-specific "Thermodynamics of Energy Transformation" within which the Input-Process-Output Mechanism is being fulfilled.

Wherever there is Mass, there is Electromagnetism; from Electromagnetism emerges Gravity. Where Electromagnetism and Gravity "engender" Fields, their cause-and-effect "mechanisms of operation," will yield Motion-force. In the Universe, Rotary Motion-force, such as Rotation and Revolution, are a priori Motion-patterns prevailing in Continuum Space-Time Curvature.

And, because the "property-of-Force" comes from or "belongs to" Mass-in-Motion, then, it is Mass-in-Motion that controls "trajectories of Force expression," which "allows" Gravity-Field "mechanisms of operation," to make manifest, other Forms of Motion-Force-trajectories, e.g., acrobatic aircrafts "moving" in all directions once in the air, can "carve" or "loop-around" a "360-degree Flight circum-sphere!"

"Overlapping Frames Transactions" engendered by gravimetric Field-Curvature Tensor-Stress Pressure-Force Energetics engineer "Force-interactions" in all their "differentiated Field-Forms," — e.g., "the strong force," "the weak force," Electromagnetism, and Gravity.

Those "Force-interactions" thus impinge upon operations of the Mass-specific Input-Process-Output Mechanism, respectively, with an ensemble of "thermodynamic entanglements" that generate cycling Differentials and force-Opposites, even as Energy Transformation patterns-of-organization "intersect" where "regions of perturbation" foster "rates-of-change" in "cycling variables and parameters," to co-determine "functional responses" that climax into "Relativity-effects," e.g., such that Planet Mercury "responds" with a Perihelion Shift.

From a mathematical Theoretics paradigm, it is clear then, that these overlapping reference-Frames necessitate interposition of alpha-numeric Constants that will operate to "normalize" cycling Differentials and force-Opposites, into Continuum operations that "reconcile all Frames" within the same Continuum Universal Space-Time Curvature Complex!

Hence, the formulation of a "Unified Field Theory of Continuum Curvature Pressure Force-Motion" is to be "built" as a General Theory of Force that "embodies" all "interactions-Forms" that characterize "Mass-Frames relationships."

"Gravitational Solar System tensor-stress tensile-Elasticity" is expressed from exertions of "the strong force," "the weak force," Electromagnetism and Gravity, for "integration" of all Differentials and Opposites, by "negotiating rates-of-change" in Cycling co-determinants, in order to "co-ordinate contemporaneous Continuum" within the Frame of Thermodynamic Time, from "Relativity-Interactions" between "all Mass-in-Motion Frames" that analogously, iteratively, and contiguously simulate, "quasi-similar replications" of Atomic Frame Dynamics for fulfillment of "Ubiquitous Universal Cycling Equilibrium."

****

HOW DO ALPHA-NUMERIC CONSTANTS "NORMALIZE" OR "RECONCILE" DIFFERENTIATED RATES/PROPPORTIONS/RATIOS PECULIAR TO MASS-IN-MOTION FRAME-CYCLING, GEARED FOR DYNAMIC UNIVERSAL EQUILIBRIUM?

The numerous structures formed by "patterns-of-Organization" by which the Input-Process-Output Mechanism finds fulfillment within each "Mass-in-Motion-Energy-Force Complex," cause Thermodynamic Frame-cycling Differentials and Motion-Force Opposites from whose complex interactions arises the necessity for alpha-numeric Constants.

In the Solar System, there is always "cybernetic feedback" between the Sun and all the Planets, respectively, and amongst the Planets themselves, for "re-calibration of Cycling variables" as each Planet, as a Unit of Mass-in-Motion-Frame, "continuously re-negotiates" each other's variegated "Mass-sensitive responses" to Gravity-induced perturbations and Field-sourced turbulence — due to solar gravitational-and-magnetic-Field-generated "Thermodynamic Curvature Pressure-Force Energetics."

Because the Solar System Frame's gravimetric parameters cycle as an open, but self-contained System, as "held-and-kept together" by the Sun's massive Magnetic Field and Gravitational Field Energetics, then, there is also solar ellipsoid plane tensor-stress-pressure as well as consistently radiated temperature gradients that cause the Plane's boundaries to be "transmuted" into "a theater of Motion-Force Influence" constituting in quasi-equivalence, "the confines of a reaction chamber," e.g., The Perihelion Shift of Planet Mercury can be predictably timed and observed with relative accuracy due to the consistent range-specific variables ratios and parametric rates-of-change that are accountable for its planetary routines.

The Earth also operates as an open, but self-contained thermodynamic System, even as its sufficiently powerful Magnetic Field externally repels, excess solar radiation energy.

Why must elements react within a closed chamber in controlled temperature and pressure Forces that contain or confine their interacting processes?

Within the Sun's nucleated Plasma-chain-reactions projections inheres Solar Gravitational Field Energetics, which are confined within an extremely potent magnetic field, in the same manner that Earth ecological processes are "bounded" by its own magnetic field, even as Energy Transformation via the Input-Process-Output Mechanism yields "measures of Entropy," in the forms of "heat loss," dipole oblation, wobbling, equatorial bulging and precession.

Chemical reactions must be "confined" to a closed chamber, because, when Frames overlap under pressurized and heated reactive conditions conducive to Transformation, "regions of turbulence" emerge as the "heat-and-pressure Complex" is thermodynamically "transubstantiated" into "Relativity conditions" tantamount to an inwardly-driven, "implosively contracted Motion-Force."

Chemical reactions continue, until all variables, parameters, and co-determinants driving "recombinant molecular dynamics" have "exhausted" their "formative energies," via achievement of "dynamic eco-equilibrium," e.g., all Electrons possessing "affinity for bonding," have fulfilled their chemical bonds, after which, "newly created elemental Forms" or "substrate compounds" obtain relative fixity, to then become the equivalent of "Steady-state systems."

Thus, though the Universe is in "constant thermodynamic Motion," there is relative constancy or reliable determinacy of iterative Forms, the Flux and Reflux of which, establishing Continuum, even while thermodynamic changes are taking place within their respective ranges-of-effectiveness and ranges-of-influence, for replenishment of resources and Transformation of Energy, via the Input-Process-Output Mechanism that animates "the Organizing Principle" initially "encoded" within the "Universe Entity Complex" ever since the Time of Creation.

Each frame-of-reference or steady-state entity, from the quantum to the planetary, from the Atom to the Solar System, is engaged in its respective Forms of Motion. The atomic frame is a steady-state structure that also seeks dynamic System equilibrium via gravimetric Motion patterns that "reconcile" the various Forces acting upon particulate constituents. Due to absence of a Neutron in its Nucleus to "anchor" its positively charged Proton, Hydrogen, possessing one negatively charged Electron that revolves around its nucleic center, is the first and lightest Element, and also, the primary solar nuclear plasma fuel.

The "complete/fulfilled particle" within the Atomic Frame," — no particle "forced into isolation" for purposes of experimentation — follows "a pre-existing program" that gives rise to perplexing complexities such as those apparently related to mass, velocity, charge, position, momentum, gravity, Curvature, and field, the complex sum of which, when "externally stimulated" via research-and-experiment instruments and apparati, undergo "a repertoire of System changes," in accordance with "specific computational re-calibration parameters," that are as imperceptible as those arising from "the simple tossing of a coin."

Though, their organizational structure of rotary Motion is quasi-similar, particulate routines are of a Form different from planetary routines — but the common platform is anchored in a "Field-and-Gravity Complex," that activates centri-vectored or pro-centric Motion dynamics, as engendered by quasi-similarly structured "Curvature tensing-stressing Motion-pressure-force Energetics."

Respective to their differentiated scalar magnitudes, as Planets are "caught" within massive solar Magnetic Field and Gravity Field boundaries, so are Electrons "trapped" within the Atom's micro-Field-and-Gravity generated boundaries.

Electrons change orbital levels in molecular change reactions as controlled by pressure-force Opposites and temperature-gradients Differentials. Planets revolve in predictable orbital patterns with rare changes in direction and magnitude, except for variations allowable within orbital Motion ranges, such as the Perihelion shift of Mercury. Though Planet Pluto at the outer fringes of the Solar System has a "peculiar orbit," still its 17-degree inclination from the ecliptic, allows for relatively accurate predictions of its trajectory routines.

Within the Atomic Frame, motion patterns can be said to be an emergent property of the complex interplay of oppositely configured Electro-motive-Magnetic-charges that cause micro-Mass-in-Motion displacements in Space, as originating from "Relativity interactions" engineered by micro-Curvature centers-of-Force, micro-centers-of-Field, micro-centers-of-Mass, micro-equivalents of Gravity-force, and micro-tensor-pressure-Forces bounded by "a micro-revolutionary plane of influence" engendering Orbitals that can "hold" only a specific number of Electrons, e.g., the first Orbital and the second Orbital, can each, only "allow" two electrons, respectively, within the "bandwidth of their range-of-Motion."

Thus, Motion-activity is generic to Quantum Mechanics and germane to Newtonian Mechanics. That all Mass-Frames "must move," from the smallest to the greatest, is unavoidable, for, Properties of Electromagnetism are relatively operating in redundant Forms of thermodynamic Energy-cycling, in fulfillment of the Input-Process-Output Mechanism geared to sustain Conservation, operational processes of which, co-laboring to temporarily thwart the onset of "Terminal Entropy." Consequently, Motion is to the Universe what sunlight radiation Energy is to plant growth.

The Input-Process-Output Mechanism operates in every "Mass-energy-Motion-force Complex" to form a Cycling-Frame possessing specific Energy-transformation patterns whose proportional and ratio quantities engender overlapping Differentials and counter-vectored force-Opposites for which relative equivalency must be established via alpha-numeric Constants that "normalize" intersecting "regions of turbulence." The perihelion shift of Mercury is in contemporaneous Continuum Motion-force, along with other apparent "anomalies" peculiar to our Solar System, such as, the periodic "Matrix of Moon-phases" accountable for engendering ocean tide surges on Planet Earth.

This is a universal Rule directly deriving from cosmic operations of the Input-Process-Output Mechanism of which must partake ALL thermodynamically cycling Mass-in-Motion-Frames: Given that every thing in the Universe occurs according to specific scientific Laws that, respectively, "plot their course," then, That every physical event, phenomenon, or process has "a physical source" and thus must be explained in accordance with physical-material cause-and-effect mechanisms and relationships is a "tautology-in-Fact."

Cycling Mass-in-Motion-Frames are "pre-programmed" from initial conditions of Creation that "orchestrate" their routines and "direct" them to "respond" to solar radiation emissions, via "mechanisms of Force" that respectively engineer "a repertoire of Motion-forms" that is pertinently designed for sustaining dynamic System equilibrium.

There is a gravimetric thermodynamic affinity between solar radiation, and "mechanisms of Force," that activate cycling-Motion, due to the Electromagnetic Properties of Matter that "re-translate" atomic frame Energy-cycling dynamics.

Within the Solar System, exo-planetary Motion occurs due to projection of "stellar Curvature Field-Mechanisms" inducing displacement of Bodies-with-Mass in Space. Curvature-Motion-force is thus "embodied" in "Momentum Field -Force Energetics" that get "re-translated" into equivalents of the Gravity Force, as "attuned" to "properties peculiar to Moving-Mass," giving rise to on-the-Surface planetary Motion-forms. From the complex interplay of sun-originated Magnetic-field Force(s) and Gravity-field Force(s) and their "inner-planetary equivalents," emerge the "triggering Mechanisms" that cause "corresponding reaction-routines" from cycling planetary Mass-Frames, which have already been "encoded" or "pre-conditioned" to process "Motion-information," e.g., Planet Earth possesses an electro-conductive molten iron magma Core, "armed" with gravimetric positioning for being centrally located within the Planet's spheroid volume to receive atmospheric Inputs such as "lightning strikes," the complex interactions of which, are "coupled" with Solar Gravitational Radiation Energetics to give rise to Earth own magnetic field as well as to its atmospheric-meteorological phenomena.

According to General Theory of Relativity, Space-curving motion-Force mechanisms are accountable for "bending" trajectories of traveling Mass within areas-of-manifestation of Gravity. Replication of "Gravity-conditions" are analogous to the dynamics present in the example featuring an elevator moving to the left due to an external Force impinging Motion upon its Frame, thus engendering "a Gravity-substitute Force," pegging a man's feet to the elevator's right side, that is, in the direction opposite to the leftward vector of the Gravity-velocity Motion-force; and this counter-Force accountable for compelling the man's feet against the right side of the elevator opposite to the leftward Motion-force vector, "becomes" the "equivalent of Gravity."

The leftward velocity pressure Motion-force engenders the relative Gravity-equivalent vectored to the elevator's right-side that serves as a substitute-equivalent for the elevator's real bottom floor-of-Gravity to which real Gravity ought to be vectored.

Likewise, the Earth rotates counter-clockwise and revolves clockwise as pro-centric forces (centripetal forces) prevail over counter-centric forces (centrifugal forces,) thus engendering a complex atmospheric convection dynamic between the Poles and the Equator. The "northern jet stream" proceeding from the Arctic follows a rotary path "from west to east" with a tendency towards the Equator to intersect with "the gulf stream" coming up from the Antarctic also moving towards the Equator, whereby the cold polar air-mass from the Arctic and the "tempered-moderated Antarctic air-mass" meets-intersects with the hot equatorial air-mass to engineer "regions of turbulence," often characterized as the locus of storm systems.

In Thermodynamic Motion-Time, "every second of arc-angle of Rotation," (1-hour Arc-of-Rotation = 15-degrees), has a corresponding "second of revolutionary arc-angle," with which it intersects, and by that, engineering "an eco-meteorological dynamic" accountable for regional variability in atmospheric phenomena, species of fauna, and species of flora.

In short, the ecliptic of Revolution "intersects" the ecliptic of Rotation at every second of Thermodynamic Motion-Time, e.g., Recurring monsoon rains of Asia and the diluvial rainfalls of the Amazon forests that account for their prolific vegetation are due to those intersecting parameters and co-determinants of rotational and revolutionary dynamics that interact to conserve Earth angular momentum.

The Universe displays an apparent self-directed, "pseudo-information processing System," so that Energy cycling Differentials and Motion-force Opposites can be gravitationally "computed" from Frame-to-Frame, in specific fulfillment of the Input-Process-Output Mechanism, for specialized functions respective to each reference-Frame, e.g., the Earth processes solar Inputs for dynamic ecological System equilibrium.

Fulfillment of the Input-Process-Output Mechanism means that there are operant variables, co-determinants, and factors that are "emitted-projected-radiated" by the Sun, to be "received" by each Energy-force-motion Frame, under specialized "Mass cycling-conditions of processing," in order that certain "Outputs" can result, as controlled by Continuum Space-Time Curvature pressure-tensor-stress-Force Energetics.

This pattern: Emission-Inputs, Reception-Processing, and Operant-Outputs, encapsulates all transactional-Steps-and-Stages espoused by the Input-Process-Output Mechanism, in order to fulfill dynamic System equilibrium for each Frame: From the Atomic to the Molecular; from "congealed Energy" to "excited Energy;" from Earth Planetary Ecological Processes to Solar System Ellipsoid Plane of Revolution Phenomena.

The Creator, Almighty God, "set things in Order" from the Beginning of Universal Genesis, with an Organizing Principle that functions to maintain universal modus operandi geared for contemporaneous and contiguous synchronized processes, by infusing the Universe with all the powerful physical Laws necessary to activate Continuum Curvature Space-Time Motion-Force for "re-translation" of the Electromagnetic Properties of Matter, into "transubstantiated cycling patterns of structured Energy Transformation," e.g., Laws for "pseudo-information processing;" the Input-Process-Output Algorithm; mechanisms for computational differentiation and specialization; relative proportions and ratio equivalencies; redundancy of thermodynamic Forms; iterative symmetry in harmonious gravimetric synchrony; the proton-neutron Force having pre-eminence as "the strong force" for Conservation of Energy predominance over Entropy Processes; centripetal Force prevalence as Gravitational Attraction responsible for molecular change; centrifugal Force opposites complementing attractive-Forces with repulsive-Forces, and thus, facilitating variegated Motion-forms etc . . .

As cycling Frames "process gravitational information" from counter-vectoring Motion-Forces, they engage in "Differentials computation" in order to achieve Continuum equilibrium within unified Space-Time Curvature.

Planets and Stars are embodiments of the same Energy but in different forms. The Earth is a "congealed" steady-state entity, a "dependent or open system" but a self-contained system receiving Energy from the Sun, which engenders various physical and ecological sub-systems designed for differentiated but interdependent functional purposes, e.g., an atmospheric system, an oceanic system, a geological system, a bio-sphere system, etc . . . All processes, phenomena and events extant within the solar inferno for solar system integration will find relative expression within the "congealed" planetary Frame.

As the Sun possesses an electric dynamo from plasma condensate molecular friction, Earth atmosphere is "electrified" whereby cloud formations electro-statically react to generate lightning and thunder, for rain and snow, and multi-factorial storm systems like typhoons, hurricanes, tornadoes, cyclones, etc . . .

The hot iron molten magma core of the Earth is electro-conductive and is therefore "receptive" to specific spectral solar emissions as processed by the magnetosphere and ionosphere, in order to generate the magnetic field. Lightning is attracted by the iron core, substituting for Earth center-of-Mass and center-of-Gravity.

Redundancy of structures for iterative Energy-Forms and Motion-Patterns ensures Continuum cycling, even as gravimetric Differentials and Opposites engender specialized fulfillment of the Input-Process-Output Mechanism into distinctively organized arrangements, configurations, and constellations aimed at establishing Frame-functionality that climaxes into dynamic System equilibrium.

As lightning is driven to seek the iron core, it engenders many chemo-geological reactions in the hydrosphere, soil, atmosphere, and landmass in order to activate certain-specific processes that lead to Oxygenation, electrolysis, and formation of metals and of fossil fuels, given respective gradients of temperature and pressure.

Electric fields are accompanied by magnetic fields; magnetic fields are accompanied by electric fields. The magnetic compass naturally seeks alignment with the North or positive pole of earth magnetic field. Quasi-similarly, this "Motion-Form pattern" is replicated in the electric fan: The electric fan is composed of a blade-stem assembly placed within an electromagnetic field generated by an electrical copper coil as a cord is plugged into an electrical socket. In like manner, Planets "caught" within the effective range-boundaries impacted by the Sun's powerful magnetic Field will rotate and revolve.

Atoms will bond at outer electron energy orbits in order to form specific molecules, as pressure and temperature controls are applied. These are inherent electromagnetic properties of Matter-mass-Energy within specific Frames where "Curvature-attuned relativity variables" operate to effect different Forms of cycling Motion.

Infinite void vacuum Space cannot give rise to the spontaneous emergence of the Universe as "a Matter-Mass-Energy complex" in Continuum Curvature Force-Motion, "ex nihilo," or "from nothing," due to the absence of dimensions — Space, void of Matter, has no physical boundaries or material dimensions! Thus, every physical object or phenomenon must have contemporaneous-contiguous relative dimensions in both Time-and-Space in order to "fill the void" for Continuum, e.g., Given that Matter must be present in order to yield Energy, the material or "singularity" responsible for the so-called "big bang" out of which the Universe Space-Time was supposedly "created," how could it have existed prior to the formation of the Universe when it also would have had to "embody Matter" that did not exist yet; and, what "contained it," given there was no Space to speak of yet for it "to fill?"

Energy Transformation from Matter with which it has "relative Equivalency," needs Space in order to cycle in "Temporal Time" or "Thermodynamic Time;" and temporal energy cycling needs Force with the inherent impetus for Mass-in-Motion — hence, given that there was no Motion-force to "trigger the singularity big bang explosion," then "the big bang" remains a conjectured imaginative fiction.

For Continuum Space-Time, must first possess, the presence of Curvature pressure-Force properties that then engender rotary Motion-Forms, the sum of which, emerging from the pre-existence of Matter as "transformable Energy."

But Matter cannot "self-create" — the Laws of Thermodynamics clearly instruct us that every physical thing in the Universe must have a Beginning, a Life-Span, and an Ending as it "engages" the Input-Process-Output Mechanism for operational functionality. Thus, there has to be a Creator to have had "set His created things in orderly Motion."

Time is eternal and Space in infinite. The necessity for the "Time factor" means there has to be a specific starting or beginning point in Space-Time to activate Continuum — a Genesis, so to speak. Given that the Universe cannot "self-start," as it is totally inanimate and devoid of Mind, then only a Creator could have begun the Universe as we know it. No Human

Being can refute this self-evident Truth by mere verbal assertions — No one witnessed the Beginnings of the Universe! And no laboratory or field experiment can prove where or when Matter appeared within Continuum Space-Time, given that experimenters would have to rely upon what already exists — upon this extant Universe which already exists with all the variegated Forms and categories of things that can ever exist. Thus, we are only discovering that which we find as logically-scientifically comprehensible by the workings of the Human Mind, given that all Humans are born as little babies who grow up to mature into learned adults with a life-span duration, the ultimate end of which is the fulfillment of Mortality!

Conclusively then, it is God, the Creator, who provided the fundamental substance for the reality of Time-and-Space by providing the "starter force" for energy cycling via the material Energetics of Curvature pressure-force-Motion. Matter cannot self-create, and due to the Equivalency Principle, Energy cannot exist without Matter! Thus, Matter must have been created for transformation into Energy by a supernatural Being; and this Being is Almighty God!

By creating cycling-Frames with iterative redundant periodicity in order that the Mass-energy-force Complex "fill the void" of eternal Time and infinite Space with Continuum Curvature motion, God initiated the beginning of Temporal Time or Thermodynamic Time in the fashion that Human Beings understand, calculate, and measure Time, e.g., as "clock time."

All these unfathomable categories of things, species of things, and Forms of things that exist in order to animate or "put-meat-on" or "flesh-out" those cycling-Frames as "impelled" by thermodynamics — As "Matter-Energy Frames-with-Motion-Force-Complex" — for effectively "thriving" within Continuum Space-Time Curvature, can be quantified as variables, parameters, and co-determinants of the Momentum-Electromagnetic-Mass Properties-of-Thermodynamic Energy-Cycling, as calculated by use of the CGS system (centimeter, gram, second) and the MKS system (meter, kilogram, second).

Thermodynamic cycling "leaves" operational indicators or "transactional footprints" that can be quantifiably measured via mathematical formulation that faithfully records the on-going interactive physical processes of Earth-ecological conditions, that climax into the natural phenomena being observed, e.g., How does Solar Radiation Energy Inputs — at noon on a certain day of the Year during a particular Season, — impact a specific swath of land, located in the Equatorial Zone, but within the boundaries of a particular area having those longitude-latitude coordinates: 60-degrees Longitude and 15-degrees latitude?

Light is emphasized here for both spiritual reasons and material connections. Spiritually, and Biblically, Light means understanding, knowledge, clarity, exposed purity, revelation, discernment, and perfect moral righteousness. And materially, Light is necessary for sight, plant growth, Earth life support systems, bio-sphere functionality, and crucial processes, products and events linked to the progressive development of the Sciences. Without understanding, nothing creative takes place; and without sunlight during the day and artificial light at night, no creative or productive activity can occur.

"In the beginning God created the heavens and the earth. The earth was formless and void, and darkness was over the surface of the deep; and the Spirit of God was moving over the surface of the waters. And God said, 'Let there be light'; and there was light. And God saw that the light was good; and God separated the light from the darkness. And God called the light

Day, and the darkness he called Night. And there was evening and there was morning, one day." (Genesis 1:1-5, NASB, Holy Bible)

The "pre-Creation stillness" of infinite Space and eternal Time in simultaneous contemporality with the existence of God as an invisible Creator Spirit-Mind, corresponds to the logic of Eternity or Eternal Time "striving in hope" with Infinite void-Space, in the absence of physical Matter, for "the becoming" of Mass-Energy-Motion Frames to "fill its void," hence, as biblically revealed, God's Spirit "moving over the surface of the waters."

The void of Time-and-Space in the absence of Matter had no beginning or end — No physical indicators for "processing" operations of Thermodynamics, and thus, no verifiable "footprints of Continuum." But only God's immanent Nature is omnipotent, omniscient, and self-sufficient. Thus, "prevailing void-ness" necessitated a yearning for movement of the Mind of the Spirit of God: God, our Creator, had to decide to Will universal reality into being.

God in His infinite goodness and by His mighty power said "Let there be light, and there was light; and God saw that the light was good." And by that, the Universe — the heavens and the Earth — had a beginning! And with that Creation Beginning, came "The Genesis of Matter:" Matter's Electromagnetic Properties giving rise to Thermodynamic Processes of Energy Cycling: Namely, Transformation, from which, ensue Conservation and Entropy.

Why should we be surprised that creative thinking gives rise to objective reality? Do we not stretch our arm with one thought from our Mind in order to grasp a book with our hand? In the same manner, God with "His omnipotent and omniscient thought-life" spoke the Universe — the heavens and the Earth — into being! God's character-qualities of His own Being: The immaterial substance and abstract reality of His own personal Spirit-Being "became" the initial Creative Wellsprings of the Universe "coming-into-being." But with Matter came Entropy thermodynamics: — "No physical-material thing lasts forever!"

God imbued the Continuum Space-Time Universe with all the physical Laws ("information Input") necessary for Electromagnetic Frames embodying Matter-mass-Force to thermodynamically cycle (Energy processing) in order to discharge the gravitational functions that sustain Earth life-support systems ("Curvature Motion-pressure-Force Outputs").

As human spiritual understanding factors into our society's creative technologies and productive activities, the Speed of Light in a Vacuum is the standard that controls and limits the range of Motion for all universal phenomena, cosmic events, and steady-state processes.

No particle or object can travel faster than 3.0×10^8 meters per second. Even particles accelerated to a fraction of the Speed of Light must "shed their mass-Energy gains" by radiating cosmic energy units. Matter has Electromagnetic Properties, in that it is constituted of Atoms whose fundamental particles have opposite electro-motive-magnetic-charges from which emerge "gravimetric Field-interactions" that engender "Field-induced Mass-displacements" in Continuum Space-Time micro-Curvature, e.g., Hence, Electrons must revolve around the Atomic Nucleus.

There is redundancy of Forms in natural multi-duplication(s) of "Electromagnetic Motion" in the various cycling Mass-Frames which "the Mass-Energy-Motion-Force Complex" embodies, due to the inherent "encoded programming" in the Universe, for structural equivalence in "the pursuit of analog Forms" possessing "functional quasi-similarity," e.g., Electrons must revolve around the atomic nucleus as Planets must revolve around the Sun.

How does Continuum-Cycling Curvature-Motion-Force "interface" with processes of Energy Conservation and Entropy for derivation of alpha-numeric mathematical Constants?

There is the need for a new understanding of "Qualitative Conservation," as each Frame undergoes iterative patterns of structured Energy Transformation that embody both Conservation and Entropy. Due to the pursuit of dynamic System equilibrium, Functional Operational Entropy (resource production and utilization) engenders patterns of Qualitative Conservation (resource accumulation for abundant replenishment) in prevention of the onset of Terminal Entropy (final decay, resource exhaustion or "death").

Conservation sustains resource preservation and replenishment (e.g., Youthfulness, Healthiness, and a Biological lifespan) while Entropy counters Conservation with consumptive utilization and the tendency to decay due to "wear-and-tear" (e.g., Aging, Disease, and then Death).

The strong Proton-Neutron force is fundamentally Conservation-directed in character whereas Electron propensity for molecular change as particulate-charge Outer-Orbital bonding, contributes to Entropy thermodynamics — thus, metaphorically, Conservation from "the strong force" gathers and aggregates whereas Entropy as engendered by "the weak force" disperses and scatters.

After molecular change crystallizes into formation of new elements or compounds, the process begins again, as in chemical chain-reactions, for Conservation dynamics to overcome Entropy processes via application of "the strong force" and pro-centric or centri-vectored Forms-of-Motion, until all Electrons at outer Energy-Orbitals have engaged in all possible bonding configurations with other Atoms — whereby they have an intrinsic reaction-affinity with other Atoms' Electrons, respectively, e.g., H_2O; H_2SO_4; HCL, respectively.

Because the Speed of Light in a Vacuum has a fixed value that constitutes a limiting standard or restrictive factor that impacts all universal phenomena, events, and processes, alpha-numeric Constants must also be fixed in value.

Regardless of how "elegant" is a mathematical equation, Constants cannot be ascribed "arbitrary values" that "do not fit" into the true operational constructs of a particular Frame's "thermodynamic in-Reality Cycling."

But Constants must have alpha-numeric mathematical values that act as "catalysts" for lubricating the gears of Frame-Relativity-Interactions, even as Cycling-Frames thermodynamically vary within their inherent operational ranges-of-effectiveness, the sum of which, depending upon "cybernetic re-calibration" of parameters and co-determinants that "engineer reconciliation" of Energy cycling-Differentials and Force-Opposites, so that all cycling-Frames continue to "self-configure" for fulfillment of dynamic System equilibrium.

A Constant implies "a relatively fixed value" whose invariable range-of-applicability must extend to all Frames under its paradigmatic structure, and hence, under its "normalizing properties," e.g., 9.8 meters/second/second is applicable to all falling Objects, whether a feather or an aircraft.

On the Earth, conservation and entropy at the scale of seasonal variations are marked by different geo-atmospheric events and hydrosphere processes that impinge upon Human living for creative productivity and resource utilization. In the laboratory, entropy has been defined in terms of "heat loss" or "unusable heat" stemming from chemical reactions during formation of elements and compounds and other substances.

Some astrophysicists believe that the Universe will undergo "Terminal Entropy" as "heat death," at which time there will be tremendous releases of Infrared Energy before "final de-thermodynamicization."

There are natural instances of Infrared release and mechanical technologies that demonstrate, under certain conditions, that specific Entropy processes climax into release of infrared heat that is then constructively channeled into beneficent utilization by Human Beings in order to prevent freezing in cold temperatures during winter months, e.g., an automobile's antifreeze-and-coolant heating system.

The Human body operates in homeostatic equilibrium due to circulating oxygenated blood, as well as due to systemic hydration whereby all organs inherently possess a colloidal-water-based composition, the sum of which, estimated in constitution, at more than 75 percent water, even as electrolytic processes sustain a steady 98.6 degrees Fahrenheit temperature, hence, causing "de-hydration," temperature elevation that is countered by "sweating," and exhalation of Carbon Dioxide accompanied by water-moisture or "a Form of steam."

"Aging" or "biological electromagnetic fatigue," might be said to be an ultimate Form of de-hydration that eventually climaxes into "cellular desiccation," due to a lifetime of organically sustaining electro-magnetically-driven, electro-motive, electro-conductive, electrolytic processes, for purposes of "Living one's Life to the fullest extent of duration bio-thermodynamically possible!"

An automobile operating with alternating current and explosive gas combustion engines must also have the operations of a liquid, as water or "antifreeze" in order to temper effects of "friction heat" that would lock all parts into a "molten meld." Heat triggers evaporation in the Human Body as "sweat," because it is "colloidal-electrolytic," whereas in the automobile, liquids must be periodically replenished with "refills" in order to maintain dynamic-Temperature-equilibrium within the mechanical-electrical System.

Combustive gas explosions generate a lot of heat that is then captured by the "antifreeze" and by that, the engine system undergoes "relative cooling" or maintains temperatures below a certain maximal limit, beyond which the engine would be said to be "overheating."

The "captured heat" encapsulated-contained-carried in the circulating antifreeze is then transferred to "a radiating core" that "traps the Heat" from the antifreeze. As a fan ventilates the heated antifreeze-filled core, the heat traverses into the vehicle's driver compartment or passenger cabin, via a process that allows drivers to control the temperature within the vehicle — the "antifreeze" gets hot, and with control mechanisms, its heat is "transferred" into the passenger compartment for the comfort of Human Beings.

Such examples demonstrating manifestations of Energy transformation via thermodynamic cycling, embody, from a mathematical standpoint, cycling-Differentials and Force-Opposites that necessitate the utilization of alpha-numeric Constants, in order to "re-normalize" Continuum processing, amenable to "an equation of reconciliation" that accounts for direct or inverse proportionality, and/or relative equivalency.

Cold prevails in the absence of heat. Ultimate cold prevails in the total absence of any heat source at all — and heat is a property of "Electromagnetic Mass;" that is why empty, vacuum Mass-less Space is cold. Without Stars to radiate heat, light, and cosmic rays, all of which constituting the Electromagnetic Energy Spectrum, the whole Universe would be "a deep freeze."

Heat and cold ratios and Motion-force proportions facilitating "thermodynamic rates-of-change" drive Earth temperature Opposites and pressure-Force Differentials that control Energy Transformation, for thermodynamic Motion-cycling Equilibrium.

What is a Planet but a greater agglomeration of Atoms in "congealed Energy Forms"? What is a Star but a greater aggregate of Atoms in "excited Matter-Forms"? Conservation of Energy also implies preservation of organizational Forms and of structural patterns-of-operation, as engendered by these prevailing Forms.

Different Frames-of-reference, e.g., Earth Atmosphere, will present "Functional Entropy Differentials" due to interactions with Frames having Opposite Forms of Motion-force Cycling, e.g., Plant Photosynthesis from Sunlight, Soil-variables and Hydro-meteorological determinants, the "complex interactions Matrix" of which, must be "buffered" via alpha-numeric Constants "intervening" in the conceptualization, "operationalization," and applicable quantification of "Continuum Unification."

Cycling Differentials are engendered by Motion-force Opposites emerging from Electromagnetism and Gravity, relative to interacting Mass-proportions, ratios, and equivalencies. Thus, Motion-force drives thermodynamic Frame-cycling amenable to sustaining Continuum by fulfilling dynamic System equilibrium via operations of the Input-Process-Output Mechanism(s).

Each frame of reference, e.g. Quantum, Molecular, Planetary, Solar System etc... already has "a pre-existing information-package" or "a pre-determined Cycling encoding," that in effect constitutes "steady-state programming," due to eventual "thermodynamic stability" from which ensues a respective repertoire of "routine system changes," e.g., Seasonal Cycles, within allowable ranges of Motion-specific "degrees of variance," hence, the necessity that alpha-numeric Constants "regulate" interconnected Continuum-relationships between interactive-overlapping cycling-Frames for "reconciliation" or "renormalization" of Differentials and Opposites.

There is no such occurrence as an abrupt seasonal change; each specific Season "oozes-and-blends" into the next on-coming Season as it "negotiates" transiting, from a previous "Revolution-Rotation interaction" where respective and common variables overlap-intersect, as the whole Planet "transposes" to "new points-of-intersection" at every instant of rotary Motion, whereby thermodynamic cycling variables, parameters, and co-determinants are simultaneously-consecutively being "recalibrated per pre-assigned valuations," as "befits" each "Revolution-Rotation intersect/overlap."

Each "Revolution-Rotation intersect" presents its own "overlap-intersect dynamic" that forecasts Earth ecological-meteorological "cybernetic responses" to co-determinants values-changes, variables values-adjustments, and parameters values-modifications.

In Nature and the Universe, these calculative, calibrated, and "negotiated" co-determinants, variables, and parameters that compel Scientists into discovery of "working Constants," e.g., each combination of cycling Frames necessitating its own "kind of Constant," are "computed automatically" in "degrees-of-variant-change" within Frame-specific range-allowances or "valuation limits," e.g., Gravity-frames necessitate a rate of acceleration for every Object "falling" within Earth circum-Sphere, evaluated at: 9.88 meters/second/second.

This "computation process" from pre-existing "information package(s)," is the equivalent of "Functional Operational Thermodynamics" — for contemporaneous Conservation and Entropy processes taking place in each Mass-Frame, respectively.

Even the Universe and Nature, — because they are "encoded" with "instructions-for-Orderly-Operation(s)" as pre-determined by "Thermo-dynamics of the Organizing Principle" whose modus operandi is encapsulated in the Input-Process-Output Mechanism designed to fulfill dynamic System equilibrium, — "cling to" procedures, processes, principles, and methods answering to specific physical Laws, the sum of which, at face-value, might appear "random" or "un-directed" to Human Beings. Nevertheless, given the predictability, periodicity, iterativeness, and fidelity of natural processes to quasi-similar patterning structures of functional operation(s), e.g., the Atom and the Solar System, then, scientific discoveries can only prove to us the Intelligent Design that is intrinsically "embedded" within the very "inanimate occurrences" that continue to arouse awe and wonder in Human curiosity and inquiry, e.g., The Weather offers a high degree of predictability to meteorologists who on our evening Network News broadcasts daily forecast our country's regional and local atmospheric phenomena, including expectations of flooding events, hurricanes, storms, etc…, as far as six to seven days in advance; forecasting the timing-occurrence of earthquakes and eruption-timing of volcanoes, is also advancing to greater degrees of accuracy in warning populations that would be affected or impacted so as to prevent catastrophic loss of Human lives.

In the inanimate Universe and insentient Nature, "Terminal Entropy," or final resource(s) exhaustion that would bring "death" to the Cycling Mass-Frame(s) upon which Human Civilization relies for reproducing life and for perpetuation of Society, proceeds more slowly than for the lifespan of biological life-Forms.

Inanimate slow-proceeding Universe and Nature "Terminal Entropy" is being "constricted" or "thwarted," even as cycling-Differentials and Force-Opposites "interface" to preserve angular momentum pertaining to "Patterns-of-Processing" respective to each Cycling-Frame, while maintaining "Organizational Pattern-integrity" during "overlaps" and "intersects" peculiar to each structured Mass-in-Motion Unit or Frame, respectively, as geared for "Structured-Orderly Proceedings of Energy Conservation," e.g., The Atom, as Matter's substructure upon which greater mass-Frames are erected, is reliably stable, as relatively stable as our Solar System, which has been "operating" for thousands of years while sustaining Planet Earth life-support systems as well as the lives of many Human generations-past.

Organized Patterns of Energy Conservation "encased" or "embedded" within Frame-units operations that factor in countering "Terminal Entropy," transform apparently "stochastic" cycling processes peculiar to Mass-in-Motion Frames, respectively, into relatively stable and reliably predictable "Steady-state structures."

As steady-state Frames, each Unit of Mass-in-Motion must develop its own "respective thriving gravimetric routines" geared for achieving, attaining, accomplishing, and fulfilling dynamic Cycling-System equilibrium within "the confines" established as "set-thermodynamic boundaries" for its Cycling-processes by Continuum Space-Time Curvature, yet, without "radical alteration of programmed information-instructions," while simultaneously, its own structure, content, organization, intrinsic Conservation mechanisms, and functional purposes are inherently retained, respectively, e.g., Each morning Human Beings wake-up from sleep, they do expect the ground of the Earth beneath their feet to "remain solid" as they walk into their kitchens to cook breakfast!

The Earth must continue to revolve around the Sun and rotate upon its own axis, regardless of disagreeable solar activity, or internal geo-atmospheric disruptions.

Thus, Planet Earth will "wobble" a little bit; its poles will experience some "flattening;" it will periodically "bulge" at the Equator; and as the magnetosphere absorbs excess solar radiation for ionosphere processing, Aurora Borealis will intermittently occur.

In addition, when the hot iron molten magma core of Planet Earth is so excessively "stimulated" by lightning strikes that increase its internal pressure determinants, a "safety valve" is opened for "lava pillows" to ooze out unto the bottom-floor of the oceanic hydrosphere; for a volcano to erupt; for the seas to rage in tidal surf-and-surge tempest; or for "plate tectonics" to cause earthquakes: All of which indicating core limitations in electro-radiant absorption rates as "innervated" by disruptions in planetary Energy-Cycling-System dynamic equilibrium, thus causing "re-calibration" of geo-ecological parameters for "returning to regular planetary routines."

Disruptions in dynamic equilibrium of Cycling-Mass Systems, must be "re-negotiated" until equilibrium is again restored for specifically pre-designed functional activities, so to speak — for the Earth must operate as it was "originally created to proceed" for sustaining its "Human-friendly" life-support structures.

All steady-state systems that are in "relatively constant Continuum-Motion" for Energy cycling must experience "re-computation," "re-calibration," "re-adjustment," or "re-calculation" of Relativity parameters, yet within their own specific allowable ranges-of-manifestation, in order to sustain dynamic equilibrium for Continuum intrinsic activities — There is Continuum, as for a ship navigating upon the Seas, only when dynamic System equilibrium is as "steady-as-she-goes."

As "Functional Operational Entropy" — as opposed to the onset of "Terminal Entropy — unfolds, systemic changes affecting proportional relationships between cycling-Differentials and Force-Opposites, must remain faithful to the character and purpose of the steady-state entity –

functional fidelity and structural integrity, e.g., Earth, as a Water-Planet, must have life-support systems that possess cause-and-effect processes and organic bio-mechanisms that are "Human-Life friendly."

Qualitative Conservation overcomes the tendency to decay via replacement of utilized resources consumed in functional cycling. Atoms must continue to bond at outer electron orbital levels with elements for which they have a "reaction-affinity," respectively, while undergoing chemical reactions for molecular change, e.g., Hydrogen and Oxygen must react to form water, as Iron Atoms must bond with Oxygen to form Ferrous Oxide (FeO) which is also a strong base with properties for absorbing Carbon Dioxide.

With respective adjustments in pertinent temperature and pressure controls attuned to elemental reaction-affinity for chemical or molecular change, Continuum Energy Transformation is insured.

Iron rusts; aluminum suffers corrosion. There is also "metal fatigue" which may cause cracks or fissures that weaken bonding-structures. When rust sets into metals, they can be said to be in a state of "Terminal Entropy:" Rusty metals are very poor conductors of electricity.

"Terminal Entropy changes," such as rust, are analogous to "aging-and-death" or "qualitative decay" in Human Beings — to the exclusion of all other factors such as fatal diseases and accidents. Given that Human blood is iron-based and Human respiration is Oxygen-based while the Human organism is constituted of more than 75% Water-composition, then, "aging-and-death" as a Form of "bio-organic rusting," can be analogously compared to rust-in-metals.

Metals and water, due to Oxygen therein, contribute to the formation of rust. Rust, as an extreme form of oxidization, annuls the electro-motive capacity for electro-static molecular bonding in a metal — Electrons can no longer bond or participate in reciprocal exchanges via electro-conductivity in order to form molecules.

Because electro-magnetic charges cannot flow to "attach" to atomic nuclei in response to gravimetric variables and Field-force determinants that compel attraction-repulsion binding Energetics, these activities between distinct particulate-atomic charges that attract and repel are no longer operative.

Opposite charges "attract" and like-charges "repel" — thus, positively charged Protons and negatively charged Electrons are "attracted" to each other; but, at the same time, Protons are "repelled" by each other due to like-electrical charges. And due to the presence of the charge-less Neutron "actively grounding" the Atom by preventing "atomic short-circuiting," micro-mass-Electrons are "kept" away from the Nucleus, at "a respectable distance."

Since negatively charged electrons are "repelled" by negatively charged electrons, then, relatively speaking they cannot "bond." However, due to the strong presence of positive Protons that attract Electrons, Electron-exchanges take place between elements having chemical reaction-affinity for mutually-reciprocating molecular change. Hence, the scientific understanding that: Electrons "bond" at the outer-Orbitals, where they are "as far as possible" from their Atomic Nucleus.

Consequently, greater Mass agglomerations are formed by the attractive force(s) between atomic nuclei by having Protons "attract" each other's revolving Electrons as particular Elements with reciprocal reaction-affinity combine for molecular change. However, it is the "recombinant activities" of the strong Proton-Neutron force "in consort with" Protons "repelling" each other, that together compel Electrons to "keep a safe distance" from each other, as well from atomic nuclei, even as they "jump from one Energy-Orbital level to another" during the time when Atoms are aggregating to form greater Mass frames.

Electrons involved in orbital energy exchanges for molecular bonding keep from each other "at a safe non-short-circuiting distance" because like-charges "repel," while at the same time, due to interactions with the nucleic strong force to which they are also attracted — the complex interplays of which, "holding" compounds, Elements, substrates, and substances together, and thus, by that, countering, disintegration and dispersal. Hence, reasons why, due to cause-and-effect mechanisms that both counter-disintegration (centripetal Forces) through "attraction," and simultaneously prevent mutual absorption or nucleic short-circuiting (centrifugal Forces) through "repulsion," that it is very difficult, and hence, requiring tremendously high-temperature Energy, to either "fuse atoms" (nuclear fusion), or "separate atomic particles" (nuclear fission).

It is due to the positively charged nuclei as "grounded" by the Neutron, that Electrons can "safely clump or cluster together" while keeping "a safe distance" from elemental nuclei and from each other, as "attraction" is tempered by "repulsion," and as Forces-and-Mass properties "within their thermodynamic Volume" are "proportionally distributed" in ratios "chemically befitting each other," for fulfillment of dynamic System equilibrium.

Given that Elements remain as initially formed via chemical reactions for molecular change, then, Conservation does temporarily prevail over Entropy — There is Continuum Stability in elemental Forms and their organizational Structures.

Therefore, elements, compounds and substances are stable, not due to "static balance," but rather, due to "dynamic System equilibrium" by which "attractive Forces" and "repulsive Forces" are "cybernetically distributed" in "befitting ratios" and "right proportions" throughout the "thermodynamic Volume(s)," of these Mass-Frames-in-Motion, respectively.

For, were "attractive Forces" to be equally distributed as "repulsive Forces," then they would "cancel each other out," and no elemental formation or molecular bonding could then be possible, let alone "take place."

Therefore, in a Biological Organism, Conservation must temporarily overcome tendencies to Entropy; or aggregating-attractive centripetal Forces must temporarily overcome disintegrating-repulsive centrifugal Forces, in the same manner that a mortal Human Being can "take his time" to grow, develop, mature, and "slowly age" until "wear-and-tear" takes its toll, climaxing into "extreme old age," and then "ultimate death."

During molecular change, Electron exchange takes place because the atomic nucleus is animated by "the strong force," as activated by the positively charged Proton(s) and the charge-less Neutron(s). But when a metal has rust throughout its molecular structure, Protons and Electrons are no longer "charge-potentiated" for counter-vectored Force-Motion Energetics, and hence, no electromagnetic Field can "deploy;" and the Neutron's "grounding role" is thereby "muted" as the metal becomes "electromagnetically inert:" Due to severe loss of electro-motive-conductivity, thus, due to poor "electro-magnetic circulation," Electrons can no longer "process Energy" for rising to higher Orbitals or for descending to lower Orbitals, during operations of which, Energy is released.

In the same vein, "Terminal Entropy" results within a Star when Atoms can no longer "fuse" or "fission" under extreme-heat-intensity conditions, as the nuclear material's capacity to "ignite" for "bonding" is weakened, muted, or annulled, due to gradual exhaustion of fuel(s), e.g., Hydrogen, that generate and sustain Corona-vectored counter-centric and Core-vectored pro-centric polarized nucleated–ionization Forces that are accountable for nuclear-chain reactions between plasma condensate molecules, the complex interplay of which, weakening the potency and strength of the Star's magnetic Field.

With the demise of "the strong force" being imminent, the Star's incapacity to pro-centrically aggregate and coalesce, "triggers" predominance of "the weak force" to govern post-thermodynamic death denouement.

In the absence of an extremely strong Electro-magnetic-Field that "confines" nucleated plasma condensate chain-reactions for simultaneous-contemporaneous fusion and fission and thus, for contraction and expansion Energetics that exert "compression-Pressure" within the boundaries of its "containment circum-Sphere," the Star will ultimately experience "Cryo-Death:" Nucleated processes are extinguished, causing centripetal Forces to cease in exerting pro-centric tensing-Pressure-Force, and bringing Gravity-field-Forces to a thermodynamic end, which triggers "Star-material" to be centrifugally expelled and dispersed outwards throughout the empty-void of cosmic universal Space, as only a "dead-Star-hulk" composed of "heavier atoms" and "denser Elements" is "left behind" with the cold-stillness of a "carbonized leviathan."

The Star under such "de-magnetized conditions" can no longer "electro-graphically" maintain an inner-Core as its center-of-Mass and center-of-Gravity, because, due to impaired Electro-magnetic Field strength, "routine differentiated distance(s)" between its molecular, atomic, and plasma condensate components is "disrupted," which puts them all "in a state of disconnect:" From its Core to its radiations emission zone, and from its convection zone to its coronal circum-periphery, there is no more "chain-reaction affinity."

A Star with an impotent Magnetic Field has no Core-centric processes that can sustain nucleated chain-reactions. It burns up its Hydrogen fuel in fast-cycling patterns of "Time-compressed" Energy transformation that results in a dead-end. The Star expands as it burns Helium, to collapse again as Helium is exhausted, after which it switches to "the next lightest Element." If temperature and pressure gradients and Force-Differentials due to altered Magnetic Field strength cannot sustain nucleation of heavier elements, the Star radiates most of its plasma fuel condensate as ejected Energy emissions leading to "Star Death."

The Magnetic Field that caused implosion of plasma molecules towards the Core will have weakened so that only explosive fission is occurring during which excessive amounts of radiated positrons, neutrinos, cosmic rays, light, and heat will be released until only the heavy particles, like "dead proton-neutron nuclei," remain.

If "Star Death" is gradual, the "explosive fission" will occur during a specific period of Time; if "Star Death" is sudden, then the "explosive fission" will take place as a major cataclysm. The Star will thus become a spherical Mass analogous in appearance in some parts, to "a mound of gray ashes" at the coronal circum-Sphere comparable to charcoal-residues after a fire is extinguished, but having a lighter textured appearance due to high-temperature surges and bursts of radiation emissions exiting the Corona. And in other parts, especially in the direction of the inner-Core, it will have the appearance of "singed timber," or appear as "incompletely burnt wood materials" displaying a dark-texture hulk of partially nucleated elemental residues.

During "routine Star operations," as opposite charges attract, the extremely powerful Magnetic Field keeps core particles imploded inwardly, as nuclear chain reactions are compressed within the inner-Star region. An opposite reaction occurs, due to like-charges repelling as explosive fission-Forces that cause the rush of core particles towards the convection region. Corona-vectored Forces then engage spectral radiations, heat, and light, released as solar system gravimetric, Field and Curvature binding-Energetics, the complex sum of which, activating the helio-Spheric quasi-ellipsoid plane of planetary revolution.

Since Energy Conservation prevails via redundancy of iteratively analogous quasi-similar Forms, fusion climaxes into plasma condensate from the attractive force between opposite charges, as atomic fission emerges from the repulsive force between like-charges. "The strong force" between Protons and Neutrons is the prevailing force. Thus, fusion is sustained as centri-vectored Motion-Forms overcome counter-centric dissociative Force-patterns having an affinity for fission. In that manner, dynamic "inner-Star System" equilibrium is achieved.

The solar Frame as "Mass-Energy-Radiation-in-motion-Force Complex" is a steady-state System in equilibrated Energy Transformation: Nucleated Plasma Condensate Thermodynamics!

Star nuclear plasma convection dynamics are possible only because the extremely powerful, "exponentially-Energized," Magnetic Field, prevents the whole Complex from completely exploding outwards in a huge cataclysmic event, as happens on the Earth due to absence of self-initiated powerful Magnetic Fields "to confine" or "hold-contain" the atomic explosion.

In view of all necessary Continuum activities that preoccupy a Star, it undergoes specific "Functional thermo-Operational Entropy-dynamics" as changes or permutations that take place, yet without endangering its "pre-existing programmed purposes," e.g., every 11 years, the Sun undergoes "polarity displacement;" every 22 years, magnetic polarity change. These cycling Differentials are part-and-parcel of Star characteristics that sustain an ellipsoid plane "pregnant with" tensing-stressing pressure-Force, within whose strength-boundaries thrive all Forms of planetary Motion for consistently "holding" nine Planets revolving around Solar System's Center-of-Mass, Center-of-Gravity, and Center-of-Field. Cycling changes, or thermodynamic permutations, when taken holistically, constitute "Entropy negotiating mechanisms" via which the Sun resists and battles the "tug-of-war" with radioactive decay and Magnetic Field degradation, so that Entropy processes are not "driven to their logical Terminal conclusion:" which is, "Star Death."

Thus, in order to formulate the Unified Field Theory of Continuum Curvature Pressure Motion-Force, it would be necessary to devise a "joint Constant," so to speak, that embodies all the "gravimetric Differentials and Opposites" that characterize both Force and Cycling, respective to each particular Frame's unique patterns of structured Energy Transformation, in specialized functional fulfillment of the Input-Process-Output Mechanism.

A joint Constant, or common Constant for both Motion-Force and Thermodynamic-Cycling, denoting overlapping operations of "Field-and-Gravity Intersect(s)" would "reconcile" all "binding Energetics" accountable for "specific Mass-in-Motion routines," in order to "renormalize" differentiated expressions of "the strong force" and "the weak force" as "sub-categories of Electromagnetism," with Gravity-field interactions, for dispensation of Continuum Unification.

CONTINUUM FIELD-CURVATURE MOTION-FORCE ENERGETICS

Electromagnetism and Gravity "move Matter" as "Mass-in-Motion Frames" to give rise to: The Continuum Principle, the Equivalency Principle, the Thermodynamic Cycling Principle, the Curvature Principle, The Conservation Principle, The Entropy Principle, The weak force (Counter-centric Force Principle) and the strong force ("Pro-Centric-Force Principle.")

When combined in operation, these Principles, as embodied in specific physical Laws that then account for "triggering" particularly coordinated "Energy Cycling routines" and "Motion-force repertoires," giving rise to fulfillment of functional dynamic System equilibrium, then, redundancy of organizational Forms prevails for reiterating quasi-similar patterns of structured Energy Cycling, via "quasi-isomorphic embodiments" that evoke symmetry and synchronicity, e.g., The Atomic Organizational Structure evoking quasi-similarity with The Solar System Organizational Structure.

Redundancy of organizational Forms and quasi-isopmorphic iterative structural Patterns, predominate in Nature and the Universe, because, — in the light of the Law of Transformation of Energy which states: (a) "Energy is never created nor destroyed but always transformed;" and (b)

When specific amounts of Energy are utilized to do work, respectively, there results an unusable portion that can no longer produce work, this process being called Entropy, — they interact to yield the following Principles of Universal Energetics:

(1)   Every category of measurable entity or variable is rooted in the Electromagnetic Properties of Matter as it is "moving" within Continuum Space-Time Curvature Energetics due to Electromagnetic-Field Force and Gravity-field Force Interactions.

(2)   The universal limiting factor or restricting Standard being the Speed of Light in a Vacuum, all unfolding Mass-Frames must necessarily be compelled to espouse quasi-repetitive patterns of structured Organization.

(3)   All unfolding Mass-Frames must fulfill functional dynamic System equilibrium via operations of the universal thermodynamic cycling INPUT-PROCESS-OUTPUT Mechanism.

(4)   And because, as stated by Dr. Albert Einstein, there are only a limited number of valid Physical Scientific Laws for proven-verified reproducible applicability within the Vast Expanse of the Universal Complex.

Why does the following pattern repeat itself throughout the Universe from Quantum Mechanics to Newtonian Mechanics: Things revolve around a Center-of-Mass during which Energy Transformation cycles within differentiated Frames with Mass-in-Motion-Force, as animated by opposite electro-motive charges?

All Motion-Force Dynamics, e.g., as Magnetic Field and Gravity Field, active within the Atomic Frame, are also active within the macro-planetary Frame, but in differentiated yet analogous Forms that sustain Continuum Space-Time Curvature Energetics.

The Atom is a complete whole Energy system displaying operations of micro-Curvature, Field, Gravity, Electromagnetism, "the strong force," and "the weak force" in cycling-fulfillment of the Input-Process-Output Mechanism, whilst in Forms that are "recapitulated" or "iteratively transubstantiated" within greater Mass-Frames such as the Solar System wherein Planets and Star "locked into Relativity-interactions" are "Motion-force synchronized" for specialized operational functionality, respectively.

A "joint constant" embodying-accounting for stress, tension, and elasticity metrics would "renormalize Curvature Motion-pressure Force," as necessary for all cycling Frames that "respond" to specific cycling patterns that are activated by differentiated rates of Motion-force, as projected by the Gravity-field, Electromagnetism, "the strong force" and "the weak force" within their "respective spheres of field influence," — The Gravity-field Motion-Force not being "particle-based" as it is not activated by electro-motive dipole opposite charges, is more "elastic" in "range-of-manifestation" and "diversity of Motion-Forms" than the Electromagnetic Field Motion Force from which it derives; whereas Electromagnetic Field Motion-Force is "triggered" by dipolar opposite electro-motive charges that Relatively-interact for yielding rotary Forms of pro-centric Motion — in a clockwise vectoring direction.

{Depending on the location of the Observer from Earth surface, from the North Pole it appears that the Earth is rotating "clockwise" whereas from the South Pole it appears to be rotating "counter-clockwise." However, "clockwise" is the correct rotational vectoring direction as the Observer is "facing" the Galactic Center from the North Pole, whereas at the South Pole, the Observer is "looking away from" the Milky Way Galactic Center. It is the positional location of the Observer vis-à-vis the Galactic Center of the Milky Way that determines the correct vector for Earth Revolution around the Sun, which is "clockwise" as the Observer is "standing at the North Pole" while "facing" the Galactic Center; or if the Observer is peering at our Solar System from the Milky Way Galactic Center itself. Thus, for determining the accurate vectoring direction for a revolving or rotating Space-body, the Center around which it is revolving or rotating must remain the primary Focus of the Observer's point-of-reference. If the Observer is "standing at the Milky Way Galactic Center" while looking at our Solar System, Earth Revolution "keeps" its "clockwise vectoring direction," even as Earth Rotation also "keeps" its "counter-clockwise" vectoring direction.}

A "gravimetric Motion-force Constant" that operates to "interface" all rates-of-change in Differentials and Opposites displayed by co-determinants of Electromagnetism and Gravity, would facilitate "Motion-force lubrication" between "intersecting/overlapping Mass-Frames," for "re-normalization" within "the complex theater" of differentiated cycling Outputs, to "link/connect/unify" all "Mass-Energy-Structures" as activated by the Cosmic Reality of Continuum Space-Time Curvature Energetics.

Continuum Energy-Cycling by all Mass-Frames within the Universe is the key to Motion-Force unification within Space-Time Curvature. "Congealed Frames" must be "reconciled" with "excited Frames," even as solar mass-ejected "radiation Energetics" exerts its Gravity-and-Field-induced Continuum Curvature pressure-Force binding Energies upon the Planets, the sum of which, exerted upon Mass-in-motion Frames that apparently "respond" to "stochastic Inputs" for thermodynamic cycling, within the context of climaxing into "Continuum Outputs."

Motion and Force accompany each other as electro-motive charges "transmute into" Magnetic-field force, while Gravity-field force emerges from rotary Forms of centri-vectored or pro-centric Motion that is directly caused by Electromagnetism.

Pro-centric Motion-patterns in all cycling Frames promote the prevalence of "the strong force" as Energy Conservation overcomes Entropic tendencies to exhaust resources. These dynamics factoring into mathematical formalization will allow computational relative equivalency between thermodynamic cycling Differentials of all Mass-in-Motion-Frames that unify Continuum Space-Time Curvature Motion-Force Energetics. Specialized functionality within boundaries of greater Mass-Frames, "recapitulates" Atomic Frame motion-force dynamics, as centri-vectored Motion is ubiquitously duplicated in its various Forms in sustenance of Energy Conservation.

Entropy is a powerful process ensuing from thermodynamic Energy cycling. However, Conservation temporarily overrides Entropy's "urgent calls" for decay, degradation, and termination.

Iterative patterns of redundant structured organization in analogously replicated Forms characterize Input-Process-Output cycling in Frames of Mass-in-Motion for functional specificity.

"A General Grand Unification Theory of Motion-Force" that mathematically formalizes all "spheres of Field-influence" engendered by their common source: Electromagnetism, would also "interface" "the strong force and "the weak force" with Gravity, so as to account for all Forms of centri-vectored or pro-centric Motion-force manifestations.

It is noteworthy to state that, though "the weak force" evokes the tendency to radioactive decay as it operates as a dispersive-force, Electrons maintain "Orbital fidelity," as "the strong force" temporarily overpowers "the weak force," except when necessary for molecular change whereby Electrons do engage in bonding at their outer Orbital levels, at which time, their complex interactions with all atomic nuclei comprised within the molecules being formed, take over to determine "Force-configurations" for the greater Element or other Space-body being formed as a result of all chemical bonding chain reactions, respectively.

For, all Forms of Motion, e.g., rotation, revolution, electron orbital motion, earth core-centric line-of-radius motion, molecular bonding motion-force, plasma fusion, ellipsoidal tensing-stressing elastic Energetics, the perihelion shift of planet Mercury, accelerated particles Energy gains, earth "wobbling," Earth bi-polar oblation, etc . . . , are composite, recombinant, and/or "convoluted renditions" of the same quantifiable Entity-category: They are tantamount to "negotiated" pro-centric Motion-force as engendered by Electromagnetic-Field-induced Gravity-force "equivalents" acting together for Continuum Curvature Space-Time Energetics.

Due to momentum Conservation and "resistance to motion," cycling mass responses to "spheres of field influence" projected by pro-centric motion-forces akin to "the strong force," electromagnetism, Gravity and "the weak force," are "negotiated" via recombinant Forms that espouse rotary, curvilinear, and rectilinear patterns, e.g., an aircraft flying continuously in earth atmosphere would display a rotary "spiraling pattern" of Motion-force around earth circum-sphere as it makes "periodic corrections to its flight-course," until it lands as it follows a core-centric path along Earth line-of-radius in direction towards its Core center-of-Mass and Core center-of-Gravity.

As the Law of Gravity and the Laws of Motion imply the exertion of a Force, the Theory of Relativity implies the exertion of a Force. "Relativity" implies relationships, interactions, inter-exchanges, relations, connections, interdependence, synergy, symbiosis, etc.., all of which being present, in the way the whole Solar System is held together via an ellipsoid plane of combined Revolution-rotation Motion-force. Planets are simultaneously revolving and rotating — revolving around the Sun and rotating upon their own axes, as respectively indicated for the Earth.

In Einstein's equation, $E = mc^2$, the "conversion factor," "excited energy" as the gravitational radiation exponential function, became the equivalent of "acceleration" (as in $f = ma$) in a Form that comprises all gravi-metric electro-magnetic determinants of mass cycling, as effected by the exponential value of the Speed of Light to the second power.

According to relative proportions derived from this Equation embodying the Equivalency Principle between Matter and Energy, due to Entropy, m cannot have an infinite value; however, if m equals zero, then no Energy can be formed and/or there is no Energy present, in any Form whatsoever. But, Energy is indeed present in the Universe! Therefore, because c^2 has a fixed-limiting value of 186,000 miles/second x 186,000 miles/second, neither can m ever equal zero, nor can m ever have an ever-increasing infinite value. For, the closer m gets to approaching the Speed of Light limit, the less Mass it must possess! Hence, reasons why only Electromagnetic Spectrum-based nearly mass-less particulates, can travel as rays, but still, below the Speed of Light in a Vacuum!

On the Earth, as "excited Matter," the smaller-in-Mass is m, the greater Energy output that can be obtained from it, and hence, the tremendous yields of nuclear explosions. But as "congealed Energy," the more massive is m, the greater Energy yields can be obtained from it.

Consequently, m must always have a positive value so that there is always a certain determinate amount of Energy as a return from Transformation cycling. Hence, why "Energy can never be created nor destroyed but always transformed!" Thus, as long as the Sun is radiating, there will always be Energy to be transformed/converted into other Forms.

Stars can be really massive, but still, because c^2 has a limiting value, there are limits to be observed in the size of Stars. The size or Mass of a Star cannot be infinite; nor can it ever be zero in Mass. Therefore, since a Star or any Form of "excited Matter" can never have m equal zero, then there is no such thing as "pure Energy" or "Energy without presence of Matter:" Which means that, there cannot be a so-called "point-of-singularity" existing as "pure Energy" to then explode and thereafter giving material physicality to the vast extant Space-bodies of great Mass "moving" in the Universe that we have come to know!

Quantifiable categories as measurable phenomena such as motion, pressure, temperature, force, velocity, mass, field, gravity, Curvature, acceleration, momentum, tensor, stress, "gravitational elasticity," "gravitational lensing," "Electromagnetic lensing," Gravity-fields and Electromagnetic-fields are all active in dynamic-cycling(s) of Energy Transformation, from a "congealed state" to an "excited state," as bounded by the fixed value of the standard limiting factor, the Speed of Light in a vacuum.

Things have "relative equivalence" in relation to "a fixed Standard." Energy is in "relative equivalency" with Matter-Mass, due to the Speed of Light standard to which their proportions and ratios must conform.

In $E = mc^2$, Einstein's equation to express the relative equivalence between Energy and Mass, they are assumed to be directly proportional, as the Speed of Light becomes the relative Constant or standard parameter that delineates boundaries for the ratio of Mass to Energy. What is accomplished by the exponential function, c^2, though not fully explained conceptually, has been demonstrated in invention of atomic bombs and creation of Hydrogen bombs.

Due to the Speed of Light as a limiting factor, Mass is inversely proportional to c^2, as c^2 is inversely proportional to Mass. Only the ratio of Energy to Mass can vary, even as their quantities are finite due to the Law of Energy Conservation, but within certain limits as bounded by the Speed of Light standard: For c^2 , as embodying the complex totality of Solar Radiation Energetics, has a fixed value.

Thus, possibilities are not endless in formalizing physical Laws that explain universal phenomena. The enumeration of valid physical Laws, representing "the information package" of the Universe, is bounded by Energy-cycling limits that cannot be breached as predetermined by the Electromagnetic Properties of Matter — the Speed of Light in a Vacuum whose application to Mass as an exponential function to the second power, transforms "congealed energy" into its "excited Form." Particles accelerated even to a fraction of the Speed of Light must "shed their Mass" as Mass-Energy Gains are "re-translated" into radiated cosmic energy.

As a "conversion factor" utilized in the computational equation of relative Mass-Energy equivalency, the Speed of Light to the second exponent power is "transubstantiated" into "Gravity-and-Field Accelerated Momentum traveling-tensing-stressing binding radiation-Mass Energetics," as the fixed constant values embodying all "Curvature-Binding Energy-Force-Motion Mechanisms" contained in emitted "Solar Gravitational Radiation" are "re-translated" into "spheres of interactive Field influence," i.e., "Every thing is revolving around every thing:" Electrons around Atomic Nuclei; Planets around Stars; Star Systems around Galactic Centers; Galaxies around Other Greater Galactic Centers-of-Star-Mass(es)!

The ultimate Form of "excited Energy" is Star nuclear plasma Energy, the complete Electromagnetic Spectrum of which is well-known, as ranging from Cosmic rays to Gamma rays, from X-rays Beta-particles, from Visible Light to low-frequency long-wave/radio-waves.

Cause-and-effect mechanisms of emitted solar gravitational radiation arising from "re-translation" of Gravity-and-Field Curvature Energetics, are primarily expressed as Motion-force in the Forms of Revolution and Rotation, which in turn effect on-Earth equivalents of the Gravity-Force as its Mass-dependent variables and co-determinants climax into composite rotary, curvilinear, and rectilinear patterns of Motion-force along Earth line-of-radius towards its Core center-of-Mass.

Extreme difficulty arises in differentiating between Curvature Tensor-pressure-force, Magnetic-Field Motion-force Energetics and Gravity-Field Mition-force equivalents, both at the micro-Atomic Frame and at the Solar System planetary–Frame, due to various Motion-force Opposites that effect "cycling-patterns-Differentials, consistent with thermodynamic System equilibrium, e.g., Earth is a Life-Planet or a Water-Planet with 1-G Gravity acceleration-rate for any Object within its circum-Sphere, and thus, Earth cycling-Routines are bound to be very different from Martian-Repertoires, given that Mars can "barely hold" any atmospheric gases at all, as its Gravity-gradient-equivalent is only about one third of that of the Earth: 9.8 m/s^2 (meters/second/second) as opposed to 3.7 m/s^2 (meters/second/second.)

At the Atomic Frame, particulate opposite-charge displacements as tensing-stressing exertions effect/co-exist with micro-Field and micro-Gravity Forces, respectively, that "trigger" Electron-motion around the Atomic Nucleus, as well as "Atomic-Nucleus Dynamics," because of Newton's Third Law of Motion, such that the whole Atom is still stable even while "the strong nuclear force" is greater than "the weak nuclear force, because of the presence of the Neutron in the Nucleus, the complex sum-of-interactions of which, allowing for preservation of atomic angular momentum or "spin."

The Solar System ellipsoid plane of planetary Motion displays "gravitational-plasticity" or "Mass-Momentum-force tensility," due to the complex interplay of Magnetic-field and Gravity-field interactions.

As solar gravimetric Field-inducing Radiation is "projected," it engineers "a complex Curvature Force-structure Dynamic" comprised of tensing and stressing, contracting and distending, pushing and pulling, compressing and dilating, — not excluding Earth-Moon and other "interaction-dynamics" that account for specific gravimetric Curvature effects, — that climaxes into inter-planetary centers-of-Mass, centers-of-Field, and centers-of-Gravity "interacting/intersecting/and overlapping" to engineer "exo-planetary perturbations" that effect, for examples: Earth "wobbling," bulging-at-the-Equator, and bi-polar oblation(s); and the Perihelion shift of Planet Mercury.

In addition, many Earth-bound or inner-Earth motion-patterns emerge, via "gravimetric Curvature entanglements" that engender geo-Core-centric rotary, rectilinear, linear, and curvilinear Motion-Forms that are to be considered as "emergent properties of Revolution and Rotation, the complex sum of which, "triggering" many well-known meteorological phenomena and geo-hydrosphere events: e.g., Atmospheric storm systems and geo-hydrosphere turbulences consequent to such "Composite-Forms of Entangled-Motion-Force Energetics," as "exponentially-potentiated" to be made manifest through hurricanes, cyclones, tornadoes, tsunamis, Core-initiated tumultuous events such as volcanic eruptions, and dynamic plate tectonics activity "triggering" earthquakes.

"Energy is never created nor destroyed but always transformed!" And the Speed of Light in a vacuum, due to its relatively fixed value, engineers: (a) A boundary that cannot be breached by events, processes and phenomena, or "traveling Objects;" (b) Specific limitations regarding which Forms-of-Energy that Matter-Mass can be converted into; and (c) The proportions of Energy which a specific quantity of Mass can climax into from molecular rates-of-change.

In the Special Relativity equation, $E = mc^2$, $m = E/c^2$ and $c^2 = E/m$, Mass is inversely proportional to c^2, as c^2 is inversely proportional to Mass. The proportion of Energy to Mass as denoted in the energy/mass ratio is already predetermined by transformative limits imposed by the fixed value of the Speed of Light, which is utilized as a "conversion factor" to the second exponent power.

Therefore, in the Universe as a whole, — (even as Space is infinite and Time is eternal,) — Mass-Energy cycling is quantifiable within Continuum Space-Time in apparently iterative but "stochastic temporal periods," due to the Electromagnetic Properties of Matter, "particulate routines" of which, embodying "duality of Form" as wave or particle that must also cycle "as pre-programmed" for "Equilibrium Conservation."

Reality speaks for itself: As in $E = mc^2$, "Symmetrical Equivalencies" do prevail in the Universal Continuum Curvature Complex, and hence, why astro-physicist-mathematicians are "always looking for equations." And it is "no random accident" that all proven-valid physical Laws governing "Energy-cycling routines" for any Space-body extant in the material Universe as we have come-to-know it, must be reproducibly expressed in the Form of Equation(s). The Universe displays, to say the least, "Mathematical Precision."

Energy Transformation takes place in "bounded Steady-states," e.g., Planet Earth, thermodynamic cycling properties of which, being "confined" within finite Mass-Energy-Motion-Force Dynamics. Electromagnetic properties of cycling mass-Frames engender "equivalency-Differentials" and "equivalency-Opposites" that must be "computed" in accordance with "the pre-existing information package" initially "created for activating" the "Universal Input-Process-Output Algorithm."

End-products prevailing from Mass-cycling Differentials, in utilizable forms or as terminal processes, cannot increase or decrease the quantity of Matter and the quantity of Energy in the Universe — for there can only be Energy Transformation. However, cycling-Mass-Frames-in-Motion undergoing "Mass-energy gains" interact with their environment by engaging in "Energy transfers" in Forms that are "thermodynamically-attuned" to the "phenomenon-type," e.g., An accelerated particle in a cyclotron "experiences" Mass-energy gains that are then "re-transferred" to its environment as cosmic radiation emissions.

Because Matter and Energy can be converted-transformed or transubstantiated-transformed into each other, all Conservation and Entropy changes have to be accounted for as "transfers" or "accommodations" between the two states-of-Matter — either as "congealed ground-State-Energy" or as "excited-nucleated Matter." Only particles accelerated to a fraction of the Speed of Light gain momentum Mass-Energy for transformation into cosmic radiation emissions due to their conversion from "a congealed State" to an "excited energy Form." However, a spaceship as an Object in "congealed bounded-State" propelled into momentum acceleration that approximates Speed-of-Light velocities would not radiate cosmic emissions in "excited energy-Forms" from "Mass-Energy gains." Rather, such a spacecraft would "shed its Mass" to merely disintegrate in the process. For no Object can travel, and no process can unfold in velocities that are faster than the Speed of Light.

When ongoing dynamic Motion is accelerated by a certain external-Force to transform Matter-weightiness into momentum-Mass-force, then, "Mass-Energy-gains" are processed as "thermodynamic Inputs" that remain within the specific context of the cycling-Frame.

For, it is still extant initial quantities of Matter that gain Mass-energy, and it is still extant quantities of accelerated Mass that gain Energy — there is Conservation of Energy, and therefore, Conservation of Mass — thus, the remaining Mass and the "transmuted Mass" from Mass-energy gains must "complement each other" so as to equate the same initial amount or quantity of Matter that "ignited the chain-reactions."

Consequently, accelerated Objects such as a Spacecraft in "congealed bounded State" would "shed its Mass" as it "approaches the Speed of Light," whereas a particle of minute micro-mass would "transmute" its Mass-energy gains into cosmic radiation emissions.

"Relative inter-exchanges" or "transmutations" between Matter, momentum-Mass, and Energy during momentum-accelerated Motion are differentiated for "congealed bounded-states" and "moving-Frames" in "excited Energy Form." The Sun's radiation emissions are consonant with "nucleated plasma condensate thermodynamics," as Entropy processes pertaining to plasma condensate particulates cause them to "shed their mass energy gains" as "excited cosmic radiation" while they move/transfer from one Energy-level to another — from the inner-Core to the "convection zone" to the "emissions zone," to the Corona during which they undergo "minimal graduated cooling" before expulsion from the solar nuclear inferno.

An Earth-bounded atomic bomb exploded for weapons research, begun with initial quantities of nucleated Matter-Mass-Energy fissionable materials, will explode and disperse the "transformed Energy Mass-gains" as: Shock or sudden impact, the shock wave, heat, Light, initial thermal nuclear radiation, and residual nuclear radiation from weapons materials residues; and these resultant Forms, as emissions, rays, particles, effects, end-products, and end-processes can be quantifiably examined as to "the total disposition" of the whole explosion — how fissionable materials underwent Energy Transformation, its explosive dispersal rate, the ratio of initial materials to residual radiation, etc . . . , as well as, measurements of proportions in resultant radiation and explosive materials "processed via dispersal rates" as "Transferred Nucleated Energetics" to the atmosphere, hydrosphere, geo-sphere, eco-sphere, and bio-sphere.

When "congealed-Energy" contained in "ground-State Matter-Mass" is transformed into "excited Matter-Mass Energy," as important are: Mathematical ramifications, inferences, implications, and deductions ensuing from the equation $E = mc^2$ that also accounts for Motion-Energy-Forces, e.g., how much initial Energy was needed in order to "start a chain-reaction sequence" that "triggers ignition" for a fusion bomb, the complex interactions-sum(s) of which, accountable for causing "transformations" or "transubstantiations" within cycling-Mass-Frames undergoing such radical changes from "congealed" to "excited."

The Sun is as much "an imploded fission" as it is an "exploded fusion" whose convection processes are confined by its extremely powerful Magnetic Field.

Earth-bounded nuclear explosions in its Atmosphere, also contain "Motion-Force-actions" that respond to "Curvature-Energetics," as imposed by Earth circum-Sphere Curvature pressure-Force(s), Gravity-stress, Field tensor-metrics, and particulate electro-motive interactions, opposite-charge(s)-induced radioactive Mass displacements, especially in the absence of a Sun-like potent magnetic field "to confine" the explosion, and hence, intense rates of particulate and plasma dispersal as "buffeted-restrained" by Earth atmospheric gases, which in turn, possess "rates-of-ignition" for nuclear chain-reactions. Consequently, the nuclear explosion is short-lived due to pre-existing cycling-Differentials prevailing in Earth atmosphere, hydrosphere, and landmass whose complex Frame-composition properties operate to put boundaries, limits, and restrictions upon nucleated Continuum chain-reactions.

All thermodynamically cycling-Frames possess a "Relativity energy-field of interactions" in Forms generated by its constituent variables, as embodied in patterns delineated by "internal binding Energies" and "external binding Energies." "Externally," the Earth "binds" the orbiting Moon, its natural satellite, into "performance of periodic phases," the sum of which, characterizing its prescribed geo-synchronous rate-of-Rotation and revolutionary trajectory.

The Earth has "internal binding Energies" engendered by its center-of-Fravity, center-of-Mass and Magnetic-Field properties, the complex interplay of which, "bonding" together, each in its proper befitting parameters, the atmosphere, the landmass and the oceanic hydrosphere. The Planet also has "external binding Energies" that define its relationships with the Sun, the Moon, and other Planets with which it "shares" the Solar System's ellipsoid plane-of-Revolution.

In the same vein, the atomic nucleus has "internal binding energies" that form "mechanisms of attraction" between Protons and Neutrons, while at the same time possessing "external binding Energies" that form "mechanisms of repulsion" that together sustain the revolution of electrons around its centers-of-Mass, centers-of-charge, centers-of-Gravity, and centers-of-Field-Curvature. These inherent "binding energies," put-into-effect the capacities of Electrons to bond during molecular change with other electrons at their outer orbital levels, in specific chain-reactions pertaining to "elemental affinity."

These inherent properties of Matter, as engineered by particulate electro-motive magnetic charges and their interactions, preserve the "Mass-in-Motion-Frame" with its own Force-Energetics Complex from impending molecular disintegration and atomic dissociation.

"Interactive Fields of Force" — Opposite-charges dynamics, Gravity, Electromagnetism, pro-centric/attractive motion-Forces, counter-centric/repulsive motion-Forces, "the strong nucleic Force" and "the weak nuclear Force" — active within "congealed cycling-Frames" constitute a "gravitating-to-Center-dynamic" analogous to the internal plasma-condensate convection Energetics "holding the solar nuclear chamber" as a Whole Energy-System via its tremendously powerful outer-Magnetic Field, even as it continuously emits tremendous amounts of radiation from its turbulent internal nuclear processes.

Interdependent and overlapping "Energy-Fields Dynamics" thriving in cycling-Mass Frames, such as from the atomic to the molecular, and from the planetary to the solar, are emergent properties of "Field-and-Gravity-induced" Curvature binding-pressure-tensor-Force-Energetics that synergistically, synchronously, and symbiotically operate, to form "Holistic fields-of-force Systems," e.g., The Atom, the Molecule, the Planet, and the Sun; the biological Cell, the Organ, the Member, the Body and the Organism!

Thus, from the micro-Structure to the macro-Superstructure, are "recapitulated," increasing "levels of complexity" that re-iterate "pre-programmed" or "pre-encoded" quantifiable categories/entities/variables, the complex interactions of which, sustaining "authentically integrated Frame-functionality" for dynamic System equilibrium!

"The Unified Field Theory of Continuum Curvature-Pressure Motion-Force," as engendered by both Electromagnetism and Gravity, conceptually formalizes the ubiquitous occurrence of "Symmetrical Equivalencies" typifying diverse but interdependent Energy cycling-Frames found-to-be in "Overlapping/intersecting thermodynamic entanglements." The scientific surmise is that "a uniformly self-iterative relationship" exists between all cycling Mass-structures, via "interactive Fields of binding-Energetics" concurrently operating within all Frames in quasi-similar iterative analogous Forms, as they interdependently overlap/intersect for equilibrated functional operation, respectively.

The Sun is an "open Steady-State Energy-Cycling System," (not isolated, but self-contained) whose specific Star properties cause the emergence of "interactive Curvature binding Energetics" so powerful as to "bind" nine Planets around it, revolving and rotating in predictable patterns of Motion, within ellipsoid plane tensing-and-stressing Revolution-Motion-Force action.

As solar system "binding Energetics" cause Curvature or "bending of Space" due to "electromagnetic-Field Matter-mass Differentials" and "Gravity-Field Motion-Force-Opposites" operating to climax into synchronized inter-planetary angular momentum velocities, the greater the Mass of the cosmic Space-body, the greater "theater-of-manifestation" for its bending or curving Field activities, within the bounded-Space around it that its "Mass-in-Motion Frame" controls.

Solar and inter-planetary "interactive Fields-of-force" or "spheres of influence" or "theaters of manifestation," are co-determinant variables accountable for "planetary routines." Planet Mercury demonstrates a perihelion shift in a periodic repertoire of Revolutionary Cycles, in response to these gravimetric Differentials and Motion-force Opposites, as caused by "cycling changes" in its "moving-Frame" as it gets closer to the Sun, due to "recalibration" of the rate at which "angular-momentum-Curvature-binding-Energies" must flow for sustaining dynamic System equilibrium. Likewise, The Moon is a quarter of the Earth Mass — but "big enough" to cause oceanic hydrosphere tidal-surge activity.

The First Law of Energy Transformation says that: "Energy is never created nor destroyed, but always transformed." "Congealed Energy," as Electro-static Matter-mass with inertial-Energy, also "contains" or is "pregnant with" kinetic Energy designed for chemical reactions that drive molecular change.

"Congealed Energy" can also be transubstantiated into "excited Energy kinematics" geared for nuclear plasma-condensate binding Energetics that effect gravitational convection dynamics.

Star, or "Magnetic-Field-restrained" atomic-explosion Matter, is "excited" electromagnetic Mass being transmuted into nuclear plasma condensate, to generate "radiation pressure-Force Energetics" that engender centri-vectored Curvature Motion-Forms. Hydrogen, the lightest element is fused into Helium while the heavier elements like Oxygen, Nitrogen and Carbon and Iron undergo fusion-fission cycles, in proportions related to "Star-thermodynamic mechanisms" that climax into nucleated-fuel Energy Conservation.

Lighter elements are "driven to fuse" while the heavier elements are driven to fusion-fission cycling patterns for Neutron(s) release and Positron(s) formation. Elemental Protons and Neutrons are engaged in the formation of various recombinant atomic nuclei destined as plasma fuel, as particulate charges transubstantiate into Motion-force Plasma-displacements that activate convection Energetics.

"The strong force" between Protons and Neutrons sustains Conservation Dynamics while "the weak force" allows for molecular transmutation, particulate transubstantiation, and radiation emission releases, as plasma condensate is cycling between atomic fuel and nucleated radiation for Energy production, accumulation, abundance, utilization, consumption, release, and cycles of Core replenishment.

Particulate charges transmute in omni-vectored and multi-directional polarized Force-patterns that cause molecular-Energies to "bind-and-release" for sustained nucleated chain-reaction Continuum, via a dynamic process, the sources of which, being anchored in cycling-Differentials and Force-Opposites, as "the wellsprings of Plasma-Condensate Molecular Change."

The quantitative dynamics of plasma binding Energies are first inwardly compressed towards the inner-Core, and then violently dilated away from the Core into the radiative zone. "Curvature-flows" oscillate between fusion and fission, contraction and expansion, compression and dilation, inner-Core stressing and convection-activated tensing, to pressurize plasma-condensate Motion-forces into convoluted charge-Opposites that "trigger" Energy cycling-Differentials.

Centripetal Motion-force Energetics, at first, "implodes" nuclear processes towards the inner-Core, to then be countered by centrifugal Motion-force Energetics that instantly "explodes/expels" radiating plasma condensate towards the Corona. And it is during these "quantum-transport moments" that radiation emissions are released into the heliosphere for ellipsoid plane Curvature Motion-force Energetics, due to "temperature-and-pressure Differentials" consequent to "Energy-levels jumps" from "higher-Orbital(s) to a lower-Orbital(s)." Ionized-particulate-radiation "absorbs" Energy when "excited" towards "higher-Orbitals" and "releases" Energy when "returning" to "lower-Orbitals."

Core-centric and Corona-vectored, magnetic Field-induced Curvature Forces engage in "counter-compressing radial patterns" that cycle in "quantum-Time entanglements" as atomic Energy is "transformed/transmuted" into gravitational Curvature radiation Energetics.

There has to be "Continuum" in "Radiation Energy Emissions," and hence, the Electromagnetic Spectrum," because "stochastic Inputs-patterning" flows into "condensate Streams" due to "Time-compression-entanglements" that "quicken" particulate interactions into "instantaneous creation" of plasma condensate molecules — There is "no-Time to cool-off" though there are "temperature-and-pressure-Differentials," "allowing" cycling-Motion-force Opposites to engineer "Orbital(s) jumps" conducive to radiation emissions release.

As plasma condensate molecules are propelled into the Convection Zone, magnetic field-induced convection dynamics impels them towards the Corona wherein interface polarized electro-motive magnetic vortices that re-direct them towards the Core — this "momentum Force-transport" from one level to another, from the Core to the Corona, is the equivalent of "cycles of heating-and-cooling," during which radiation Energy emissions are released.

Convection dynamics caused by the extremely potent "external magnetic-Field," engender counter-vectored polarized electro-motive Fields from the Sun's "nucleated electro-plasma-hydrolytic dynamo." Particulate Energy cycling from fused nuclei to condensate radiation are core-heated at first and then convection-cooled, their Energy-level "quantum leaps" being accountable for the release of the Electromagnetic Radiation Spectrum.

The Sun's inner-inferno also possesses Input-Process-Output operations that engineer micro-Magnetic Field-vortices," within its massive plasma-sphere condensate Entity, — nano-Magnetic-Field(s) "contractions" upon nano-Magnetic-field(s) contractions — as powerfully-exponentially-compressed by the Sun's "externally polarized Magnetic-Field."

The Sun "negotiates Entropy" in ways akin to "Star-type Space-bodies" that have great "kinetic electromagnetic radiating-Mass," the "particulate Kinematics" of which, "transubstantiating" into plasma-condensate "Hydrolytic Molecules" that inter-act, as energized by "pluri-potent" inner-Core-centric polarized Magnetic-fields — the plasma-condensate itself possesses "polarized magnetic-field packets," whose composite cumulative polarity "competes" with the Sun's "external magnetic-Field," hence, the magnetic polarity displacement every 11 years, and polarity change every 22 years.

The "fluidity of plasma condensate molecular Streams" magnify the Electromagnetic Properties of Nucleated Radiation Mass, the sum of which, engendering the Sun's extremely potent "external" bi-polar magnetic Field, which in turn "climaxes" or "culminates" into maximally-pressurized nucleated processes, with "Space-displacements" taking place in a "quantum-compressed Instantaneous-Time-Frame."

Solar winds, solar flares, plasma rains, and coronal mass ejections are produced during "negotiation(s)" of "Functional Operational Entropy" as the Sun initiates corrective measures against the onset of Terminal Entropy or "Star Death." The Sun must process fuel, radiate Energy, and emit the Electromagnetic Spectrum, along with Continuum Field-and-Gravity Induced Curvature Motion-Force Energetics.

The Sun's Energy-Cycles have to insure maintenance of fuel processing needs, radiating Energy requirements, and Continuum-flow-patterns of gravitational spectral emissions.

"Qualitative Conservation" is triggered thereby when the rate of Core-directed "inner-vectored implosive Force" is countered by an increase in the rate of the "Corona-directed outer-vectored explosive Force" — as the excess pressurized radio-active "excited-transubstantiated fuel Mass" cannot be "contained/confined," it escapes into outer-Space from the "Coronal circum-spheric region." In that manner, dynamic Sun-System equilibrium is sustained for Energy production and utilization, as solar nuclear fuel quantities, radiating photosphere needs, and gravitational emission requirements are "balanced" for "internal" and "external" Star-type Nuclear-Thermodynamic Energy-cycling Continuum.

Though on the earth the two known nuclear processes, fusion and fission, have been separated in the pursuit of nuclear research for the manufacture of bombs, in the Sun these two phenomena thrive within the confines of the bi-polar magnetic field whose exponentially maximized magnitude compresses nucleated chain-reactions in Curvature-convection flux. Hence, reasons why, in order to fuse Hydrogen in an Earth-bounded explosion, in the absence of such great Magnetic Field strength, a fissionable material (fission-bomb) is utilized due to the tremendous degrees in heat-temperature(s) and pressure-gradients required for nuclear fusion.

Continuum Space-Time Curvature is suffused with redundant Forms of the same Energy, redundant patterns of the same kind(s) of Motion, redundant Forms of the same type(s) of Force, and redundant Forms of the same type(s) of Matter.

Thus Matter-mass, cycles, as the Input-Process-Output Mechanism fulfills its measurable quantities of Energy-Motion-Force, framed in functional specialization. Iterative symmetry of Forms framed into complex cycling structures of Mass-energy-in-motion-Force operates in relative equivalence, as intermediated by "conversion factors" and/or alpha-numeric Constants that "renormalize" proportional Differentials.

Interdependent whole Energy systems in steady-state equilibrium functionally operate via distinctly structured, but variant Forms of the same Energy, and with variant Forms of the same kind(s) of Motion, in response to analogous quasi-similar structures of the same kind(s) of Force.

Energy is constituted of electromagnetic Mass; Motion is patterned after gravitational Curvature tensing-stressing Energetics; Force is in the Forms of Magnetic-field and Gravity-field equivalents that effect Motion.

Curvature force dynamics is Mass-energy dependent; Field force tensing is Mass-energy dependent; Gravity force-stressing is Mass-energy dependent; centri-vectored rotary Motion is Mass-energy dependent. Given that the Universe is only composed of Matter and Continuum Space-Time, then, all categories of measurable Entities or all types of quantifiable dimensions originate from the Electromagnetic Properties of Matter "transubstantiated" as Mass-in-Motion!

Curvature-Frames in Relativity-interactions experience Mass-energy gains that are then "re-translated" into an Energy-motion-force complex; a particle accelerated to a fraction of the Speed of Light emits cosmic radiation in "re-translation" of mass energy gains.

Earth "congealed" magnetic field Energy is dependent upon solar "excited" electromagnetic gravitational field Energy. As solar gravitational magnetic field is dependent upon a nucleated core-engendered plasma-generated electric dynamo, Earth magnetic field is dependent upon an electro-conductive hot iron molten magma core.

Solar power is "processed" by plants via the mechanism of "photosynthesis" and then converted into plant fiber, glucose, cellulose etc . . . Solar radiation is "processed" by the magneto-sphere and the ionosphere from which derive electro-conductive ionized processes for cloud formations and "binding charges" interacting to generate lightning that engenders rain, snow and atmospheric storm systems.

Biological Species are designed with innate "genetic development encoding(s)" that insure the procreative reproduction of organisms of the same kind. Biological organisms possess a corresponding physiological mechanism, metabolism, via which "solar and environmental Inputs" are "processed" for constructive life-affirming existence with prosperous activities-of-living — The Sun's rays may even prompt "manufacture" of vitamin D through Human skin exposure to solar radiation Energy.

Inner-spirit determines perception and understanding. Building a storm-worthy house requires designing a blue-print for a firm foundation. The spirit a person owns determines his thought-life.

Does the Universe function as believed by astro-evolutionists? "Steady-state programs" are different and distinct per each Frame-of-reference, but the uniform "Input-Process-Output" Mechanism operates in all Frames via Relativity-variables that interconnect them for overlapping cycling-Continuum, e.g., organic and inorganic, (Nitrogen-fixing soil bacteria); "excited" and "congealed," (Aurora Borealis; lightning-and-precipitation from solar radiation Energetics,) etc . . .

The "mc^2 exponential function to the second power" is a "fundamental Input" in determining specific Frame "excited kinematics" constituent characteristics embodied in Mass, Energy, and Motion, e.g., the Sun, as plasma-molecular condensate-flux momentum-Force cycles in "compressed Time-frames," akin to "excited Energy."

Because the Speed of Light is the limiting fixed value for all things in the Universe, only Mass and Energy can co-vary, as c^2 remains constant. The greater the Energy of a Space-body or Mass-in-Motion Frame, in relation to its Mass, the more "excited" it would be; and the smaller its energy in relation to its Mass, the more "congealed" it would be. The Sun has great Mass and great Energy displaying tremendous "gravitational-elastic-tensor Curvature stress-power" as Field-force and equivalents of Gravity-force "effect momentous torque" in Forms of planetary rotary Motions, such as Revolution and Rotation.

The Sun is a "fused explosion" as well as an "imploded fission" — counter-vectoring Forces are at play to engineer the greatest "pressurized nuclear-reaction chamber" in the Universe in the Form of an exponentially potent "external Magnetic Field."

Pro-centric and counter-centric force Energies "agitate" nucleated mass into violent currents of fused plasma quanta that then flow into streams of condensate fission. With tremendous momentum force, nucleated energy is expelled from the inner-Core to only be compressed back again by convection Energetics that resist dilation. As condensate molecules and plasma particulates "change energy levels," they emit/release radiation energy.

As the equivalent of "radiation pressure force," the propelled nucleated mass-in-motion engenders Field, Curvature and Gravity energies that climax into planetary centri-vectored Forms of rotary motion operating as "re-translated" quasi-analogous "mass Energy gains."

Cause-and-effect relationships that characterize the Earth as "congealed Energy" flow from the equation of Mass-Energy equivalence — the Earth does not project any radiation akin to Star emissions, but, rather, receives, absorbs, reflects, processes, and reacts to solar emitted electromagnetic gravitational Radiation, in various ways that identify specific patterns of fulfillment for the universal "Input-Process-Output Mechanism."

The Earth as a whole— atmosphere, hydrosphere and land mass— is in "suspended animation" as it travels through vacuum Space. There are no "buffers" or "barriers" to absorb "cosmic shocks" from solar gravitational disturbances and magnetic storms. The oceans absorb heat, light, shock, vibrations, equivalent G-force effects from land mass activity, as the electro-conductive core is "stimulated" by Earth magnetic field and Ionosphere that co-labor to engineer lightning and storm systems as indirectly induced via "magnetosphere super-plasticity activities" and ionosphere-dependent interactions.

The ability of a massive earthquake to cause a "tsunami" or "extreme high surf surge" upon the coastal areas is directly amenable to the fact of "land mass suspension" as "connected" to the hydrosphere — shock waves have no "other place" to go, nor is there any other component to impact, except the oceans. A motor vehicle has the land surface under its tires to absorb the friction of stopped traction when the brakes are applied, and human passengers know how to "negotiate" the various "functional Entropy changes," such as equivalent G-force motions that impinge upon their bodies as the driver operates the vehicle in accordance with the "rules of the road."

The Earth is, however, "suspended in Space;" it will always be in constant Motion — the Atmosphere will always produce winds and storm systems; the Oceans will always have waves and tidal activity, including raging tempests; the Landmass will always engage in earthquake-producing "plate tectonics processes" and core-erupting volcanic activity.

Conservation is paramount to natural processes and universal events — proton decay or "proton release" is a rare occurrence in the Universe. Even when atoms are nucleated, particles other than protons constitute the mainstay of emissive materials, the sum of which, embodying the variegated ways in which Entropy is made manifest.

In a nucleated Energy system: Gravity, Field, Curvature, Motion, and Mass interact to engender "interactive spheres of influence" or "intersecting theaters of manifestation(s)," that are in "Relativity-entanglements" for re-translating these gravimetric Field-interactions, as tensing-stressing pressure-Force Energetics that then reiterate these Forms within related-overlapping cycling-Frames, e.g., the Sun causes the Earth and other Planets to revolve and rotate.

When approaching the expression of Gravity from the standpoint of Relativity, the relationship is no longer a "direct-physical-solid interconnection" akin to the direct connection between an Electromagnetic Field and the rotary Motion Forms it causes. But rather, Gravity is expressed as an "indirect-Field-of-Force interconnection." No Gravity-Field can exist apart from an Electromagnetic Field Motion-Force System. According to Newtonian Mechanics, Gravity is expressed as "Mass attract Mass." However, to support this assertion, Newton posited three Laws of Motion, the complex interactive operations of which, not independent from Gravity-Forms of expression.

Centers-of-Mass-in-Motion, thus, get "transformed" into Centers-of-Field as the Input-Process-Output Mechanism re-translates the Electromagnetic Properties of Matter into its various re-convertible states — gravitational-and-magnetic-Field(s)-Energetics generates Curvature pressure-force dynamics, among which are, "Motion-force-binding-Energies" that cause Planets to not escape from the Solar System while yet remaining in rotary Forms of Motion-force Systems, i.e., Rotation and Revolution.

When there is "acceleration" upon an already-moving System of cycling-Mass at a rate of Force projected by a gravitational tensor-stress mechanism that is greater than the Planet's rate-of-Motion-Force for its "regular Motion-routines," — as in the solar System ellipsoid plane, — then, there is a "surge" of Motion-Force that is, hence, accountable for a deviation from the regular-routines, such as in the perihelion shift of planet Mercury, which is co-dependent upon equilibrium variables controlled by solar gravitational and planetary Centers-of-mass, Centers-of-gravity, and Centers-of-field.

$E = mc^2$, $m = E/c^2$, and $c^2 = E/m$. In the inverse relationships, what is the relative proportion of the Sun's energy to its mass — is it in "perfect equivalence?" In other words, is the Sun the most efficient Energy system ever created — its mass being the exact equivalent of its Energy, as controlled by its core-centric pluri-potent electro-motive polarized magnetic Fields?

Is released energy, "excess energy," or is the Sun engaged in "shedding its mass" due to Energy gains engendered by nuclear chain-reactions dynamics?

How does Conservation sustain the Sun's Field-force properties consonant with the ratio of its Energy to its Mass? What are its "proportional rates of Mass Energy conversion?" How does the Sun "negotiate" the unfolding of terminal Entropy?

What are "Differentials" between operational dynamic Energy consumption and its rates of emitted gravitational Energetics?

In what ways is emitted-projected-radiated Energy as rays, particles and waves, "differentiated" from inner-Core plasma-condensate constituents? Are these emissions also manifestations of inevitable Entropy processes: The Sun is "dying" even as it's radiating gravitational Energetics that sustains Earth life-support Systems?

The Earth inner-Core is relatively solid as the outer-Core possesses "elastic tensility" to "flow in rotary motion" or "move around" as the Earth is rotating and revolving, in accordance with the level of intensity that strokes-of-Lightning engender throughout the electro-conductive hot-iron molten Magma outer-Core, as necessary for generating Earth Magnetic Field.

Does the Sun also have a solid inner-Core quasi-iterative of Earth-core arrangements, or does it have a "plasma-fluid" inner-Core that reflect Differentials between "congealed electro-conductive Core-processes" and "excited electro-conductive Core fluid-Nucleated Plasma processes?

Temperatures in the Sun are too high for the Core to be "solid" as we understand "solid," such as Earth inner-Core, or such as a stone or rock on Earth surface. We expect greater pressures at the inner –Core of the Sun, however, in an albeit quasi-fluid state with great thickness-of=flow having immense capacities for "superelasticity" in response to the mega-tremendous strength of the Sun's "external Magnetic Field" that climaxes into "compressing" and "compacting" the solar nucleated infernal plasma-fluid Mass.

Given the immensity of the "pull-of-Gravity equivalents" towards the Sun's centers-of-Mass, Gravity, and Magnetic-Field, coalescing within its inner-Core, and given that Hydrogen is the lightest elemental nuclear fuel in its plasma condensate Mass, it is expected that the heavier particles remain within the Core as "recycled nuclear fuel"— Hydrogen cycles into Helium as heavier elements undergo fusion-and-fission from atomic Energy to radiation Energy, with protons always sustaining Energy Conservation as they engage in neutron capture.

Spectral radiations-releases come from "particulate shedding of Mass-energy gains" as cosmic radiation, due to "compressive momentum-acceleration" of nucleated chain-reactions.

Inner-solar Field-force(s) with "binding-Energy Properties," are "transubstantiated" into the equivalents of gravitationally-induced "torque-Force, tensor-Force, and stress-Force, that together, "work in compelling" solar nuclear operations into "confinement" within the "greatest stellar pressurized chamber-cooker."

The Sun, as the Star-type, "must shine all-the-time anyway." As it observes the Law of Transformation of Energy, it emits visible Light and infrared heat in the same manner that "regular Entropy mechanisms" common to Earth-bounded combustive processes engender these releases. But the release of X-rays, ultraviolet rays, and low frequency electro-motive radio waves, etc . . . , comes from accelerated particles "transformed into high-frequency-electricity" nearing Speed of Light boundaries from the Sun's nucleated plasma condensate dynamo. As complex nano-Masses-in-Motion, they must "shed their momentum-Mass energy-gains" as radiated cosmic Energy.

Solar Energy Plasma Nucleation, within the solar-dynamo inferno, is engaged in "thermonuclear cycling entanglements" that, in operations, "recapitulate" all the Forms in which Matter-mass-Energy-in-Motion can "ignite for burning," from the "combustive-transformations" to the "nucleated transusbstantiation(s)," the complex entity of which, emitting not only Visible Light and Infrared Heat, but also radioactive Energy units, such as X-rays, Alpha particles, Beta particles, and Gamma rays.

Both "the strong force" and "the weak force" simultaneously "trans-agitate" nuclear processes as they complement each other in cumulatively engendering Motion-Force Energetics constituting the "interactions-dynamic" giving rise to the "super-elastic revolutionary-Field-plane of influence." This revolutionary ellipsoid plane has for its Sources the complex Relativity-interactions geared for the Sun's dynamic System equilibrium, as characterized by, not only tensor-stress-torque-metrics akin to the Quantum Mechanical Atomic Frame, but also by the molecular-chain-reactions-dynamics that pertain to Solar System Frame of Star-type-nucleated-Force-Energetics — the fundamental substructure of Matter-mass entanglements from atoms to molecules to planets to Stars is "maintained" or "recapitulated" but in nucleated-molecular-chain-reactions-Forms.

Momentum field-force binding-properties tend to reside in Energy Conservation as engineered by "the strong force," whereas emitted radiation is an emergent property of "negotiated Entropy" from operational dynamics of "the weak force" — plasma-condensate materials must "cool off" via convection processes "to be re-cycled" by the Core as nuclear fuel, and by that, release Energy as they are internally "transported" from one level to another towards the coronal circum-periphery for ejection.

The heavy nucleated protons would tend to remain as plasma-fuel within the inner plasma-condensate solar-Core as they "gravitate" towards the Sun's center-of-Mass; whereas, it would be neutrons and electrons that are re-transmuted into their radioactive Forms, to be ejected as high-velocity, high-energy, propelled discharged-emissions; thus, only a very minute quantity of the nucleated Atomic Protons "would show decay," if any; and hence, the greater duration of the Sun's lifespan, in terms of billions-of-years, akin to a "Star-type Form of Universal Energy."

For, if the Sun emitted all the radiation it operationally produced, it would then quickly "die out" or "extinguish itself." Therefore, there must exist a "dynamic Differential" between operationally produced Energy, convection processed Energy, and emitted radiation Energy, hence, Input-Process-Output cause-and-effect mechanisms explaining why "the strong force prevails," why Conservation of Energy predominates, and why pro-centric or centripetal forces are controlling.

Are not tornado dynamics driven by temperature, pressure, and electro-motive-magnetic force Differentials, hence their tremendously destructive powers?

Wherever Differentials and Opposites exist, a relatively constant ultimate Standard must regulate intersecting relationships so as to climax into dynamic System equilibrium. Therein are revealed the reasons why the Speed of Light in a Vacuum is 186,000 miles per second or 3.0×10^8 meters per second — giving rise to prevalence of "the strong force," of centripetal forces and of Conservation of Energy, while the photon "appears" to be mass-less and charge-less as a "stream-of-waves" constituting Visible Light, which is inert, in terms of radio-activity and electro-motive charge, so that Human Life can prosper to its maximal empowering spiritual-scientific creativity — as Earth distance from the Sun is more than 93 million miles, and atmospheric layers restrict Earth surface radiation penetration to Visible Light, Infrared-heat, and ultraviolet rays. This initial spectral property being intrinsic to Star-energy steady-state systems, and residing in the ratio of the Sun's Energy to its Mass, is "confined" or "encased" or "contained" by its extremely powerful Magnetic-Field which derives from nucleated plasma-Energy chain-reactions that cannot surpass the Speed of Light limiting Standard, and must "shed their mass-Energy gains" when accelerated even at a fraction of that Standard-Speed.

What happens when a photon particle is accelerated? Things always operate within allowable ranges, and consequently, even radiation particles are bounded by the Speed of Light.

The Speed of Light in a vacuum is definitely related to "the kinematic velocities" of plasma-condensate particulates akin to solar-Core "molecular change dynamics," — while those same kinematic velocities are limited by the Speed of Light that they in turn produce.

Now then, is Star nucleated Energy the only means in Continuum Space-Time via which Visible Light can be generated? Or, are there other Energy processes that can yield Visible Light and other spectral radiations?

The Atom is the fundamental constituent of all matter-Mass. Phenomena, from the atom to the planet, and to the Star only reiterate atomic-Frame quantifiable properties-categories, in Forms configured by their functionally unique fulfillment of the Input-Process-Output Mechanism, the complex interplays of which, operating to differentiate and specialize them into either "congealed Energy" and/or "excited Matter."

There are other forms of the Star-type in the Universe but only as differentiated by their radiation emissions that qualify them as "quasars," and "pulsars," etc . . . The Speed of Light is the limiting Standard to which all natural processes and universal phenomena must defer. For, if all of the Sun's mass were converted into emitted energy, not just "Functional Operational Entropy," but also "Terminal Entropy" would be prematurely accelerated.

Thus, possibilities are not endless as to the way in which physical Laws can be conceptualized, mathematically framed, and structurally operationalized for reproducible, iterative, or semi-similar patterns of organizational applicability.

On Earth, duplication of nucleated processes occurs in bomb-making activities, or in activities geared towards building and operating electricity-generating nuclear power plants. Only a small quantity of the total Energy-to-Mass ratio in those systems, respectively, is converted into Motion, infrared, photons, cosmic rays, geared for constructive, utilizable Forms. After all, what is a bomb used for? To destroy "the enemy" and/or his "facilities," etc . . . And how is radiation exploited in an electricity-generating nuclear power plant? Rods of enriched Plutonium emit cosmic rays, the released heat energy of which, is utilized to make hot, fast-moving steam that then powers turbines and generators. In nuclear power plant operations, "radioactive enrichment" degrades and rods must be replaced after, from between 15 to 20 years of operation, after which they must be safely and securely "disposed of."

Both nuclear bomb research methods and nuclear power plant operations are lethal to Human life as they are very inefficient in generating the Forms of utilizable Energy that engineers seek. However, nuclear weapons are commonly regarded by governments as a powerful deterrent against invasion by other nations whose military forces would be contemplating totalitarian fascism.

Condensate particulates approaching the Speed of Light must "shed momentum mass Energy gains," as dispersed cosmic radiation. Projected "rotary-Motion-Force properties" of "Field-force binding-Energy," operate, as "the equivalents" of gravitationally-induced Field-Curvature Mass-in-Motion Energetics.

Radiation emissions that accompany the "rotary-motion-Force properties" of Curvature-Field binding-Energetics also represent, in the aftermath, the "Entropy equivalents" of "solar shed mass," that find their analogs, in turn, as planetary Revolution and Rotation, along the "super-elastic Frame of tensor-stress Energetics," as exerted by the Solar System ellipsoid plane.

Some other Space body, or outer-Space itself, would be "the best place" to build a fusion nuclear power plant as great distance from the Earth would protect Human life. In the case of the nuclear bomb, the encasement for the bomb, catalytic materials, the Earth atmosphere, land or oceans, all constitute "barriers" to Energy yields, a scenario which does not exist for the Sun.

The Sun operates within the external dimensions of void, vacuum, gas-less Space. By comparison, the Sun is very efficient in its purposeful operations — for planetary and solar system functions, such as Earth revolution, rotation, atmosphere, hydrosphere, biosphere, "phyto-sphere," and geological activities, etc . . . Given the Law of Entropy, ultimately, the Sun "will run out of nuclear fuel." For, as molecular condensate particles continue to "shed their mass Energy gains:" Every ray, every wave, and every particle of emitted solar radiation, testifies to the Sun's "laborious journey" towards "Terminal Entropy."

When two Objects are in Relativity-Force interactions, "every action" will trigger "a reaction" that is co-equal, co-linear, but opposite in vector!

Thus, Relativity-mechanisms, that make the Earth a "congealed whole-Energy system" where Speed of Light limitations are processed to restrict/hamper/confine/thwart/resist certain States into which Energy can be transformed, — among which is "a state of Earth-resistance" such as, to Energy Transformation into "the nuclear-plasma-State," — ought to be deciphered from the standpoint of matter-Mass-Energy Equivalencies, as framed within operations of the Law of Transformation of Energy, "as informed by "electro-magnetic Field Energetics, Curvature pressure-Force Mechanics, Dynamics of Opposite-Gravity-Force Equivalents, and Mass-in-Motion cycling-Differentials.

In consideration of all the overlapping gravitational Frames that unite Continuum Space Time via "Curvature pressure-force Motion Energetics," each Frame presents thermodynamic cycling Differentials and Force-Opposites, within structure-specific, "Gravity equivalents," that are amenable to analogous patterns-in-transmuting Electromagnetism into its "congealed/ground-State," such as liquid-Water; solid-Aluminum; gas-Nitrogen; or in its "excited Forms," such as:

(a) Combustive: such as petroleum-fueled automobiles; diesel engines/air-compression/gas-ignition engines; fire-wood-burning; natural-gas-stove ignition; coal-burning stoves or furnaces; locomotive coal-fired steam engines; an explosive grenade; stick-of-dynamite; "conventional" bombs;

(b) Electro-Energized: such as incandescent Light-bulbs and fluorescent Light-bulbs, Electrolysis, Hydrolysis, Photo-electric Cells; spark-plugs;

(c) Nucleated-plasma condensate chain-reactions: such as the Sun's nuclear dynamo; a Hydrogen Fusion Bomb; an Atomic-fission Bomb;

(d) Radiation-induced Steam-making: as powered by radio-emissive/Radiation-Emission Energies, such as in electricity-generating operations of a nuclear power plant from radioactive rods of "enriched Uranium," or Plutonium.

Whether due to an Electromagnetic-Field or to a Gravity-force Equivalent, all Frames answer to Curvature-motion-Force parameters "triggering" pro-centric, rotary Forms of motion-force-patterns, lest there be "an external Force" that modifies the Object's course-vector for fostering Force-opposites that are counter-centric to centers-of-Mass, e.g., a jet-propelled aircraft.

Consequently, from Quantum Mechanics to Solar-planetary Dynamics: Gravity and Electromagnetism have "Relativity-interconnections" that "lie on a Continuum," the complex operations of which, being accountable for producing not only the centripetal rotary Forms of Motion witnessed in all cosmic Mass-in-Motion spheres, but also centrifugal Forms of Motion-force akin to the "Gravity-force equivalents" preventing a Hydrogen-Atom's Electron from "crashing" or "merging" into its Neutron-free Nucleus.

In addition, due to these "Relativity-interconnections" between Electromagnetism and Gravity, Human Beings are blessed with capabilities to engage in a rich, diverse, flexible, versatile, variegated repertoire of Motion-force patterns, the complex combinations of which, allowing a gymnast to poll-vault, or a rocket ship aircraft to lift-off Planet Earth in a trajectory or course pre-determined by its Human pilot.

A comprehensive conceptual theoretical framework seeking a mathematical equation resolution for the "Unified Field-Theory of Continuum Curvature Pressure-Force Motion" would have to utilize alpha-numeric Constants that "re-normalize" Relativity-relationships between all cycling mass-Frames, i.e., the Atom; the Planet; the Star; and the Solar System's Ensemble of Compositely-Combined Motion-force Energetics.

Frame-specific expressions of Electromagnetism and Gravity, as: Magnetic-Field, "the strong force" and "the weak force" constituting phenomena that complement "Gravity-force equivalents" — as induced by operations of the Gravity-field, which effect Processes, Events, Forces, and Motions prevailing within each Frame-of-reference, — will have to be "reconciled" for synergistic, holistic, interdependent, symbiotic, overlapping/intersecting, Continuum-thermodynamic-cycling Unification.

All fundamental Forces, composed-induced primarily by Electromagnetism and Gravity, project "a sphere of influence" that consists of a "Field-theater of interactions" that effect specific Forms of Mass-Energy-Force in cycling-Motion "situated" within Continuum Space-Time Curvature: (1) "The strong force" (proton-neutron); (2) "The weak force" (electron-Motion in molecular change, atomic/particulate decay, radiative emissions); (3) Electromagnetism (Magnetic Field rotary-Forces); and (4) Gravity-Field-Force Motion-Forms (rotary, curvilinear, rectilinear, linear, and composites/combinations thereof).

It is Motion-force that is uniformly consistent within Continuum Space-Time Curvature. No phenomenon, process, event, principle, Law, transformation, or mechanism takes-place or operates without Motion-Force being applied to its "parent-reference-Frame."

And it is due to "Curvature pressure-force-tensor-stress super-elasticity Energetics" (pressure-force-tensor-stress-and-torque), as encapsulated within Magnetic-field Motion-force and Gravity-field Motion-force "equivalents" that frame-control Time-elapsed cycling processes of thermodynamic Energy, transformations of which, being effected for "Ellipsoid Plane Equilibrium."

****

DISTINCTIONS BETWEEN FUNCTIONAL OPERATIONAL ENTROPY AND TERMINAL ENTROPY!

Energy Conservation —("Energy is never created nor destroyed but always transformed") — and Entropy Processes — (Energy utilization for productive work climaxes into unusable/unproductive Forms of energy-Frames) — characterize all Mass-in-Motion Frames that are engaged in fulfilling the purposes of the Input-Process-Output Mechanism, as they are transforming Energy in cycling patterns that qualify their uniquely structured, specialized operational functions.

All Energy Transformation Processes involve cycling quantities in Continuum Space-Time geared to sustain maximal-Frame-abundance for resource(s) production, utilization, self-maintenance, and consumption. Frame functional operations often require "corrective measures" that compensate for on-going operational Entropy dynamics that compromise Qualitative-Efficient Frame-Maintenance Conservation (efficient functioning in correct quantity proportions, amounts, ratios, quotients, gradients, and equivalencies).

Phenomena that cycle between Qualitative Conservation and Terminal Entropy (ultimate resource(s) exhaustion and functional death) are engaged in Functional Operational Entropy — regular operations for fulfillment of purposive functions that involve "constant corrections" that thwart eventual decay and terminal death, hence, displaying a "cybernetic characteristic" that thrives on "feedback Field-Energetics Systems" rooted in "gravimetric information processing."

The Universe, the Earth, and machines we invent, all display patterns of Energy-cycling that have a beginning, a periodic duration, and an end. Analogously, Human life begins at biological conception, goes through growth, development, and maturity, and ends at natural death.

"Seasonal cycles" are framed such that, each Season from Spring to Summer, from Autumn to Winter, and from Winter to Spring comes and goes; but individual Human Beings only have one-and-only terrestrial biological lifespan, which, eventually succumbs to Terminal Entropy or biological death. Why? Death contradicts Eternity, but not "thermodynamic or temporal Continuum."

How is Human Living on Planet Earth affected by Biological Thermodynamics as their physiological limitations "interface with" the apparent permanency of the Universe in comparison with the relatively short Human lifespan?

"Thermodynamic or Temporal Continuum" has to do with the Law of Transformation of Energy and the Law of Entropy: We cannot create the electrolytic-colloidal organic life-support Systems with which the Human Body has originally been endowed by our Creator from our very biological conception from the Male's sperm and the Female's ovum, which come from both our parents, respectively!

At the same time, because our lifespan or our limited living-duration in Thermodynamic Time is also "embedded" within Eternity, then, Biological Thermodynamics or Temporal Physiological Continuum must also reckon with the eventuality of Terminal Entropy, which climaxes into biological death!

Cause-and-effect reasons for our mortality are spiritual-moral, on the one hand, and on the other, biologically-scientific: They have to do with logical Input-Process-Output Mechanisms animating our Scientific Organic Physiological Constitution and its Electrolytic Modus Operandi.

Human Nature is activated by "spiritual-scientific principles" of living that embody both heavenly and earthly fulfillment: Sinfulness is in our biological Flesh-nature while Entropy is in our mortal physiological Body! Morally, we have a sinful spiritual constitution! Physically, we have Entropy in our Flesh or ultimate death in our Biology!

We are Spirit-Beings who live in "a thermodynamic world." Sin entered Human Nature, as Entropy entered the physical world, due to the wicked works engendered by Satan's evil spirit in the souls of our Forebears — Adam and Eve, our biological ancestors, succumbed to Satan's deception, by their disobedience of God's commandment not to eat of "the tree of the knowledge of good and evil." (Genesis 3:1-24, King James Version, Holy Bible).

We are mortal in essential or natural constitution. But we are given the blessed opportunity to live eternally beyond our mortal terrestrial biological life. However, Eternal Life is not possible within a Thermodynamic Referential-Frame! Eternal Life begins here, on the Earth, but only by "passing away" as natural death comes in order to begin anew in Heaven.

Other attempts at arriving at a conceptualization of how Human Beings could obtain fulfillment of their desires to "live forever" have driven many men to entertain the possibility of "re-incarnation." But no Human Being can experience "re-incarnation" via biological reproduction from other people or from "future parents."

We are born biologically, only once! Thus, "re-incarnation" is a debilitating myth that subverts the spiritual inspiration motivating Human Beings to treat each other with justice and magnanimity. The concept of "re-incarnation" also defuses the moral incentives that "move the Human heart" to fulfill inner-drives that impel Humankind to improve "the condition of the world" in which they live!

Time is forward-vectored, and personal Human life cannot repeat itself exactly and similarly as previously lived. Not only do personalities differ, but also experiences and environments.

As individual Persons, we have a specific life-span. The specific individual Person conceived by two parents, a Man and a Woman, respectively, is a unique "Spirit-Being" born only once to parents who will have had died and whose sperm and egg, respectively, cannot "resurface" in other Human Beings at a later time.

All Forms or categories of things that physically exist must partake of Thermodynamics. Human Beings are subject to Biological Thermodynamics. Earth seasonal duration is approximately three months as the Earth gravitationally cycles in Continuum Space-Time Curvature. In the same manner that there is duality of Form as Energy is made manifest as particle-wave, "Biological Thermodynamics" as configured by the Speed of Light, Electromagnetism, and Gravity, imposes certain physiological boundaries upon the extent to which individual Human life can endure, persist, or perpetuate. Individual personal Life does come to an end which we call "biological death."

As a "quantum-moment event," individual life complements the generational continuity of the Family of Humankind, which corresponds to a "continuum stream" or "continuum wave," akin to points constituting a line, or units of Energy constituting the Electromagnetic Spectrum. But the Family of Humankind has generational Continuum as "a life-stream touching the shores of Eternity — of Infinite Space and Eternal Time."

Individual mortal Human Beings are born at different times so that generations overlap for Continuum. However, individuals "pass away" at different times, due to many Entropy-factors that impinge upon the continuation of "thermodynamic-terrestrial Life." But the Human Family as a Whole Species enjoys living Continuum through generational procreation.

Seasons stream for sustaining Earth ecological Continuum, while incorporating dynamic changes that also uphold requirements of Earth life-support systems. Even seasonal cycles and planting-harvesting cycles appear to come to an end as configured by gravimetric variables that affect ecological processes, yet without an interruption in Earth motion patterns.

Nature is imbued with self-iterative cycling characteristics that seem to defy Terminal Entropy, because Energy Conservation has replenishment prerogatives that temporarily prevail over Whole-System resources exhaustion.

The Laws of Thermodynamics comprise two extremely intermeshed-intermingling properties of Law: Conservation and Entropy whose "entanglements" involve temporal Energy cycling — measurable as rates-of-change within a certain period of Time. Mass-in-Motion Frames are engaged in Qualitative Conservation, even as their cycling processes are combating and thwarting "Terminal Entropy."

This "State of Temporal Continuum" — such that Human Beings enjoy a relatively "long lifespan" in comparison with other living creatures on Planet Earth — constitutes the structural framework within which Biological Thermodynamic Processes are "embedded," and from which Functional Operational Entropy is launched — Energy must also be consumed by operations of Cycling-processes, even as Energy is being produced for cumulative conservation of constructive Frame-functions.

Functional Operational Entropy amounts to what a "whole-Energy system must do" in order to sustain Continuum Curvature Cycling Energetics while it produces, accumulates, utilizes, and consumes Energy for Frame-specific operations, as it is also accounting for "un-recyclable waste" or "unusable Energy that can no longer produce work."

"Corrective measures" must be taken in accordance with requirements of "structure-bounded Qualitative Conservation," the comprehensive operations of which, thwarting the acceleration of the advent of Terminal Entropy or "eventual death."

There are micro-processes and macro-processes in biological organisms, in Nature and in the Universe that are real occurrences, but which are not yet amenable to mathematical formulation.

How would the operations of the Input-Process-Output Mechanism be mathematically formulated? For example, a biological organism has an intrinsic, "genetic code" akin to its own kind: life is procreated from life of the same kind — indicating the First Law of Bio-Genesis. In addition, a biological organism is imprinted with "an immuno-hormonal metabolic-electrolytic chemo-motor mechanism" that allows for "processing" Solar Energy for the sustenance of life, e.g., respiration; digestion; conversion of proteins to Glucose, in the same manner that plants "process Solar Energy" during photosynthesis.

Likewise, in Nature, we have discovered that Electromagnetism, or The Electro-Magnetic Spectrum, amongst which, Light, "carries information" or can be "encoded with information" to be "transported," either via metallic wires or via the air waves. From such discoveries, we created technologies ranging from the telegraph, the phonograph, the telephone, radio, television, the computer, the cell phone, the iPad, the iPod, the tablet, to artificial telecommunications satellites, super-computers, and 3-D printing technologies, etc . . .

As an electric field will generate a magnetic field and a magnetic field will generate an electric field, the two fields are combined for articulating operations of the television antenna. Though "how these two fields work is known," or their cause-and-effect lines-of-interactions can be mathematically "connected" for practical applications, there are no mathematical equations for formulating the essential-fundamental reasons why an Object that falls within a magnetic field must start revolving, e.g., a Planet revolves around a Star; the blade-stem assembly of an electric fan revolves as it is placed within a magnetic field generated by a copper coil. But these fundamental phenomena must take place as they do. And so, Scientists only accept what is believed to be understood about them for fostering practical technological applications.

"Science" has no true explanation as to why these things must be, but does provide an understanding of their working operations, such as cause-and-effect connections that result in climaxing into specifically observed experimental data.

Due to Electromagnetism and Gravity, Motion is fundamental to all cycling-Mass-Frames, processes, phenomena, and events that engage in Energy Transformation. And biologically, even the male sperm cell has motility, while at the same time, the female egg is gently "moved along" the fallopian tubes via its muscular contractions, in order that the male sperm cell might reach it for fertilized conception. Likewise, a plant will seek light radiation energy as it engages in photosynthesis.

There are certain statements that, on their face, might appear to be "true" but their veracity must be anchored within the contextual Frame of their physical-material applications. Contextual Frame defines conditional parameters for cause-and-effect mechanisms relative to the frame-of-reference of the "observer," which when explained establishes validity for all Frames where conditions are similar.

The sense in which these statements are "true," then becomes generalized for equivalent Frame conditions. For example, let us consider the statement: "Human Beings are born to die." What exactly does this mean? Does it mean that Human Beings are born to simply die two seconds later? No; we have a life-span; some human beings live to be over a hundred years old.

But it means that, given the natural biological constitution of Human nature, Human Beings are mortal creatures who will eventually grow old and then die, (barring fatal diseases or accidents). Yes; Our Creator made us mortal in natural constitution; ultimate mortality is in our genes. And hence, the real-true sense of the statement: "Human Beings are born to die."

This statement presents the most mysterious and the most paradoxically perplexing generality ever conceived by the Human mind. A human baby is conceived in his or her mother's womb, after which an infant is born. The child can be said, for example, to be "an hour old," "a day old," a "week old," a month old," or "a year old," as the case happens to be. One thing we do know, from human experience, study, and revelation that within his genes are contained growth and limitation, gradual development and delayed expression, and ultimately, life and death.

Quantitative development simultaneously accompanies qualitative growth. The person is alive, developing and growing, learning and maturing, while at the same time, within "his biological DNA-core," resides a predisposition for "Terminal Entropy" or biological death. Yet, "Terminal Entropy," though potentially present, appears to be "arrested," while living growth and maturing development are occurring. At a specific point in time or "age," gradual growth reaches its pre-programmed genetic limits and stops, such that "the progress of Conservation" is equal to "the progress of Entropy" — and then, "a new kind of maturing" begins — it is called "aging." It culminates into "seniority" or "old age" whereby Entropy processes predominate those of Conservation.

Our God-endowed "will to live," as "inner-spiritual impetus" must have been "triggered" meanwhile, to effect such a wondrous miracle. Terminal Entropy is arrested as life thrives towards ultimate maturity.

This lifespan dynamic has engendered a constructive process in the Form of a "corrective mode of replenishment" called "Functional Operational Entropy," — thirst begets drinking water, hunger begets eating food, fatigue begets resting and sleeping, etc…etc… And these "variant corrective changes" continue the whole span of our lives in order that our organism reaches dynamic operational System equilibrium as it fulfills "the laws of biological thermodynamics."

Functional Operational Entropy is observed, for example, in our sleep cycles, and nourishment cycles, etc, . . . For the growing infant, "aging" is both in terms of "gaining years" quantitatively, and growing in size, developing mentation, perfecting immuno-metabolism, which is "Qualitative Conservation," as opposed to "Terminal Entropy." Human beings define these periods of life as infancy, childhood, youth and adulthood. The observable cycles as noted above, are the ways in which "Qualitative Conservation" is sustained, by engaging the beginnings of "Functional Operational Entropy," as the human body "negotiates" the "tug of war" of "Terminal Entropy." As conservation overcomes entropy, the child develops, grows and matures into adulthood.

In the material world, "recycling" has been introduced as the "third law" of Thermodynamics. Among the things that are recycled are, paper, glass, plastic and metal. This is practical "mechanism" via which attempts to sustain "Functional Operational Entropy" effect replenishment of resources in a way that mitigates the costs of "unearthing" raw materials that would have to be supplied in order to bring about totally new products.

Likewise, valid scientific laws will apply within all Frames where gravimetric conditions are similar. They will complement each other and overlap in applicability, akin to the way in which the Relativity reference-Frame complements and overlaps the Newtonian reference-Frame in explaining Energy cycling within boundaries established by Gravity, Field, and Curvature Energetics.

"Waves of probability" at the atomic frame, when considered in light of Functional Operational Entropy, reveal that particles are engaging in Qualitative Conservation of angular momentum as "external energy" is injected into the cycling constellation predetermined by micro-Curvature dynamics engendered by centers-of-mass, centers-of-gravity and centers-of-field, via experimental methods and mechanisms of measurement that cause them to engage in "shedding their mass energy gains" as cosmic radiation.

The Laws of Thermodynamics — Conservation of Energy and Entropy — operate both at the same time, in systems that are either "closed" but not isolated, or "open" but self-contained.

In the experimental laboratory environment, the experimenter determines the degree to which a system is "closed," for he or she is the one adding or subtracting factors and variables. However, in a non-artificial environment, like the Solar System, or Earth ecological system, the degree to which these systems are "closed" or "open" is already predetermined by the universal "programming code," "instruction," "algorithm," or "ordering mechanism," that structures inherent properties of functional organization and operation while "processing" the Law of Energy Transformation via the Input-Process-Output Mechanism. Earth ecological processes unfold in accordance with the Organizing Principle operating in Nature since the time of Creation as attuned to the Planets components, constituents, and their fundamental interactions, e.g., Atmosphere, Hydro-Sphere and Land-mass, in constellation with solar gravitational radiation Energy, revolutionary and rotational Curvature Motion-Force(s), electromagnetic field dynamics, and ellipsoid plane tensing-and-stressing-pressure Energetics.

No natural or universal System is "totally closed" or "totally open." When a System is considered as a whole, and external variables or factors are taken as "Constants," then, it can be said that the System is a "closed system." Still, this can occur only for a specific period of time, and only for a circumscribed set of purposes. For example, for purposes of study, many areas or branches of knowledge regarding the Human body, the Earth or the Solar System, can be compartmentalized or isolated under contrived or controlled conditions that address a specific question requiring a respective solution. Such cases are: the treatment of a disease, the replacement of a part in an automobile, or the fertilization of a plant. Calculating the average luminosity quotient of the Sun in light of its supposed electromagnetic Mass from an Earth-bounded experimental station, would present "degrees of openness" as adjusted per planetary radiation absorption and processing rates. In each case, many variables have to be factored while others are excluded depending on their applicability, but for a specific period of time during which the quest for a respective solution is finalized.

In short, the Earth is "a self-contained System" that can be considered either "closed" or "open" in accordance with the experimental Frame within which variables, parameters, and co-determinants are engaged in interactions, and then, their cause-and-effect mechanisms measured and analyzed for pertinent results and respective conclusions.

For example, in the case of an automobile, the replacement of one part in order for the vehicle to resume proper functioning is predicated or contingent upon the optimal operation of all other parts. One cannot replace only the radiator if the battery has no charge whatsoever and yet one expects the engine to run. Concerning a new car, for a short period of time, all the driver has to do is "fill up the tank," so to speak, that is, all other things being held "constant," e.g., no flat tires from an incidental nail on the road, no rocks flying into a head light from a semi-truck passing-by, etc, . . . Thus, Entropy operates in all steady-state Systems in ways that present stable functional operation, given certain specific conditions that sustain "Qualitative Conservation." It is "normal" to expect that with that new vehicle, "Terminal Entropy" is "a long way off," whereas "Functional Operational Entropy" must be continuously "negotiated" with pertinent changes that have to be made within the allowable System-specific ranges, i.e., oil change, tune-up, etc . . .

There are also System-specific boundaries that cannot be breached, just as the Speed of Light is the limiting standard that cannot be surpassed. For example, a vehicle is not a plane; though the vehicle possesses the quality of Motion, it cannot fly, but is operational only on "land surfaces;" another is that its tank of gasoline is limited to a certain use-duration or mileage — it cannot "drive forever." In this physical-material universe, the Laws of Thermodynamics will not allow for a so-called "perpetual operating machine" or "eternal working apparatus."

On the one hand, conditional parameters, relativity factors and co-determinant variables must be observed within their allowable specific Steady-state ranges; and on the other, there must be resource replenishment via Energy cycling for Continuum working-operation.

System-defined operational efficiency within allowable ranges-of-productivity affords dynamic change-repertoires that hold great social utility. Assuming controlling Force-opposites to be relatively "constant," e.g., routine explosive Forces within a combustion engine, all Relativity factors, co-determinant variables, and conditional parameters, are in Continuum System relationships of "productive equivalency" as mathematically derived from direct or inverse proportions that are determined by cycling Differentials geared for establishing dynamic System equilibrium.

But even totally new vehicles break down, as in the case of a manufacturing defect unknown in a part with "Delayed Terminal Entropy." "Delayed Terminal Entropy" is different from "Functional Operational Entropy," in the sense that "Delayed Terminal Entropy" cannot be "negotiated" "with a "temporary fulfillment," e.g., replacing one part, due to many other parts having been affected by the defective part when it completely failed. For example, a small fissure in a spark plug wire can cause discharges or "arc-ing," during sudden engine acceleration, which could then engender an electrical fire or discharge the battery due to loss of vectored Energy.

These disastrous changes could not have been caused by "aging" or "continual developmental use" or "wear and tear." If the engine is not totally ruined by the ensuing fire, then, one would think that only a new spark plug wire would be necessary. However, one part with "Delayed Terminal Entropy" could also cause a System to undergo changes so destructive as to precipitate the onset of "Terminal Entropy" for the system as a whole.

In the case of the vehicle, if the engine is totally ruined by the electrical fire, a new engine would be needed or the vehicle could be declared "totaled," if other parts or areas of the vehicle could be too damaged for mere repair — if the hood, passenger compartment, etc . . . had suffered heavy smoke, fume, and flame damage.

A state of arthritis can cause "impaired systemic Qualitative Conservation," hence causing the person to experience long periods of sustained Functional Operational Entropy with appropriate medication that temporarily restores periodic dynamic equilibrium. Another example of "Delayed Terminal Entropy" is contraction of the HIV virus which has not yet developed into full-blown AIDS disease. Both can cause death of the Human Person, for the virus is so systemically destructive and damaging, that other body organs and processes are irreversibly damaged. However, with appropriate drugs, patients could "continue to survive" while undergoing severely painful attempts at "triggering" the application of "Functional Operational Entropy." But the disease is fatal; there is no final cure and biological death or "Terminal Entropy" results.

Biological Systems have much more complexity than mechanical or electronic Systems due to Creator-endowed design, structured processes, and orderly organization. And the application of the Laws of Thermodynamics may present moral dilemmas which do not occur in physical and electrical applications.

In addition, Human Beings are spiritual-moral beings with psycho-motor affective sensibilities that have great impacts upon their biological living systems for thriving purposes — "the will to live" for which there is no cause-and-effect "Science logic."

"The will to live," is, "spiritual programming" built-into DNA. "The will to live" accompanies the "Human genetic code" for controlling species-specific functional development, instinctual adaptation, and organic growth designed for beneficent purposes under-girding Human Civilization. However, because Human Beings are endowed with "spiritual gifts" that transcend their biological constitution, abstract Human capacities such as spiritual discernment, creative inventiveness, wisdom, judgment, and moral righteousness have no limiting physical boundaries.

"Nature's clock" for on-going Continuum Energy transformation is pre-set since the time of Creation. And built-in corrective mechanisms sustain Qualitative Conservation via Functional Operational Entropy processes geared for constructive resource-replenishment and molecular change, from day to day, season to season, and year to year.

As we interface with the Universe for creative knowledge, God's Word empowers our inner-being with optimum "thought-life," "gifted speech," and moral maturity. Constructive beneficent living and comprehensive soul-prosperity must sustain moral fortitude in facing life's difficulties due operational workings of Sinfulness and physical Entropy. (Romans 8:1-11, 26-28, NASB, Holy Bible).

Particulate behavior at the quantum mechanical Frame embodies System boundaries whose parameters are already pre-determined by structure-specific conditions of Energy-motion-force cycling, due to micro-Mass-configured dynamics engendered by interactive Continuum parameters of Force, Curvature, gravity, field and charge displacement Energetics. Experimental technologies, as well as instrument-dependent or device-based Systems of measurement, impact determination of results as particles undergo change, every time external energy involves a disruption of Relativity variables that climax into dynamic System equilibrium.

Therefore, it can be deduced that no System is "completely closed" or "completely open." It is crucial to determine the extent and degree to which a System is "closed" or "open" for purposes of analytical research in order that co-dependent variables and co-determinant parameters are accorded their right proportions and ratios for relative equivalency conversion. The same analysis can be applied to the Human Organism, the Atomic Frame, the Earth-Moon Complex, and the Solar System, respectively.

"Closing a system totally" is useful when compartmentalization results in effecting provisions for problem resolution — f or example, a Person with a "new disease" or a known contagious disease can be "quarantined" or "isolated" until a cure is discovered, or until the Person has healed. But in certain specific cases, especially in the discovery of processes or events whose formalization would aspire to possess "totalizing characteristics," ignoring external variables or factors thought to be "extraneous" might be detrimental to end-results. Certain variables held "outside" of the paradigm, might have had import for conceptualization of related phenomena not under consideration.

In Physics and many other branches of knowledge, cause-and-effect mechanisms inter-link variables, parameters, and factors for determining hypothesis validity towards experimental and operational applicability. And oftentimes, this requires experimentation and testing, which involve logical understanding, scientific analysis, and "objective detailing," as well as laboratory and/or field operations.

In studying physical events, phenomena, and processes while aiming to discover comprehensive conceptual schemes or mathematical short-cuts as "in toto equations" or "totalizing formulations," such as, $f = ma$, $F_g = G\ m_1m_2/r^2$, or $E = mc^2$, the paradigm of application ought to foster the utilization of analytical Systems that are "held open" rather than "closed."

However, "a closed System" does not imply "an isolated System;" rather, it evokes "a self-contained System."

Thus, the Earth-Moon Complex is "an open System" that is also "a self-contained System." Because the Universe in One, no physical event, process, or phenomenon can emerge from operations of "a closed System." For none such exists in the Universe as we know it!

If the Earth is considered a "closed System," solar radiation would have to be factored as a "non-contributing Constant." Yet this is unrealistic when it is solar activity that is the primary parameter determining Earth capacity for Continuum functional cycling as a life-sustaining Planet.

But when the Earth is properly held to function as an "open but self-contained system," then variables, factors, and parameters externally operating upon its Mass-in-Motion Momentum-Force Complex can be "all-Universe-inclusive" for validly framed experimentation, variables testing, scientific analysis, and reproducible applicability.

The Continuum Principle demands "system openness." Though the principle of parsimony is active in scientific research and experimentation, certain endeavors require the painstaking, time-consuming commitment to open the "Input-Process-Output Mechanism" to the broadest array of co-determinants — as they physically interact within the Frame of Continuum Space-Time Curvature Energetics.

In the case of attempts at deriving a Unified Field Theory of the Universe from available scientific data, categories of Force, such as: Gravity, Electromagnetism, "the strong force," and "the weak force," have been considered apart from Motion, as they remained too "compartmentalized," as if they are independently operating from each other; which does not correspond to the real framework of universal Systems operations.

Rather, the above-noted Forces should be subjected to "transactional analysis" from within a Scientific Paradigm that factors the heavy role played by Relativity Physics in unifying their overlapping and interdependent operations, effects, and influences.

Are not various complexities triggered when a particle is isolated for acceleration in a cyclotron while apparent interpreted results are then presumed to be applicable to universal phenomena-at-large? Though distinct from each other, Electrons or particulates in general exist in conjunction with a Center-of-Mass around which they revolve as "streams," "waves," "rays," "beams," or other "consolidated Forms," whereby they "cluster" in order to form Continuum field-of-operations. The only Element possessing "a lone Electron," is the lightest and Number One Element, which is, Hydrogen; but even that "lone Electron" has the Hydrogen Atom's positively charged Proton at its Center to revolve around!

Therefore, apparent observation(s) extracted from experimentations on artificially isolated particulate environments such as in "a cyclotron," will present great difficulties for synthesizing a General Theory of Unified Field for the Universe as a Whole. The Universe is an ordered Reality with a specific general Organizing Principle Algorithm that operates to yield intrinsically specific iterative cyclical results, events, processes, and phenomena in accordance with pertinent Laws of Physics that particularly operate, respectively, to produce them. And hence, why "fission" ("separating/isolating particles via explosive force;" or "fusion" ("forcing particles to merge/agglomerate/coalesce/trans-blend,") are tantamount to "denaturing" the Universe, or "denaturing the Earth," to be more specific, as these States (fission and fusion) are being "artificiated" within the geo-Sphere of Earth natural-intrinsic-essential "Life-sustaining co-determinant components," i.e., Atmosphere, Hydrosphere, Landmass and their Ecological Transactions and Interactions.

But all things in Nature and the Universe "operate within a range" that "takes effect" according to very highly synchronized Laws of Physics, but the parameters of which, once disrupted, will "transubstantiate" dynamic System equilibrium "to force it to yield' results that are inimical to the very existence of Human Life in the Universe.

Thus, fission and fusion are designed to occur only in Stars specifically possessing nucleated plasma-flux-processes requiring very highly synchronized details and specifications, as to temperature and pressure ratios, quantities, proportions, location, position, situation, and operational distance from other Objects-in-Space such as Planets.

Those very highly synchronized "inner-Star processes parameters" operate within "ranges-of-effectiveness" that cannot be "breached" without violating intrinsic Laws of Physics that particularly sustain Earth life-support Systems, i.e., The Earth cannot "contain a Star" within its constituent components and simultaneously-consistently retain its present thermodynamic Energy-Form that's already designed for sustaining the perpetuation of Bio-Organic Life! Hence, it amounts to the final realization that the Family of Humankind has launched "a martial directive" that will climax into "slowly committing suicide," e.g., The so-called "hole in the Ozone layer" is such a consequence!

Fission and fusion, — or "inner-Star Processes" — demand very intense heat-derived phenomena to activate, ignite, or conflagrate them, but yielding fatally destructive processes, such as, the Atomic Bomb and the Hydrogen Bomb whereby "specific distances" between particles and molecules are "purposely breached" for obtaining munitions-grade yields to be utilized in warfare; the sum of which, being tantamount to attempting to build a Star or construct a Sun on the Earth.

Likewise, constricting all physics research into a tunnel-vision circumscribed by the sole feverish pursuit of "proving the big bang theory" is deleterious to theoretical progress towards mathematically formalizing universal forces unification.

Fact is, the Earth does not exist by itself — though self-contained, the Earth is not "an isolated System." The Earth is part of a Solar System that comprises a specific number of Planets "operating as a Star-Planets Network System," the complex Motion-forces of which, being harmonized, synchronized, and fine-tuned with "Operational Symmetry," to work as System-components belonging to, or consisting of, a Star (the Sun) and nine Planets. The "Network of Planets" revolving around the Sun, along with their particular "operational Motion-Forms," respectively," constitutes the complex interplay of operations, phenomena, events, and processes that co-determine the details and specifications of Solar System Dynamic Equilibrium.

The Solar System itself is within the Milky Way Galaxy, beyond which, function other Galaxies having other Stars, and possibly, other Star-planet(s) Systems, if any.

It is unlikely that the Earth could retain its Moon in the absence of solar gravitational parameters and inter-planetary field Motion-force Energetics. For, without Star projection of gravitational radiation and Field Curvature Energetics as engendered by Electromagnetism and Gravity, the Earth would not possess a magnetic field, nor engage in heliocentric rotary patterns that form its Mass-in-Motion-force iterative repertoire, such as Rotation and Revolution.

In seeking to formulate the Unified Field Theory of Continuum Curvature Pressure Force Motion, it is necessary to conceive a cycling frame like an Atom or a Planet, as a mass-in-motion-energy-force complex within which the Laws of Thermodynamics find "transformational expression" for functional operations, via "range-specific changes" that fulfill requirements and properties of the Input-Process-Output Mechanism.

This conceptualized framework of scientific principles must be applied to cycling Frames held as "open systems," but also as "self-contained Systems," such that, the whole Universe is held as an "open System" constituted of components operating as "self-contained Systems" thriving in dynamic equilibrium.

The Universe is in constant Motion where Continuum steady-state whole energy system functioning is dependent upon the "relative totalizing interaction" of each and every Space-body within Continuum Space-Time Curvature. All these "Space-bodies" operationalize the Input-

Process-Output Mechanism in specialized terms that regulate analogously "re-translated" universal variables as properties of Matter within Frame-specific conditions on which, proportions, ratios, and equivalencies depend, for achieving dynamic System equilibrium.

These cycling Differentials and Force-Opposites must be "renormalized" or "reconciled" for Continuum, via "recombinant numerical operations" produced by specific alpha-numeric Constants that embody the "hidden variables" or "unknown co-determinants," operations of which, to "smooth out the waves of turbulence" that arise as those "Space-bodies" transact, within overlapping regions of Frames-intersection.

The Solar System receives inputs from the Milky Way Galaxy's other star systems and stellar bodies, even as each planet receives inputs from the Sun and from each other — note here Mercury's perihelion shift.

Space itself cannot "move," for it is "a container of Matter" that must be filled by Matter. Time itself cannot be "curved" for its interconnectedness with Space has to do with the physical dimensions of cycling mass in Continuum Space-Time Curvature. Motion is a property of Matter-Mass, and not of "void-emptiness."

What in Space could "move?" It is accelerated moving Matter under gravitational Curvature tensing-stressing-torque pressure-Force that travels through Space; without Matter-Mass to move within it, Space only has "inertial stillness."

Consequently, quantities active in a General Theory of Force must be expressed in proportional equivalencies represented by values that inhere in Frames of "mass-Energy-in-Motion," counter-polarized Forces and cycling Differentials of which, would need "reconciliation" via alpha-numeric Constants as "conversion factors."

The totality of Matter-Mass that is moving and traveling through Space accounts for the measurable physical dimensionality of the material Universe. Continuum Space-Time obtains physical dimensionality via Curvature Energetics that engenders centri-vectored Forms of Motion within the organized structures of Mass-Frames designed to fulfill the Input-Process-Output Principle via thermodynamic Transformation of Energy in order to attain dynamic System equilibrium.

All Mass-Frames are in Motion or belong to a greater-Mass-in-Motion Frame, even as they maintain their respective distances from each other. All Mass-Frames are moving or "traveling" within Continuum Space-Time, at the same initially prescribed rate, respectively assigned to them since the beginning of the Time of Creation. And hence, both classical interpretation and Relativity understanding tend to support the following observations:

(1) The appearance of quasi-similar iterative Forms of structurally synchronized symmetry;

(2) The concept of inertia, of inertial Mass, and of inertial Energy; and

(3) The factual occurrence of observed Continuum Motion is not necessarily a Form of evidence for the hypothesis of "universal expansion." How could "invisible, barrier-

free, emptiness" expand? Space cannot expand, except as distance between Objects therein, or as Volume contained within or framed by Objects therein.

If when Scientists say that "The Universe is expanding" it is meant that "The Space between Mass-in-Motion Frames" is "getting larger;" or that the distance between Space-Objects is "expanding," then, it has not been proven that the spatial distance between Mass-Frames is "increasing" so as to connote "expansion."

Space itself cannot "expand" as it is "empty-infinite-void" with no "solid barriers" to "stop," "frame," "contain," or "enclose it."

Were Space to have had physically manageable boundaries, then "moving" such boundaries would cause the Space contained or enclosed within those physical frontiers to "expand." But only some thing that possesses material-physical substance can expand or increase.

Therefore, given that "emptiness" cannot move, the presumed hypothetical theory of universal expansion is predicated upon the relative Mass-in-Motion-Curvature-Force of extant Frames that are already "traveling" in Continuum Space-Time.

Is distance between cycling Galaxies "increasing" so as to represent "expansion?" Is the distance between Planets in our Solar System "increasing" so as to be interpreted as "expansion?"

If a "solid" or other "physical barrier" is imagined, then, what would be "behind" or "beyond" that "barrier?" Mass-Bodies occupying Space, travel through Space; but that phenomenon cannot be counted as "evidence of Space expansion."

Mass-bodies are moving and traveling but not "expanding," for that would imply that "they're moving away" from some thing presumed to have had had a specific location in Space, or to have been situated at some specific place in Space. What would the Universe be "running away" from? What would the Universe be "expanding" away from? What would the Universe be "expanding" towards?

But Space is void-empty without Matter-Mass occupying it; and vacuum Space is boundlessly infinite with no physical contours to encompass it. Space has to be infinite, for it is a form of "non-Solid invisible-Reality." Space is a Form of Reality having "a complementary objectivity" that is conjoined only with Matter contained therein. Matter and Space "complement each other" to form the Universe as we know it to exist.

The Universe as we know it, exists in the Form of Space-Time with Matter therein! Space-less Matter, or Matter in the absence of Space, is however inconceivable! Space can exist without Matter therein; but Matter cannot exist without the presence of Space! Hence, the so-called "big bang singularity" could not have "created Space" by its existence or explosion!

We all know this, for, we too, have Space around our bodies, within our bodies, even as our own bodies occupy Space — but Earthly Space is "atmosphere-filled Space!"

There are no solids in vacuum Space, except as specific Mass-bodies like stars, planets, asteroids, comets, etc . . . that move and travel through Space, in accordance with pre-set, pre-determined patterns, as espoused by rotary Forms of Motion.

Even if distance between Planets, or between Planets and Stars is shown to be "increasing," this would not constitute "universal expansion," in the sense that, given that the Universe is both Matter-and-Space, and only Matter can be said to be moving, then, it is not Space and Planets together that are "expanding" or "traveling."

Planets and other Objects are moving through Space; but the Space within which these Objects are moving, does not itself "move" with them. For void-Space itself, or "atmosphere-void Space," being naturally intangible-immaterial, cannot not be "expanding" as "within a physical bubble," or as "on the material surface of a bubble," as imagined by some astro-evolutionists.

Did not Halley's Comet "return" every 75 years? "Halley's Comet" was traveling through Space in a way that is analogous to following a "curvilinear orbital path," displaying Curvature pressure motion-Force, which appeared to cut a wide-angled swath within our Solar System, to which it faithfully adhered, every 75 years. When the comet was not visible within our Solar System, it was still traveling; it was not "parked somewhere" or "waiting behind a stage" to make its sudden appearance.

Therefore, the "pattern of momentum rotary Motion-Force" exerted upon the Earth by the Sun, is the equivalent of the gravitational Energy Field-pressure-Force necessary to move its bulky Mass in Continuum Space-Time; that is, Earth rotary Mass-in-Motion momentum wake-Force is "the equivalent" of solar gravitational Field-pressure Force.

Solar gravitational field-pressure-Force operates as a cumulative system of "Rotor-Energetics" possessing "torque-tensing-stressing effect" that is "curving" or "bending" the spaces within which the Comet is traveling, to cause it to "display a stream-of-movement routine" that characterizes its every appearing as "orbiting in a specific traveling way."

Cosmic Space-bodies have specific pre-set, predetermined, "steady-state motion patterns" that identify their types/kinds/categories, e.g., the Earth is a Planet because it revolves around the Sun and rotates upon its own axis. A Space-body having "planetary characteristics" should display rotary Forms of centri-vectored Motion identified as Revolution and Rotation.

But at the same time that the Earth is revolving and rotating, the whole Solar System to which it belongs is "traveling" through Space, revolving around the Milky Way Galactic Star System situated right-smack at the "focal centre" of the Galaxy.

The Sun itself slightly rotates upon its own magnetic axis every 27 Earth-days, as it "negotiates" the tremendous "gravi-Field forces" acting upon its nucleated cycling plasma Frame. The Sun's own "repertoire of activities" or "operational routines" are necessary for "negotiating" its internal magnetic-field-activated convection dynamics and incipient planetary mass-Differentials and gravimetric Force-effects that are impinging upon the external strength projected by its magnetic field. The Sun's external magnetic field, as affected by internal plasma condensate kinematics, impacts projected strength-levels of Gravitational Curvature binding-Energetics, the complex interactions of which, "measurably calibrated" to keep Planets in their respective places, positions, locations, and situational spheres-of-influence, e.g., The Earth has a Moon to "maintain."

An "interactive sphere-of-influence" is constituted of the complex interplay of centers-of-Mass, centers-of-Field, centers-of-Motion, and centers-of-Gravity comprising the totality of Solar System ellipsoid plane gravimetric projections, such as, tensor, in torque, pressure, and stress effects.

The Sun itself undergoes specific steady-state "nucleo-dynamic" changes within allowable ranges of cycling equilibrium, as it "negotiates" the "tugs of Entropy." The Sun is "consuming fuel," Hydrogen, even as it continually radiates! The Sun also has limits in Mass: Its internal Hydrogen supply is not inexhaustible! It operationally functions as the Solar System "primary-determinant Mover" for each Planet under its gravitational "ellipsoid-plane sphere of influence." Solar emissions comprise radiation/heat/Light, Curvature binding energies, field force dynamics, and equivalents of gravity force mechanics that together effect planetary Forms of heliocentric rotary Forms of Motion.

In return, as according to Newton's Third Law of Motion, each action causes a reaction, Planets re-adjust/recalibrate/renegotiate their gravimetric thermodynamic cycling parameters, the complex interplay of which, in effect, "bouncing back" towards the solar gravitational "sphere-of-influence" that engages its "re-calibrating" co-determinants of Curvature pressure force Energetics — planetary momentum "feeds upon" planetary momentum, as solar activity compels reciprocal "counter-force mechanisms" (pro-centric/counter-centric) "to hold" the whole solar-planetary "Frame of gravithermal operations" together as One Whole Energy System.

Gravitational "re-calibration dynamics" engage co-determinants of Field-dynamics that vary within ranges of operational System equilibrium geared for maximum system Energetics.

This "tension dynamic" involving "pro-centric compression-push" countered by "centrifugal distension-pull," yields the controlling variables/parameters/co-determinants that cause the perihelion shift of Planet Mercury.

As the Sun's tremendous magnetic field causes "imploded fission" to compress inwardly as a "fused explosion," in order to sustain the Star-type whole Energy steady-state structure together in System-Oneness Integrity, ellipsoid-plane Energetics compels each Planet to hold its respective Space-position and System-location — for example, the Earth is preceded by Mercury and Venus, holding, respectively, the First and Second place-position from the Sun's center-of-Gravity; the third position/place/location is Earth-situation from the Sun, which is 93 million miles away from Earth center-of-Mass. The Earth has "the enviable honor" of being the Third Planet from the Sun – closer to the Sun it would have been too hot and therefore non-habitable!

The whole Universe itself represents a "mega-Sphere/theater" or "interactive-Field of Mega-influence," in the sense that it is a "wide, unbroken region" or a vast limitless boundless contiguous expanse with no end-in-sight where interdependent, overlapping, and intersecting yet measurable Mass-energy-motion-Force Units, such as Planets and Stars, thrive together, with all their respective routines, repertoires, processes, operations, events, and phenomena.

Magnetism, Gravity, electro-motive Energetics, moving Mass-frame dynamics, Electromagnetic Field-stress mechanics, nucleated Energy cycling permutations, etc . . ., are all differentiated Forms of "interactive Relativity parameters" being impacted by Curvature pressure-torque Motion-force Energetics.

The Whole Complex of Solar-System Ellipsoid Plane Operational Energetics is comparable to the "tensor-stress-torque pressure-Force-dynamics" embedded within "a very powerful stretched rubber-band" that can neither contract nor expand, that can neither compress nor dilate, but upon whose super-elastic ellipsoid circumferential plane, would be located, Space-Objects analogous to Planets all along its "flowing-moving ecliptic."

Because "the strong force" temporarily prevails over other cycling-Frame properties ensuing from Motion-Force, qualitative Energy Conservation predominates over Entropy, promoting "Functional Operational Thermodynamics," rather than the onset of Terminal Entropy or "in toto" resources exhaustion.

Thus, as long as the Sun remains the primary Energy provider, catalyst, or "transubstantiator," there will always be analogs of solar energy on the Earth for transformation into utilizable Forms, but with differentiated degrees of efficiency for producing "working power," e.g., "electric cars" in "comparable efficiency" to gasoline engine powered vehicles.

Ellipsoid Curvature plane tensor-stress-pressure-Force Energetics compels each Planet "to hold" its relative Space coordinates for Continuum thermodynamic cycling, as corrective Qualitative Conservation Mechanisms thwart the onset of Terminal Entropy (or the tendency for Planets to adopt a random straight-line dispersal rate away from the Sun; and also the tendency for the ellipsoid plane of Revolution to either dilate outwards/relax or contract inwardly/compress).

Sustaining dynamic System equilibrium conservation in a state of Functional Operational Thermodynamics via the Input-Process-Output Mechanism of Energy Transformation involves "ranges of variability" or "ranges of performance" in regards to planetary centri-vectored Motion repertoires, consistent with all Mass-dependent proportional quantities/ratios/gradients, "enthralled" in "re-translating" the electromagnetic properties of Matter.

Though differentiated in duration, rotational angle of ecliptic, and distance from the Sun, all Planets within our Solar System must revolve around the Sun and rotate upon their own axes, respectively, even as "cumulative-composite-recombinant" momentum-Motion-Force dynamics remain relatively constant, and proportionally consistent with the electromagnetically-induced Field-tensor-stress-torque and Gravity-pressure-force metrics that are necessary for "relative gravitation" towards solar center-of-Mass — whether at aphelion, perihelion, or equidistant from the Sun — of all Mass-in-Motion planetary Frames within the solar ellipsoid Curvature revolutionary plane.

"Qualitative Conservation" of Energy predominates in cycling patterns of structured thermodynamics wherein thrives the Input-Process-Output Mechanism, over Motion-Force Differentials and Opposites engendering deviations caused by gravimetric perturbations arising within overlapping/intersecting/interacting "regions of turbulence." Though planet Mercury displays a perihelion shift, this "compensatory cycling mechanism" becomes part-and-parcel of its equilibrium repertoire of routines, in fulfillment of the Equivalency Principle.

****

DEFYING ENTROPY: CULTURALLY COUNTERING THE THERMODYNAMIC TENDENCY TO DECAY

As defined by Webster's Dictionary, "Entropy: A thermodynamic measure of the amount of energy unavailable for useful work in a system undergoing change; a measure of the degree of disorder in a substance or a system: entropy always increases and available energy diminishes in a closed system, as the universe; a process of degeneration marked variously by increasing degrees of uncertainty, disorder, fragmentation, chaos, etc., specifically, such a process regarded as the inevitable, terminal stage in the life of a (social) system or structure."

We observe the extension of the above definition to include "social system or structure." Are there also "laws of Social Thermodynamics?" In what ways is natural physical Thermodynamic Transformation of Energy "connected" to the political structure of Human Society?

The Law of Transformation of Energy is constituted of two entangled, intertwined, enmeshed, interrelated, and interdependent processes: Conservation and Entropy. Energy must be produced in order to be utilized. Production involves supply, accumulation, and abundance; utilization involves consumption, waste, and possible exhaustion. Processes of Conservation tend to replenish and preserve resources, while processes of Entropy tend to utilize and exhaust resources.

As there are steps and stages to replenishment and preservation, utilization of resources is sometimes accompanied by degradation in quality and decrease in quantity before the onset of final exhaustion.

Generally speaking, the Law of Conservation and the Law of Entropy concern the thermodynamic cycling of Energy in natural processes, ecological phenomena, and universal events. Forms of government embody Human representatives whose sinful nature might engender destructive will and error-prone judgment. Cultural phenomena and social activity, however, are not "physical categories," though a nation has physical infrastructure, such as roads and buildings, power plants, fire departments and hospitals, schools and armies, whose operations involve the utilization of physical resources and material things.

But because there is a "joint interface" between Human will and the quality and quantity of material resources that society utilizes for infrastructure building and operation, the Law of Conservation and the Law of Entropy have been extended to apply also to Forms of government, including their administrative structures, and legislative, executive and judicial processes. In a way, the extension of Entropy to Forms of government is also an extension of Human sinfulness to physical processes that are amenable to social control.

Physical Entropy that affects natural processes as well as Human will is the "secular analog" to moral sinfulness that affects Human behavior. However, the Holy Bible tells us also that because of our ancestors' sinful disobedience against our Creator, Almighty God, even processes of Biological Nature and phenomena of Ecological Nature have been imparted an intrinsic characteristic of transient temporality, tendency to decay, and propensity for degradation.

In what ways is it true that all physical things "tend to degrade" to eventually "pass away?" Are forms of government, cultures, societies, nations, countries, religions, beliefs, faiths, ideas, etc . . . also subject to the Law of Entropy?

To what extent could "physical Laws" be truly applicable to "social processes" having to do with the ways in which Human Beings are "moved" by their spiritual values for making decisions characterized as moral judgments?

Is not the extension of Entropy to moral Human affairs merely the affirmation of biological and moral sinfulness as the fundamental basis for Human imperfections that touch all aspects of living our lives in Civilized Society and the Physical World?

We know that physical processes, events, phenomena, and things in general have a beginning and an end; in sum, that "they cycle in temporality or "thermodynamic Time." We understand sunrise and sunset! And comprehend that Seasons come and go!

We know that a vehicle's tank of gas has a certain volume limit that requires refilling after traveling a specific number of miles.

How about biological beings? Biological organisms have a certain life-span resulting in individuals dying after living a definite number of years. Human beings, for example, begin life at conception, develop in their mothers' wombs until birth, after which they grow, mature, age, and then ultimately "pass away." Thus, physiological biology implies or begets mortality.

The Family of Humankind has Continuum from generation to generation, though specific individual Persons "pass away" at their "divinely appointed time."

In the same vein, the nation has Continuum perpetuity from century to century, from one generation of its citizenry to another generation, pursuant to Qualitative Conservation of its constitutional laws, legal processes, and administrative procedures, constitutional operations of which, climaxing into faithful adherence to fundamental principles of free government that have already secured our God-endowed inalienable rights, consistent with our "firm reliance on the protection of divine Providence."

We know that some Forms of government that prevailed in the past, like monarchies, are no longer the norm for systems of government.

What are some of the conditions that led to their disappearance as the prevailing Form of Human government? In other words, why did Entropy prevail over Conservation when it comes to monarchies?

Basically, there is a qualitative spiritual quality and moral dimension in social systems that must accompany Conservation in order for its processes to prevail or be preserved over Entropy or eventual decay. Once a social foundation is historically laid down for Posterity, its principles must continually be affirmed from one generation to the next in order to preserve the fundamental Good that has already been established.

Generally, Human Beings, from our innate spiritual conscience, have the knowledge of good and evil; meaning that, we know the difference between the two! Thus, we can't escape "believing" that "there is right and there is wrong;" nor can we avoid being instructed by religiously inspired Morality.

As Human Beings, we understand, know, and pursue that which is good, not only for ourselves, but also for our fellow Humankind, in order to establish Justice in the world, even as we live in this vast barren Universe. That, Human Beings are bent on pursuing that which is Good for themselves, is no longer a doubtful enterprise.

In the past, monarchies or rule by one king or one queen, tended to be unjust in proclaiming edicts that dispensed inequality of condition before the Law, the sum of which, contradicting intuitive Human understanding of a transcendent Source-of-Origin for their terrestrial, existential, personal, social and civil Rights to freedom, peace, and justice; and hence, giving rise to an impetus for discovering transcendent, or God-given spiritual liberty that provides guidance and instruction for moral righteousness in Human affairs and social relations.

But we cannot transfer application of the Laws of Physics to the social structure that frames our daily living. For our Form of free government is a God-centered social phenomenon creatively instituted to secure our inalienable rights; and it is not a physical process answering to fixed thermodynamic laws of Energy cycling. All natural processes tend to be eventually inevitable. But social processes having to do with the free exercise of informed Human Will are not necessarily inevitable.

Within physical Frames, Conservation dynamics outweigh the deleterious consequences of Entropy processes. Under industrialized and technological conditions, given the "progressive refinements" that can be achieved in engineering scientific technologies and commercial products, Human Beings have taken beneficial advantage of the Laws of Energy Conservation, applications of which, they utilize for the betterment of their daily lives.

However, as social systems are also activated by processes that depend upon Human will, a form of "static or inertial Conservation" could exist, in the sense that "things might not change for the better," e.g., in monarchical systems. Thus, qualitative improvement must accompany benevolent Conservation in order to be of beneficent value to Human Persons. We who are "created unto the likeness and in the image of God," are also endowed with God-given inalienable rights. Consequently, the pattern of "progressive refinement" in establishing what is good and what is just, also implies "improved corrective measures" as well as greater understanding in framing solutions within the boundaries of God-given commandments for righteous living, e.g., sexual perversion is not a so-called "right," but rather, willful engagement in immoral, unnatural, and abnormal sexual relations condemned by God because they are inimical to our general welfare and well-being.

In physical systems, "progressing refinement" concerns improving efficiency and reliability of performance as "thermo-Energy cycling," geared for Continuum conditions that sustain dynamic System equilibrium for longer periods of time. Corrective maintenance and preventive amelioration are a regular component of every Form of technology designed for optimal operations, processes, and outputs.

The process of "progressive refinement" is a creative socially-driven impetus overarching Human ingenuity, and does not inhere in physical inanimate Nature whose Forms are pre-determined by fixed physical Laws designed to climax into predictable patterns of Continuum Energy cycling.

Thus, no "natural-biological evolution" ever took place on the Earth or in the Universe as-a-whole. For "improvement" and "amelioration" are socially-rooted conscious Human undertakings that transcend repetitive patterns of information processing — In physical Nature, thermodynamics proceed in accordance with Input-Process-Output Mechanisms that are consistently predictable: Inputs are unchanging, but vary within fixed ranges in operational values; Processes are self-iterative per ingrained organizational structures established since the Time of Creation; and Outputs are predictably similar, though not identical in repetitive structural patterning.

And given that the Laws of physical Nature are fixed in permanence for predicting their operations and applications, they are not due to any "evolutionary proclivities" that would imply built-in systems of natural self-improvement. The "theory of evolution" is therefore imaginative fiction! Physical Nature cannot "self-improve!" Only Human Beings can create machines which they improve through the process of "progressive refinement!"

The Laws of Thermodynamics cannot be transferred in framework or function to any form of government by way of equivalency of direct application. Astro-evolutionists make pronouncements that contradict the Laws of Thermodynamics by positing that designed order and organized complexity can arise from "residual chaos" and "consequent anarchy" as engendered by the so-called "big bang explosion." Their assertions also contradict the Law of Energy Conservation by presupposing that "missing intermediate Species" constitute "biological links" that presumably belong to or lead to "mutated biological Forms" that we observe in the Present.

The conjectural "missing links" or "intermediate Species" could not have existed; nor could they have been Human; for they would not have preserved the "Human Species genetic algorithm" that sustains the first Law of Biogenesis. In addition, they "died," while leaving no physical-biological trace in the so-called "fossil record" of their ever having been real!

Even steering an automobile on a straight road requires periodic, if not continuous, corrective movements in order to remain within the two lines that delineate the width, of the straight road. Highly technical electronic instruments require extreme sanitary and dust-free conditions as well as frequent re-calibration of their command-and-control mechanisms. The Earth has cyclical seasons for ecological rejuvenation; our bodies must rest when tired in order to reinvigorate our bio-metabolic-hormonal immune system.

The foregoing are physical applications of the laws of Conservation and Entropy as concerns physical phenomena, technologies, and physiological processes that God instituted into our biological organism for life-giving Continuum. As the Laws of Thermodynamics are being fulfilled, "Conservation mechanisms" must counter Entropy processes in order to sustain "Qualitative Continuum."

The Laws of Thermodynamics say a resounding NO to random chance and infinite Time! Inanimate processes cannot "self-organize." They must first be framed into specialized Mass-in-

Motion-force structures for Energy cycling operations, functions of which, as predetermined by initial conditions of gravitational Creation.

God's creative endowment of Intelligent Design unto the Universe prompted the apparent deterministic pattern in the unfolding of natural phenomena and cosmic events, as perceived by Human Beings in the iterative periodicity of material occurrences and natural processes.

America's Founders also said a resounding NO to random chance and infinite Time! A deliberately written Constitution is the antithesis of chaotic development or anarchic happenstance. For, in faithful adherence to the Judeo-Christian tradition and the Prophetic heritage, and in accord with the scientific principles they knew and understood, all of which guiding their creation of this nation, they consciously and deliberately instituted representative self-government with systemic Qualitative Conservation measures that preserve the rectitude of their cause. Opposite to duplicating the monarchical system that had predominated over Human history up to the Eighteenth Century, they proceeded to thereby establish earthly Continuum between Human government and God's heavenly government for life, liberty, the pursuit of happiness, with equality and justice for all.

For, it was only by interfacing spiritual and moral conviction with historical precedence as influenced by deliberate Human will, that the apparent "continuum of political Entropy" that hitherto characterized the unfolding of Human history could be broken by conscious constructive action for nation-making!

Therein, resides, the essential difference between presupposed "spiritual thermodynamics" (or the extension of Entropy cycling to social systems and political structures), and "physical thermodynamics" (Entropy cycling common to Matter-Mass systems of Energy). Physical thermodynamics is not the equivalent of "spiritual thermodynamics." Nature has no intrinsic autonomous Will — but Human Beings do.

Nature cannot "change course of action;" but Human Beings can decide to renounce falsehood and redirect their thought-life and political habits to hold belief in "self-evident truths." For, human beings have the knowledge of good and evil, and they are equipped with spiritual discernment, moral judgment, and the rule of beneficent constitutional law.

Phenotypic DNA does not determine personal environmental performance or life-long goals for the future — DNA builds cells via RNA replication but does not control or activate Human Will. DNA only holds natural predispositions for developing personal capacities — however the course of their development and direction is driven by personal desires, acquired skills, and emergent levels of private ambition aimed at pursuing certain specific interests until their fulfillment to one's heart-felt contentment.

Likewise, the Atom qua Atom — or "a naked Atom — does not determine the relativistic Energy state of any specific cosmic body-Mass — but Atoms qua Atoms are fundamental building blocks common to all chemical Elements and other Objects in Nature and the Universe.

Thus, to declare that "every thing is genetic" closes the door to personal freedom of choice in both faith and action, in both thought and activity. The natural Human predisposition or capacity for thinking, moving, collecting information, analyzing data, and acting upon derived conclusions, are "genetically based;" but not which actions will be freely chosen by the Human

Person. Freely chosen Human activity as a Form of "macro-expression" of distinct behavior by the Human Organism, do not have "a genetic basis," e.g., opening a bottle of whiskey to drink a full glass there-from is a personal behavior choice; when a Person finds himself under the desert sun, so to speak, with parched throat and broken skin, thirsting after sweet water is "genetically based," but not thirsting after alcohol. Likewise, practicing sexual perversion instead of engaging in natural and normal sexual relations as prescribed by our Creator and Nature, is therefore a personal choice: There is no such thing as a so-called "homosexuality gene." Inventing a "homosexuality gene" without the biological-scientific basis to prove it, is a demonic lie geared to destroy religious morality in order to engineer a world in which all moral standards of right and wrong are destroyed. Godless, unnatural, and abnormal sexual behaviors cannot thrive without the divine moral and societal condemnation they truly deserve.

Astro-Evolutionists aspire to "reconstruct the Nation" according to their own internal attitudes towards the Physical Nature of the Cosmos, due to their frustrated attempts at understanding the sub-atomic world which they have chosen to perceive as being amenable to "random chance probability," the sum of which they say is "quantum probabilistic."

Thus, from such a pseudo-scientific paradigm, astro-evolutionists and bio-evolutionists have proceeded to teach our children and young people to expect a so-called "value free society" without absolute standards of right and wrong, where any godless or evil behavior can be entertained without guilt, shame, regret, remorse, or condemnation; and thus, by that, they have annulled or obliterated all vestiges of individual responsibility and personal accountability for their own actions.

As above-demonstrated, "macro-expressions" of willful Human behavior have neither "an atomic basis" nor a so-called "genetic basis." However, they engage variables, parameters, co-determinants, and factors amenable to spiritual principles and moral values that effect "thought-life," personal character, analytical judgment, and moral choice. There are "macro-expressions" of personal and social characteristics, therefore, that are not derivatives of the micro-DNA bio-sub-structure or of physical sub-atomic organization.

DNA qua DNA and Atomic sub-structure qua Atomic sub-structure, are not pre-deterministic of the contents animating Human spirituality, professional choice, personal morality, intellectual mentation, and/or freely chosen individual activity. The Human Genome presents only natural predispositions or natural capacities with capabilities for development, growth, maturity, and fulfillment, but is not pre-deterministic of their contents for any specific Person.

Spiritually and morally, we oftentimes witness "social-political Entropy" manifested in terms of "periodic punctuations" rather than as proceedings that appear to have "consistent Continuum-streaming," e.g., World War I (1914-1918); World War II (1941-1945).

But Human Beings cannot continually engage in "perpetual warfare." Peace eventually comes and tranquility returns.

God "reconciled: or "re-united" Heaven and Earth; Christ Jesus Messiah, came as "a Son of Man," and died once and for all to redeem all Humanity, but to rise from the dead as a perfect

"Son of God." We've been re-empowered with freedom to begin anew and the "glorious liberty of children of God" for performing "works of righteousness" on the Earth.

Likewise, the American Revolution (1775-1783) occurred as a "one-time historical event" that would set the patterning precedent for establishing a long-lasting Form of free government, to operate as a lawfully founded constitutionally secured representative democracy; as a republic, one nation under God, with liberty and justice for all, intrinsic principles, rights, and processes of which, being inherently based upon democratic voter-consent, substantive God-given inalienable Rights, and a foundation of Self-government anchored in constitutional due process of law and the equal protection of the laws.

Due to our innately spiritual character for moral understanding, Human Beings understand the need for inner-renewal and inner-rebirth that then overflow unto personal conduct and social behavior, emanations of which, impacting our culture, our economy, and our politics, in accordance with agreed-upon rules governing lawfully ordered proceedings from which arise accrued benefits ensuing from reasoned deliberation, analytical judgment, and constitutional fairness of due process of law.

However, the Universe and Nature are "pre-programmed" for Energy-cycling replenishment via processes of "Qualitative Conservation" yielding resource production, accumulation, abundance, and utilization, whereas Entropy proceeds in order to account for "consumptive processes" that involve resource recycling, waste, and possible depletion and exhaustion. For example, in cold climates, after budding in Spring and growing fruits and grains for post-Summer harvest, trees shed their leaves in Autumn and "go to sleep" during the Winter, after which they will be "rejuvenated" to begin anew with the thermodynamic seasonal cycle "picking-up" again in the Spring.

In Human societies, Conservation processes take a deliberate Form of applied constitutionality for supreme Continuum of substantive individual rights secured in due processes of lawfully administered justice under the equal protection of the laws. Without the self-conscious, God-commanded, spiritual rudder of deliberate moral living, "as time goes by," things will tend to degrade, degenerate, dissolve, and disintegrate.

Nature does not have self-consciousness for analytical reasoning, but Human Beings do. Thus, physical principles animating the material Cosmos cannot be analogically "transferred" for application to Human society. Nature is "pre-programmed" to undergo Conservation and Entropy processes that cycle from season to season.

However, the perpetuation of Human societies requires faithful adherence to fundamental organizing principles without whose consistent application, decadence begins to take place.

As Entropy processes would set-in, in the absence of deliberate constructively channeled moral Will, Human affairs would not improve or get better, but would degenerate into chaos and anarchy. Thus, it takes applied conscious Human Will to preserve social order and institutional organization, — "in order to form a more perfect Union," — for devoting constant attention to always restore Continuum from lawfully protected constructive secure social processes of personal and social living as anchored in freely structured self-government.

Just cook a meal and expect the dishes to wash themselves and the kitchen to re-organize itself, without conscious Human Will to perform "orderly clean-up" in the equation — it will not happen! Only roaches will move in without a complete kitchen clean-up!

Life cannot arise or emerge as an "outgrowth," by-product, or "spin-off" from essentially inanimate "quantum events," probabilistic chance, and random processes.

The spiritual intelligence of our Creator infused design, order and organization into the universal structure. Without these fundamental principles structuring ongoing Continuum, no Frame would cycle with the right proportions that effect relative equivalency between specialized gravimetric Differentials and Opposites pertinently respective to Frame-specific functions or outputs.

All these Mass-in-Motion Frames, e.g., Solar Systems, Galaxies, Planet-Moon Systems, etc..., cycling with differently structured and organized patterns of fulfillment for the Input-Process-Output Mechanism require an overlapping-intersecting-interfacing integrative Continuum Algorithm that unites their specialized processes, functions, and products into relative Equivalency.

All universal steady-state or structured whole-energy Systems are engaged in restoring "thermodynamic Conservation" to their processes by applying "corrective mechanisms" that address the "instabilities-anomalies-aberrations" that disrupt dynamic System equilibrium because their "areas-regions of intersection" are marked by "overlapping turbulence."

These "negotiated patterns of corrective maintenance" engender cycling Differentials and Force-Opposites that must be "reconciled" via alpha-numeric Constants, "hidden values of which, that will "normalize" overlapping regions of intersection marked by "thermodynamic turbulence."

As the variables, factors, parameters, controls, and co-determinants that activate Functional Operational Entropy interplay and connect for "Qualitative Continuum," the potentiality for, ultimate, final Terminal Entropy, is averted. Human Beings do have the capacity to develop an understanding of these natural processes for the preservation of the social Good that self-government structures have achieved, by applying their expressed consent and moral Will to participatory processes grounded in constructive political rejuvenation and beneficent social activity.

Functional Operational Entropy exists because of natural irregularities in Continuum Space-Time Field-Curvature, e.g., Perihelion Shift of Planet Mercury. Qualitative Conservation daily overcomes these natural irregularities.

Likewise, due to sinfulness in Human nature, errors, accidents, and misapplications of foundational principles might occur. Rectification also takes place to make amend for errors and correct mistakes. Due to the fact that variables operate in definite structure-specific allowable ranges within the boundaries of the rule of physical Laws, fundamental principles and Form-specific operational dynamics, respectively, foster the fulfillment of the Input-Process-Output Mechanism for "resource replenishment" via "corrective mechanisms" of "thermodynamic cycling," as unavoidable.

There can also be errors in Human judgment requiring amendatory proceedings of reformation, restitution, and restoration. Testing these amendatory proceedings in light of the letter, spirit, content, and inherent Design of the constitutional foundation in conjunction with wise counsel and prudent deliberation, work to resolve contradictory approaches to problem-solving.

It is also apparent that psychological principles of Human self-governance do present imperfections in execution or implementation due to error-prone macro-expressions of Human Will as influenced by the "sinful nature." This "dynamic" necessitates active self-awareness, as well as consciousness of external parameters affecting administrative proceedings and social events that impact daily life. God has not left us bereft of the good news of Human redemption by Messiah Christ Jesus for beneficent righteous moral living on Planet Earth.

When consciously transformed by our Creator's Spirit, Human will is activated and quickened for righteous love and compassionate charity, the sum of which, superimposing over Nature, a framework of lawful order and principled organization that transmute Matter and Energy into usable Forms — for example, the coal, the oil, the natural gas, or the metallic ores would remain underground with no utility for Human society, were it not for Human ingenuity, creative industry, and constructively channeled productivity for establishing methods of extraction for mechanized industrialization resulting in social utility.

Animals do not have free will; they answer to genetically pre-determined modes of existence. Animals breathe, move and do, but only as pre-determined by their Species-specific genetic "programming." Certain insects created by God, like bees, appear to "impose a Form of order" upon the inanimate, seemingly grown-at-random, nectars of the field, due to their genetic programming for honey production. Evolutionists are wrong however in labeling "honey-making" as "demonstrated intelligence." A personal computer is pre-programmed to print certain output, but that is not summed up as "intelligence," which would imply a thought process conducive to analytical reasoning for deliberate choice.

But "pre-wired functionality" in structure-specific genetic Forms, cannot be categorized as: "intelligence." Even computers can be pre-wired and pre-programmed to execute specific functions. Creative intelligence implies self-conscious application of purposeful Will. Only we, Human Beings, are blessed with such giftedness. God gifted us, Humans, with the conscious application of the Organizing Principle and with a portion of His own capacities of Mind, so that, as His created Human Family in this Universe, we might take a creative part in "the Continuum of created things," on Earth" and in the Universe at-large, that are reflections or "mirror emulations" of spiritual Realities already extant in the Heavenly Realm. Maintaining a productively functioning constitutional social order is a consciously performed Human activity demanding continuous attention.

In Psalms 8, the Holy Bible declares that we are created "a little less than God," or "a little less than the heavenly beings." God is eternal; we are mortal. And thus as God said "Let there be light, and there was light and God saw that it was good," likewise, Man said "Let there be the light bulb." God had already instituted this pattern of "intelligent Creation" in the Universe whereby things-inanimate respond to spiritual instruction from Beings created unto the image and likeness of God. For God endowed us with a portion of his own character-qualities of divine intelligence when he created us to become like Christ Jesus Messiah in the world, even though we should not be of the world.

God allows us to duplicate phenomena via deliberate replication of natural processes that simulate original events in Nature and the Universe. We invent technologies in functionally-specific structural Forms for our utilization — atmospheric lightning finds its analog in alternating electrical current that supply our homes and industries with power. Lightning strikes can kill; the Sun can burn; but a light bulb provides light to our eyes.

America's Founders defied culturally inimical historical Entropy by their inspired persistence, willful determination, and deliberate insistence on spiritual principles and moral precepts that would prove absolute significance for their establishment of a free nation that would endure for thousands of years into the future. As we read the Preamble to the Constitution of the United States of America, we observe that Founders had studied previous kingdoms and empires like Greece and the Rome, in order to discern their shortcomings and weaknesses, dysfunction and frailties.

Mid-stream through the 18th Century, America's Founders could not "botch-up the job" by re-instituting the tyranny, despotism, corruption, and other vicissitudes that had crippled the developmental transformation of previous Forms of government into free republics. And hence, they took great pains to infuse our Constitutional system with all the precautions, checks and balances, preservation measures, and safeguards necessary to ensure its security. Amendatory processes and consent-based representatively enacted legislation oversee lawful government operations for the peaceful and orderly transition of power after each election, once and for all, from one administration to another.

> "We, the People of the United States, in order to form a more perfect Union, establish Justice, insure domestic Tranquility, provide for the common Defence, promote the general Welfare, and secure the blessings of Liberty to ourselves and our Posterity, do ordain and establish this Constitution for the United States of America.................Done in Convention, by the unanimous consent of the States present, the seventeenth day of September, in the year of our Lord one thousand seven hundred and eighty-seven, and of the independence of the United States of America the twelfth. In witness whereof we have hereunto subscribed our Names." (Constitution of the United States of America)

These transcendent spiritual, moral, and political mandates as carved unto "the body of the Law," encapsulate the most miraculous understanding of Biblically proclaimed truths and moral principles for right living on the Earth. As enshrined in our representative Form of free government, these principles find their Form constitutionally patterned after God's heavenly blueprint for beneficent living that upholds our well-being and general welfare.

America's Founders intended that social-cultural Entropy be thwarted, as morally-founded and spiritually-discerned constructive change would operate to sustain lawful Continuum Conservation of the Good which we would achieve from generation to generation within the framework of Biblically-tested laws and ratified amendments to the Constitution of the United States, "in order to form a more perfect Union." As local administrative structures continue to uphold the principle self-government, individual liberty takes precedence over enlargement of federal powers beyond its enumerated boundaries. And in order to curb abuses of power and prevent the gradual centralization of bureaucratic federal encroachment upon individual rights, there must be constant rejuvenation of our Form of free government in the

Judeo-Christian foundation of liberty upon which our representative republic is based, as its free principles of local self-government proscribe remote-control federal micro-management of our lives. For, powers not delegated by the Constitution to the federal Government are reserved to the States, respectively, or to the People. (Amendment X, Constitution of the United States of America).

Why is it so important to address these issues here? Because the scientific theory that predominates an era usually makes its way into personal morality and public conscience as well. Has not the Theory of Relativity been "transferred" unto our culture as embodied in the saying "Everything is relative!" Well, is it?

The founding principle substantiated by the Declaration of Independence and the Constitution of the United States is constructively anchored in beneficent personal-moral liberty of Citizens, and not in government supremacy. Government cannot ever become the sole source of solutions but should create and frame a lawful atmosphere within which individuals exercise their God-endowed inalienable rights to life, liberty and the pursuit of happiness, for self-government, political participation, and productive initiative.

Citizens are to be motivated and encouraged to work out solutions to problems which they are already constitutionally empowered to resolve. Public officials are elected servants of the people and their pre-occupation should not be engrossed in concocting devious legal convolutions by which to increase powers of government outside of limits already prescribed by the Constitution.

The federal government, whose limited powers are constitutionally enumerated in order to secure our inalienable individual rights, should be engaged in insuring the greatest extent of lawful personal liberty for the creative expression of individual initiative, scientific activity, technological productivity, and nation-making freedom, via "the spirit of entrepreneurship."

There are no constitutional prescriptions authorizing government to dissolve, weaken, redefine, and imperil the founding principles of the nation. For, by contriving so-called "gay rights" out of thin air, public officials are destroying the nuclear family as the pillar of ordered liberty and civilized living.

Not every willful behavior qualifies as a "right." The willful engagement in sexual deviancy is not a "right" requiring nationalization by public officials who have strayed away from the biblical foundation upon which our nation is erected. Nor is there a "right" to commit infanticide by abortion and other godless methods of baby-killing. God's commandments against such sins are well known; so are the sound spiritual principles and salient moral doctrines that undergird our Constitution and laws.

Consequently, when safeguards, precautions, and checks and balances are not faithfully adhered to, carefully established boundaries that control the "sinful human tendency" to be influenced by corruptions of power are hereby made void, resulting in cynicism and destruction of hope. And, "as time goes by," temptations to extend the application of the physical law of Entropy to the processes of human government result in lawless behavior on the one hand, and power centralization, and concentration of political will into fewer and fewer hands, on the other. This scenario of extreme decadence would eventually lead to disintegration, decay, and degradation of lawfully secured individual liberty, paving the way for tyranny and fascism. But thanks be to God, America's Founders warned us regarding human nature and "the necessity of auxiliary precautions" when it comes to "a government to be administered by men over men!"

It means that, within the Constitution itself, are structured inherent processes for designing constructive, edifying, or up-building change for Continuum moral living and Continuum public administration, as framed to sustain good principles and practices, which when implemented, would consciously thwart the degenerative effects of Entropy. But standards must be observed—the inalienable substantive and due process rights of individuals have to be unswervingly secured in all administrative and public proceedings "with liberty and justice for all." The spiritual foundation of the nation circumscribes the preservation of God-endowed inalienable rights within the moral framework provided by our knowledge of God's commandments of righteous living, which is the foundation for the Declaration of Independence and the Constitution of the United States of America.

God is righteous and holy and thus, standards of right and wrong as defined by his living commandments control the definition of "what a right is." God's commandments in the Holy Bible, which constitute the Judeo-Christian foundation upon which our form of representative government is based, already define what is right and what is wrong. Sexual deviancies and behavioral abnormalities like homosexuality and lesbianism cannot be defined to be exercised as "rights" requiring constitutional nationalization by unelected judges and other employees of the American people. Nor can there be a "right" to perpetrate infanticide in an abortuary. Not every human desire, nor every behavioral propensity, can be "added as a right" to our already secured inalienable rights, lest it undermines the constitutional foundation whose spiritual and moral source is the Holy Bible.

With the Bill of Rights, legislative enactments, the amendatory process, elections, checks and balances, and the resulting Constitution and laws, the administrative system can constructively resolve problems that arise, but within God's commanded standards of right and wrong, while simultaneously protecting, preserving and defending the Constitution, and avoiding the traps of tyranny, disorder, chaos, and anarchy.

Within the framework of American principles of jurisprudence and political philosophy, is active, an array of willfully applied, self-corrective operational mechanisms. When judiciously applied in faithfulness to the Judeo-Christian foundation of spiritual logic and moral reasoning for the righteous design of time-tested constitutionally-based legal processes that preserve the dignity and lawful liberty of the individual, they ensure well-organized, productive, conscientious, and effective individual and social activity for constructive self-government.

With utmost reverence for these self-evident fundamental principles and their lawful application to our "cultural condition," we can again echo the immortal pronouncement of President Abraham Lincoln at Gettysburg in AD 1863: "That this nation, under God, shall have a new birth of freedom and that Government of the people, by the people and for the people shall not perish from the earth."

As we can see, Spirit matters. A skyscraper cannot stand without a solid comparably-great, strong foundation, the roots of which, securely reaching into the depths of the earth underneath it. Again, Americans continue to defy Entropy—this republic shall not follow the so-called "rise and fall pattern" common to all petty empires that mar the landscape of liberty-destroying human history. From our Founding, God our Creator has been our strength, our shield, our Deliverer, our Rock, and our Salvation: as a free nation, a republic under God, where citizens, and persons legally living in our midst, enjoy "the equal protection of the laws" that have been established with substantive and due process proceedings that secure "liberty and justice for all" in accordance with God's commandments that prescribe standards for right and wrong.

These events and processes are not "by accident." As we have just witnessed above, they were deliberately begun by America's Founders with the expected faithfulness of each generation to the wellspring of its living foundation.

Yes, physical matter is subject to the Law of Entropy. But our Creator made us "more than mere biology." We have a spirit and a soul created in the image and likeness of God. Social organization is not a "free fall," but proceeds in accordance with appropriate laws that sustain Continuum moral living. God's commandments, spiritual principles and moral standards determine how human laws are structured for public administration. Morality is not "relative." Good and evil do exist. Actions have consequences that are either constructive or destructive of our personal wellbeing and general welfare. Human attempts at "re-inventing the wheel of morality" only mask transient carnal interests bound to bring about personal disaster and social disintegration, and increase in human misery and hopelessness. We already have the best "blueprint" for earthly living as blessings from our Creator — the Judeo-Christian foundation which is our inheritance for moral standards of right and wrong as elucidated and fulfilled by Messiah Christ Jesus our Lord and Savior.

Organization does not sprout spontaneously in Nature — it takes pre-programmed, Creator-imprinted principles of order so that ecological, geological, and meteorological processes continue as created for dynamic System equilibrium.

But our Society is not Nature — we live by laws enacted from our given consent as we remain faithful to God's standards of right and wrong. God endowed us with inalienable constitutional rights, the blessings of which have been secured since our Founding, even since "before the foundation of the world."

It is necessary for us not to lose sight of our innate spiritual constitution and our intrinsic moral character as we "individually embody" those ideals for lawful and conscientious practice, which makes for living prosperously in this blessed land. These are not make-believe fantasies on a movie screen, but a reality; based not merely on Law, but living within the hearts, minds, souls, and spirits of our people, from conscientious moral apprehension and spiritual discernment of the Good which impels us from deep within our inner-being to constructively "will to live" as we enjoy God's blessings as "self-aware Americans." Therein are treasures untold and prophecies foreknown.

****

UNIFYING ALL BINDING-ENERGY FORMS IN THE UNIVERSE AS CONTINUUM CURVATURE-PRESSURE-MOTION-FORCE!

Electromagnetism, "the strong force," "the weak force," and Gravity, all of which being Properties of Mass-in-Motion, together, are in Continuum Motion-Pressure-Force, as induced from "Field- Curvature Energetics."

All Mass-in-Motion Frames process Energy via the Thermodynamic Input-Process-Output Mechanism in order to achieve dynamic System equilibrium as they participate in the structured Complex engineered by certain fundamental principles ensuing from Continuum Curvature Field Energetics.

"Energy is never created nor destroyed but always transformed;" an electric field is accompanied by a magnetic field as a magnetic field is accompanied by an electric field; centers-of-Mass "attract" centers-of-mass, and centers-of-Mass "bend" or "curve" Space by projecting tensor-stress-pressure Motion-force. "The strong force" prevails via centri-vectored or pro-centric rotary Forms of Motion. Energy Conservation "temporarily" triumphs over Entropy tendencies towards decay, via inherent corrective mechanisms of Qualitative Continuum.

The Speed of Light is the standard limiting factor for all universal and natural phenomena, processes, operations, and events. Motion is generic to all Space-bodies with Mass, in response to Curvature pressure-Force, as all Frames cycle in Continuum Space-Time for functional Energy transformation.

At the planetary level, the Earth participates in rotary motion forms: Revolution is helio-centric, while Rotation is geo-centric. Objects on Earth surface participate in Gravity-induced Forms of Motion: Curvilinear, Rotary, Linear, and Rectilinear patterns of Motion, and combinations thereof, along Earth line-of-radius. Vehicles and objects only appear to have a non-rotary Form of Motion not directed towards Earth center-of-Mass due to "obstacles" in their path, such as, the surface of the Earth, and trajectory-direction or vector of roads. However, vehicles and Objects "negotiate obstacles" in their path, vectors of which, always pointing towards Earth-center-of-Gravity and Earth-center-of-Mass.

From Earth atmospheric Frame, passenger aircrafts and flying Objects follow centri-vectored or pro-centric Motion-patterns along Earth line-of-radius as directed towards its center-of-Mass and center-of-Gravity, thus, necessitating flight-path corrections that keep such passenger aircrafts and flying Objects within the boundaries of Earth Atmosphere, so as to prevent such aircrafts and flying Objects from venturing into outer-void vacuum-Space.

Thus, on the Earth, all Motion Forms are core-geocentric. Vehicles, aircrafts and Objects follow rectilinear, rotary, linear, and curvilinear trajectories that are composite Forms of radial Motion towards Earth center(s) of Gravimetric Mass, even as they "navigate" through or "negotiate" apparent geological, surface, oceanic, and atmospheric "obstacles" in their path.

In the quantum mechanical Frame, particle Motion is estimated as "waves of probability" because Curvature pressure-force encapsulates "composite" or "recombinant Forms" of rotary, curvilinear, linear, and rectilinear patterns of Motion that are compelled to "move" in trajectories that are centri-vectored or pro-centric in magnitude, vector, and direction.

Because Electrons are operationally simultaneously "attracted to" and "repelled by" the atomic nucleus — Like charges "repel" (Each Orbit can only "accommodate" a certain number of negatively charged Electrons); but opposite charges "attract" (Positively charged protons "attract" negatively charged Electrons) — hence, Field Curvature Energetics (Electromagnetism and Gravity) impinging upon Electrons that "travel in rotary Motion-Forms," or that "revolve" around the nucleus of the specific Atom.

Atomic Frame cycling is relatively activated by all Curvature-induced pressure-Force motion Forms. Trajectories climaxing into nucleus-vectored patterns of Motion are attuned to function-specialized Frame properties of Electromagnetism. Particulate charge-displacement Force-mechanisms activate Curvature Energetics, as induced by both "attraction" and "repulsion," the complex relations of which, being "re-translated" into equivalents of the Magnetic-Field type-of-force and equivalents of the Gravity-Field type-of-force.

Electrons have "revolution trajectories" that are nucleus-centric as formed by rotary, curvilinear, linear, and rectilinear composite-recombinant Forces activated by "charge tensing-stressing-pressure-torque micro-Energetics" peculiar to the Atomic Frame. There is "Form fidelity" from the micro-Scale (Quantum Mechanics) to the macro-Scale (Solar-Planetary Dynamics) as the Universe perpetuates "Relativity-preservation" through iteratively synchronized and fine-tuned patterns of structured Motion from Scale to Scale, i.e., Electrons "revolve" around the atomic nucleus as they also display "spin;" Planets also "revolve" around the Sun as they similarly display "rotation" on their own axes.

Preservation of Scalar Fidelity of Motion, such as from the Atom to the Solar System, marks the most mysterious pattern of iterative redundancy, yet most logically scientific, of all universal redundancies.

Redundancy of iterative Forms permeates the Universe, from the smallest Mass to the greatest Mass, via progressive accumulation of aggregate Matter into cycling Frames of "Energy-Motion-Force Processing." Rotary motion-Forms follow the pro-centric line-of-Force. Linear and rectilinear Forms of motion-Force trajectories occur only upon the imposition of an external Force countering that of Mass-induced centers-of-Gravity, or only in the apparent absence of prevalence of Gravity-field-force, e.g., A jet-engine powered aircraft following "a straight-line flight-path" across the sky. Centri-vectored or pro-centric motion-Force patterns prevail in the Universe due to the predominance of "the strong force" that "resonates" from the micro-Scale to the macro-Scale, as it relatively operates to preserve Conservation of Energy and conservation of angular momentum, over Entropy processes that could yield to atomic decay, on the one hand, and on the other, to planetary dispersal from the Ecliptic.

Within the solar core plasma condensate Frame, — Star inner-radiometric processes — particulate Curvature Energetics unfolds in extremely "compressed Time," so to speak, due to an extremely potent Magnetic Field that engenders inner-core-centric, convoluted Forms of motion. Gravimetric Field-Curvature conditions climax into "motion-force-Opposites with trajectory-pressure-tensor-torque Differentials," as controlled by molecular plasma condensate centers-of-Mass, centers-of-Gravity, and centers-of-Field.

The Continuum Principle necessitates that "mathematical theoretics" aspiring to "go beyond" all known physical laws by exploring yet-to-be-explained or yet-to-be-understood phenomena, events, processes, and principles, must begin to address phenomena falling outside of the applicable limits of both Quantum Mechanics and Relativity Theory, e.g., Is there a mega-unifying Force holding the whole Universe together as One? Could "a multi-verse reality" mathematically and realistically be "justified as true?" What does it mean that the Universe is "expanding? Did the Universe really start by a so-called "big bang singularity explosion?"

A totalizing comprehensive General Theory of Force must be launched from the starting point of the Law of Energy Transformation, as induced by Curvature Energetics animating cycling-Mass operations ranging from quantum mechanics to Newtonian mechanics; as well as from Relativity dynamics to Electromagnetism and Gravity, understood as "Fields of Interactivity."

The Mass-Radiation-Motion Frame of nucleated plasma condensate, as a whole, e.g., the Sun, constitutes the initial platform from which the electromagnetic properties of Matter get "re-translated" into various structured energy-processing Forms subjected to "elastic dynamics" of Curvature pressure-force-tensor-stress-torque Energetics. In fulfillment of the Input-Process-Output Mechanism whose operations give rise to Cycling Differentials and Motion Opposites, Energy Transformation pegged to Mass-specificity, finds functional expression under relatively redundant conditions of Thermodynamics, e.g., revolving and rotating planets within the Solar System ellipsoid tensing plane.

Because the same Energy-form is being transmuted into many convertible states, Quantum Mechanics and Newtonian Mechanics incorporate all force-Forms such as Gravity-field and electromagnetic-field, as engendered by solar system Curvature projections operating under Relativity conditions.

And it is through overlapping Mass-cycling Differentials and Motion-force Opposites creating "regions-of-interactivity" that all "processing Frames" achieve Energy Transformation for "Curvature Equilibrium." The Atomic Frame displays "a sphere of Field-force interactions" climaxing into "an elastic plane of Revolution" characterized by tensor, stress, and torque motion-pressure-Force Energetics that activates "Electron routines." Likewise, the Sun-Planetary Frame-of-Reference displays an ellipsoid sphere of Field-force interactions characterized by an interplay of Motion-force Energetics that is quasi-similar to that of the Quantum Mechanical Frame, but complex operations of which, holding Planets in their respective ordering, as to their places, positions, and locations in the Solar System.

The Electromagnetic Properties of Matter, as embodied in Mass-in-Motion Frames, give rise, via radiometric electro-motive dynamics of gravitational radiation, to gravimetric conditions bounded within the Solar-System Frame, that "re-translate" particulate and molecular binding-energies, into "Curvature pressure-Force motion Energetics," taking the Forms of Magnetic-field Force and equivalents of the Gravity-field Force.

"Re-translated moving-electromagnetic-Mass" as preserved by "angular momentum fidelity," "bends" or "curves" the Space within which it moves. Only a Force that has "push-and-pull characteristics," or "attraction-and-repulsion qualities," can engender "tensing-stressing pressure" to cause Matter-mass to "bend" or "curve" Space.

Thus, "Curvature Energetics" stands for: The projection of Forces that induce Motion from "tensor-stress-torque-pressure dynamics," as generated by magnetic-field and gravity-field forces, the complex interplays of which, engineering predictably iterative heliocentric Forms of rotary motion, such as Revolution and Rotation.

The universal complex has specialized functional and operational "identification-of-types," such as Stars as opposed to Planets, only due inherent thermodynamic properties of electromagnetically-substantiated Matter-mass that fill the void of Continuum Space-Time. Matter-mass is "electromagnetically motive" and Curvature pressure-force Motion is its expression. Due to pro-centric motion-force-Forms prevailing over "counter-centric dynamics," "the strong force" predominates for Energy Conservation.

What are the various Forms of Curvature-pressure motion-Force? Electromagnetism and its derivatives: "the strong force" and "the weak force," and Gravity, — as proceed from atomic frame motion-Force dynamics, — are generated by particulate charge-displacement Energetics that engender "spheres of interactive field-force influence" whose structures are analogously "replicated" via greater body-mass agglomerations, like planets and stars.

Gravity, electromagnetism, (and "the strong force," and "the weak force," display patterns of Motion that are but only "converted states" of Curvature pressure force as embodied in mass-frame characteristics that are anchored into thermodynamic cycling of the electromagnetic properties of matter encapsulated in the atom.

From Newtonian Mechanics within "the Framework of remote-Gravity," to the fall of a ripe apple from a tree due to local center-of-Earth Gravity; from a motorized fan driven by an electric coil embodying electro-magnetic "rotary energy," to internal solar core nucleated mass dynamics; from solar convection region dynamics, to corona-magnetic field interactions; from solar-interplanetary frame dynamics, to inter-galactic frame "telemechanics:" Motion-Force is generic to all cycling thermodynamic structures comprised of Mass-and-Energy.

Thus, the fundamental forces, e.g., Electromagnetism, Gravity, "the strong force," and "the weak force," constitute cycling Frames wherein is encapsulated, a structured complex composed of "Mass-in-Motion-with-Momentum-Force-Energy" manifested in differentiated Forms, e.g., the Atom made up of protons, neutrons and electrons; the solar system made up of a Sun and nine planets.

"Field Continuum Curvature Gravi-Energetics" embodies the tensing-stressing-torque-pressure-force "Energies" projected by radiating solar Mass, as Planets respond via "externally exerted" heliocentric Motion patterns that model the rotary form, the sum of which, engendering within planetary Frames, "gravitational-Fields of interactions," e.g., Planet Earth must tilt its rotational axis to a 23-degree-angle in order to "accommodate" the Moon's Mass, the sum of which, being a quarter of Earth-Mass.

And inner-planetary exertions of Gravity force, such as on the Earth, climax into Core-centric Forms of Motion along its line-of-radius that model composite-recombinant rotary, rectilinear, linear, and curvilinear patterns of Motion, e.g., A jet-powered aircraft or a turbo-jet powered helicopter can perform many acrobatic feats-and-loops and twists-and-turns in the air, movements of which, "competing with," countering, or converging with Earth line-of-radius gravimetric Core-centric Forms-of-Motion.

Equations such as f = ma, which is the equation for force; $F_g = G\, m_1 m_2 / r^2$, which is the equation for universal gravitation; and $E = mc^2$, which is the equation for Transformation of Energy, together, explain or "operationalize" the Solar System, as Continuum Space-Time Field Curvature Motion-Force, which is constituted of electromagnetic field-forces and equivalents of gravity-field force, complex interactions of which, holding its ellipsoid plane of Revolution in tension-and-stress pressure-force equilibrium, as required for all-planets gravimetric and radiometric thermodynamic cycling.

The first two equations are structured to work together in modeling accelerated gravity-induced momentum Mass-in-motion-Force for "congealed Forms," explaining how the Earth travels through Space, with a displacement-Force that is the equivalent of its momentum-Mass; as well as illustrating how the Moon maintains its orbital trajectory around the circle of the Earth. The third equation explains how a Hydrogen atom is transformed into nucleated momentum plasma condensate radiation-Mass under specialized high-temperature and extreme-pressure conditions.

Given that Electromagnetism (Magnetic field force, "the strong force" and "the weak force") and Gravity and their equivalents involve tensing, stressing, torque, pressure, and Force, that may cause disruptions in dynamic System equilibrium, e.g. Perihelion shift of Planet Mercury, then, a general theory of force aspiring to integrate all universal forces into the Unified Field Theory of Continuum Curvature-Pressure Force-Motion, would have to "transform" all cycling-Differentials and force-Opposites into "Thermodynamic Equivalencies," indicators of which, being qualitatively characterized as "tensor-stress,-pressure-torque metrics" that amount to gravimetric expressions of Field-force.

Field equations that account for Electromagnetism and Gravity as "fields of interactive influence," e.g., yielding the Solar System ellipsoid Revolutionary plane, must encompass a framework of "Mathematical Theoretics" that establishes their inseparable contiguous simultaneous operations as "inevitable confluent equivalencies" accountable for keeping momentum Mass-in-Motion efficiently performing its functions in accordance with the universal Input-Process-Output Mechanism that inheres in the Law of Thermodynamic Energy Transformation.

Electromagnetism and Gravity operate simultaneously to sustain functions of Mass-frames "caught" in thermodynamic Motion while experiencing rates-of-change to their cycling parameters within the span/period/duration of Thermodynamic Time.

Thus, Electromagnetism and Gravity are "Field-forces," in the sense of eliciting "operational theaters of influence" upon Space-bodies with Mass, such that they relatively "trigger" momentum-Motion and torque-tensor-stress-Pressure, configurations of which, embodied in overlapping Frames-of-Reference constituted of components such as Mass-Motion-Energy-and-Force, complex interactions of which, being linked for relational interface, by alpha-numeric Constants that "re-normalize" their cycling-Differentials and "reconcile" their force-Opposites, together with "conversion factors" that establish "relative equivalencies" via direct proportionality or "indirect dimensionality."

Newtonian Mechanics emphasized how centers-of-Mass "attract" centers-of-Mass. General Relativity explains how centers-of-Mass "curve" or "bend" the Space(s) within which they are moving. Though the Universe is physically infinite in the sense of possessing no physical barriers or material boundaries that restrict its spatial expanse, (but possibly "finite in the quantities or amounts of Matter-Mass" that "fill its void-vacuum Volume"), its physical Laws operate within resolutely defined boundaries, respective to their "natural" scientifically demonstrable principles, parametric ranges of essential definition, and "allowable fields of application," e.g., The Law of Gravity as propounded in Classical Newtonian Mechanics could not extend in application to provide a mathematically definable, measurable, applicable, proven relative explanation of the cause-and-effect mechanisms that actuate the Perihelion Shift of Planet Mercury.

Thus, both Newtonian Mechanics and Relativity Dynamics must reckon with operational physical Laws that inherently possess scientific ranges of bounded applicability.

"Curvature Torque-Force Relationships" permeate-and-perfuse all Mass-in-Motion Frames from Quantum Mechanics to Newtonian Mechanics, from Relativity Dynamics to projected solar nucleated plasma condensate Energetics.

Only needed is a mathematical formulation accounting for "all-Force equivalents" in the totality of their thermodynamic cycling, as interactive Frames that are naturally and consistently encountering "Continuum Operational Principles" that compel them into Field-unified Relational Integration. The utilization of "conversion factors" would enhance "accelerated quantification of mathematical operations" via "re-normalization" of cycling-Differentials that "reconcile" with Motion-force-Opposites, interfacing of which, mediated by alpha-numeric Constants.

Gravity is Mass-in-Motion dependent, with a "centripetal bias" towards a center-of-Mass. Electromagnetism is particulate-charge-displacement-Motion dependent, with a uniformly applicable Form of rotary, but also pro-centric, Torque-Motion-Force.

As binding-Energy Forms, "the strong force" and "the weak force" find their equivalents within the frameworks of Electromagnetic Properties of Matter. Field-force and Gravity-force operations, manifested as analog quantifiable expressions of the Electromagnetic Properties of Matter, remain pertinent to each Frame within which they are "exerting their influence," respectively.

"Energy is never created nor destroyed but always transformed" — thus, Motion-Force, as an embodiment of energy-cycling Force, also "gets re-transformed," in the same manner that an automobile's engine torque is "re-transmuted" as "traction-Motion-Force" transferred to wheels-with-tires, via the intermediacy of an automatic or manual transmission. It is the cycling RPM Motion performed by the automobile's combustion engine (torque) that gets "transmuted" into road-traction, as mediated by the differentiated speed-gears of the transmission device (horse power).

Motion-Force as "transformed Energetics" finds various analog expressions in different Mass-energy-in-Motion Frames, structured for specific gravimetric functions. Opposite charges attract as like-charges repel: Thus, proton-neutron strong force as mediated by revolving negatively charged electrons, in toto, engenders within the whole Atomic Frame, a strong yet predictable "sphere of Motion-force interactions" that holds the Atom together, as a complete Unit of Matter-Mass endowed with an intrinsic "Concentric Field-of-Revolution-plane."

When "stand-alone" particulate motion-Force patterns are interpreted as "waves of probability," it is from experiments that isolate particles or separate them from the Atomic Frame to which they naturally belong.

Consequently, a magnetic field, having electromagnetic properties that engender rotary motion, is "a transmuted analog" of the proton-neutron strong force that "triggers" electron revolution.

Gravity, as an emergent property of pro-centric Forms of rotary Motion, is "the transmuted analog" of "the weak force" that allows electron bonding for molecular change and projection of radiation emissions, comparable to the way in which a ripe apple will fall upon Earth surface in a motion pattern that flows along its line-of-radius but towards its center-of-Mass and center-of-Gravity.

The Neutron serves as a "ground" or "buffer" or "neutralizer" between the positive proton and the negative electron, allowing them to "approach each other," at a safe orbital level distance, without "short-circuiting."

It can be said that the Sun is a self-confined nuclear explosion that is forced to inwardly implode by its extremely powerful external Magnetic Field; whereas in Earth-bound Hydrogen or atomic bomb explosions, there is no magnetic field to contain chain reactions as a self-perpetuating plasma condensate inferno.

However, the absence of an extreme magnetic field is compensated for by "congealed" atmospheric gases that restrict nucleated expansion, in order to protect life-support systems that make the Earth a habitable planet — lest a greater explosive nuclear force causes the whole atmosphere to be "converted" into a gigantic ball of nucleated plasma fuel.

Mass Frames "clump together," as "binding Energetics" keep particles of Matter or of other Space-bodies such as Planets, from randomly dispersing, e.g. The atomic Frame, the solar system Frame. Curvature binding Energetics cause particles or other Space-bodies to aggregate and move, in ways that are consistent with "gravitational bonding." Curvature pressure-force binding-strength depends upon Field-induced Frame-conditions of thermodynamic energy cycling, which are Mass-in-Motion dependent.

The Sun projects an electromagnetic field that prompts planetary rotary motion Forms such as revolution and rotation, e.g., The Earth; the Earth in turn projects its own magnetic field that causes satellite orbital motion, e.g., the Moon.

Both magnetic-Field force and gravity-Field force are equivalents of Curvature pressure-force Motion "condensed" in all matter-Mass-in-Motion Frames or Space-bodies as "Thermodynamic Units of Energy Transformation," — due to the electromagnetic properties of Matter that get "re-translated" into "Frames of thermodynamic cycling." Matter has Mass with intrinsic Rotary-Pressure-Force born-out-of its particulate constitution that can't help but "trigger Motion-displacement in Space" by "curving" or "bending" the Space within which it is moving while Gravity-field Force and Electromagnetic-field Force operations insure conservation of angular momentum.

All these reference-Frames, — regardless of the motion-Form(s) induced by "rotary binding Energetics" caused by gravity-field force equivalents and magnetic field-force indicators, — engage activation of "electromagnetically sensitive spheres of influence," complex interactive relations of which, keeping such Frames together as "Units of Motion-Force" that surge forth with "momentum Curvature pressure-tensor-stress-torque action."

In the absence of nearly spherical Space-bodies such as Planets within Continuum Space-Time possessing "an intrinsic natural affinity" for thermodynamic cycling causing them to respond to field-Curvature momentum motion-Force due to their fundamentally electromagnetic constitution, Star-caused gravimetric Motion(s) would attempt to follow a "straight-line pattern." However, a Space-body, — or properly speaking, whether a particle at the quantum mechanical level, or a planet in the solar-planetary frame, or even an artificial satellite such as "a Space-Station" — "caught" within the sphere-of-influence of an electro-magnetic Field, will naturally display rotary pro-centric Forms of Motion.

How is the Force of Gravity at the quantum mechanical Frame to be differentiated from the Force of Electromagnetism operating therein as well, when they are so intertwined-recombinant and composite-entangled with the dynamics of Electron rotary Motion that occurs within the compass of "compressed Time?" Electrons revolve much faster than Planets: Can micro-Gravity-field be measurably differentiated from micro-Electromagnetic-field in the quantum mechanical Frame?

The quantum mechanical level presents great atomic-Frame complexities when attempts are made to understand the differentiated exertions by both, an electromagnetic field and the force of gravity, as to their respective effects upon constituent particles, and upon the atomic-Frame complex. Electromagnetic-Field and Gravity-Field spheres of Motion-Force influence, are configured by micro-centers-of-Mass, micro-Curvature Energetics, and particulate opposite-charge-displacement motion-Forces taking place "in compressed Time," representing "cycling-Forms" that compound force upon force, momentum upon momentum, and motion upon motion.

Electrons are simultaneously attracted and repelled by the atomic nucleus where magnetic-field force and gravity-field force equivalents take centri-vectored/pro-centric Forms of motion as displayed in rotary, rectilinear, linear, and curvilinear patterns, from applications of the Continuum-micro-Curvature principle.

Given that the Universe is constituted only of Matter and Continuum Space-Time, then, all Field-properties triggering every Motion-Form are inevitably Mass-dependent.

Thus, every category of phenomena, every Form of events, or every manifestation of a Unit of Motion-Force being observed in the Universe, is necessarily Mass-dependent, respective to the kind, Species, or type of Matter-Mass Frame under consideration, e.g., A Star v. a Planet and the natural Forces acting upon them.

When Electrons are "agitated" by many external Force-vectors that cause disruptions of atomic-Frame equilibrium, their trajectory routines tend to appear as "waves of probability," because they are no longer amenable to the Researcher's experimental controls that make them consistently and reproducibly predictable.

As Electrons revolve around the atomic nucleus, they "bend" and "curve" in path whose trajectories have patterns, with gravimetric indications that their negative charge is being attracted to the proton's positive charge, while at the same time, demonstrating that they are also repelled by the atomic nucleus due to the "grounding properties" of the charge-less neutron: This micro-Mass-dependent gravimetric electro-motive Complex, being then, interpreted, as "waves of probability."

Thus, particulate motion(s) in the quantum mechanical Frame would be constrained by "tendencies-for-deviation" elicited with multi-vectored and omni-directional Force(s), — as produced by high-intensity Electron-beam microscopes — which, have to be "negotiated" via pro-centric or centri-vectored Forms of rotary Motion, e.g., displayed as rotary, rectilinear, linear, and/or curvilinear patterns and their composite/recombinant Forms. such that they, together, operate to sustain "Atom-wholeness integrity" as a Cycling Energy Transformation System in continuous pursuit of dynamic equilibrium.

Thus, it is "redundant" to speak of "the strong force" and "the weak force" apart from Electromagnetism, or in ways that fail to factor, in a "Unification Equation," the common source of origin for their binding Energetics properties, i.e., as anchored in their particulate opposite-charge displacement foundation.

No scientific experiment has proven that Gravity has a particulate charge basis or foundation. As "the strong force" concerns the positively charged proton in relation to the charge-less but comparably massive neutron, and the weak force concerns the negatively charged electron in relation to atomic decay and emissive release, these Relativity variables address the electromagnetic properties of Matter that get "re-translated" into Curvature pressure-Force Motion-Energetics, the complex sum(s) of which, accounting for magnetic-field projection(s) and operation(s) of gravity-field equivalents.

Within the atomic frame, there is a "charge attraction" between the proton and the electron. However, due to the presence of the comparatively massive charge-less neutron, the negative electron is prevented from "crashing" into the positive proton at the atomic nucleus, which, if it did, would "short-circuit" the Atom. Even within micro-Frame Curvature dynamics, Gravity-field and Electromagnetic-field co-determinants remain micro-Mass(es)-dependent. Thus, Electromagnetism and Gravity are still present, even within the micro-sized Quantum Mechanical Frame.

These counter-vectoring Forces: attraction and repulsion, contraction and expansion, centrifugal counter-centric energy and centripetal pro-centric energy, pro-centric compression and counter-centric dilation, complex interactions of which, keeping electrons "at a safe distance" away from the atomic nucleus, induce Motion-Force-Opposites that are also micro-Mass-dependent, not only to keep the Atom as a "congealed frame," but also to constitute the Curvature-Energetics Differentials that work therewith to allow molecular change to take place under specific pressure and specific temperature controls, on the one hand, and on the other hand, and to fulfill requirements of universal Continuum as well as achieve dynamic System equilibrium.

Curvature Motion-Force Energetics must sustain both Continuum and Equilibrium within thermodynamically cycling Mass-Frames that operate "through sharing" their overlapping-interdependent co-determinants of functional operations, respectively, in order to collaboratively engender, produce or yield Universal Wholeness Integrity within the boundless expanse of immanent Space-Time.

Particulate forms of motion expressed as patterns of rotary, curvilinear, linear, and rectilinear "quantum movements" are "sensitive to mass energy gains," hence, the mathematical difficulty causing interpretation of momentum particulate trajectory parameters as "waves of probability."

For example, in a nucleated state, atoms "shed" their neutrons and electrons which escape from the nucleus. It is in that manner that a "congealed atom" of Hydrogen gains positrons to be transformed into Deuterium; and then, through successive, respectively applied "Neutron-gains," into Tritium, to climax into Helium.

Thus, even as "the weak force" is being applied for neutron escape and electron release, "the strong force" is being applied for positron gains, hence, amounting to achieving dynamic frame-System equilibrium for Conservation of Energy. "The strong force" prevails, albeit temporarily, depending on "the half-life" of the Chemical Element, for pro-nucleus-centric motion-Forms, thus, sustaining Energy Conservation.

During an Earth-bounded atomic explosion, it is "the strong nuclear force" of "congealed" atmospheric gases that restricts expansion of nuclear-chain-reactions due composition of the Atmosphere, being 80 percent Nitrogen, 16 percent Oxygen, and 4 percent inert gases, among which Hydrogen. These atmospheric gases resist neutron release and electron escape, subject to the relative intensity of "the Heat-force" or "temperature-gradient" involved in the nucleated plasma condensate reactions.

Hydrogen is the most abundant Star nuclear plasma condensate fuel in the Sun, but only constitutes less than 2 per cent of Earth atmosphere. And these structural atmospheric proportional Differentials, in addition to temperature and pressure Differentials, keep the Earth as a "congealed Frame" (non-nucleated) as opposed to a Star-type "excited Frame" (nucleated).

The neutron's role is analogous to vacuum space distance "insulating" the Sun from Planets in consort with its extremely powerful magnetic Field-force acting "to implode" its Thermodynamic Energetics back into its massive internal inferno.

The Sun stands at 93 million miles away from the Earth. Hydrogen is the lightest element. Relatively speaking, due to "elemental density" or "particulate distance" that allows the positively charged proton in the Hydrogen atom to thrive with the negatively charged electron without short-circuiting each other, Hydrogen exists in stable dynamic atomic System equilibrium even in the absence of a "grounding Neutron."

Within the atomic Frame, "distance is compressed" in fast-Time energy-cycling patterns, hence, the necessity of "neutron grounding." Distance separates the Sun from the planets. Deep, void, vacuum Space acts as a form of "ground" or "buffer" between the Sun and the planets in "trapped" in macro-Curvature-Motion, with cycling patterns that allow Planets to interface within overlapping Field-projection regions, via binding energies imposed upon the Solar System ellipsoid revolutionary plane by centers-of-Gravity, centers-of-Mass, centers-of-Field, and centers-of-Force.

Were it not for the Moon's imposition of the 23-degree axial tilt upon the Earth, life-support systems would not be possible. In addition, were the Planets to be any closer to the Sun, they could become "nuclear fuel" and "burn up," or their gravimetric Differentials would cause greater interference, collision, and perturbations at their "points of intersection," the whole complex of which, to then operate to prevent dynamic thermo-cycling equilibrium.

In the presence of an electric Field that generates a magnetic Field for rotary field-force Motion-Forms, "insulation-mimicking void-Space-distance" must be present in order to prevent "short-circuiting" or "premature merging" between the Source of the magnetic field (the Sun) and the Object(s) undergoing pro-centric rotary Motion, (a Planet) e.g., preventing Planets from becoming "Star-fuel."

Likewise, as the void of vacuum-Space-distance acts as a "universal ground" or "quasi-insulation" between the Sun and the Planets, and in the same manner that the charge-less neutron "grounds" the Atom, a "ground wire" is utilized in structures, machines, houses, or buildings endowed/equipped with electrical systems, in order to have electrical current flow alternately without "short-circuiting," e.g., Positive and negative electrical charges must be "kept apart," or away from each other.

Still, Electromagnetism and Gravity do interface as Fields that overlap, intersect, interact, and interconnect by engendering "common Frames of Entanglement," even as Relativity-factors makes it possible for "operational equivalencies" to emerge, thus allowing all Frames to thermodynamically cycle in Continuum Space-Time Curvature while pursuing dynamic System equilibrium.

The quantum mechanical Frame embodies simultaneous expressions of many Force equivalents, such as electromagnetism, gravity, Curvature pressure, mass-in-motion dynamics, opposite-electro-motive-charge(s) Space-displacements, "the strong force," "the weak force," "potential Energetics" for eventually producing molecular-change binding Forces, "particulate kinematics" within allowable orbital levels, etc . . . whose "pressure-Force manifestations" effect particulate trajectories as composite-recombinant patterns of centri-vectored or pro-centric Forms of Motion.

Thus, "a totalizing theory" being sought is one that integrates all "equivalents of tensor-stress-torque-pressure-force-Motion Energetics" into "a general theory of Continuum Curvature-Motion-Force," as embedded in centri-vectored or pro-centric Motion Forms. These Motion Forms are said to be "centri-vectored" or pro-centric, because their patterns are directed towards the center of the particular reference-Frame within which they are moving, in the same manner that it is said that the Solar System is heliocentric, i.e., all Planets revolve around the Sun.

For example, revolution and rotation are rotary patterns of pro-centric Motion; an apple falling towards Earth center-of-gravity and center-of-mass demonstrates the rectilinear pattern of motion-force along Earth line-of-radius. A vehicle traveling upon Earth surface displays the curvilinear pattern of motion as it "negotiates obstacles" or road layouts in its path, including the inevitable presence of "Earth-tablature" or "natural Earth-Surface substance" under its wheels-tires.

Complexities arising from the dynamic interplay of centri-vectored Forms of motion displaying variant expressions of rotary, rectilinear, linear, and curvilinear patterns that emerge from micro-centers of Mass, Gravity, and Field made manifest as "Continuum micro-Curvature Motion-Force(s) within the quantum mechanical frame, account for explaining why particulate trajectories appear to resist simultaneous measurements of position and momentum; and also why particles accelerated even to a fraction of the Speed of Light "shed their mass-Energy gains" as cosmic radiation, within boundaries fixed by Speed of Light limits.

Undoubtedly, the micro-motion-pressure-Force-Energetics extant in "the electron beam," exerted by electronic instruments controlled by the experimenter, "becomes" the "equivalents of the strong force," but in a way that does not correspond to pro-centric parameters demanded by the atomic strong force; hence, particulate trajectory displacements that "veer away" from a desired Curvature-path when the magnitude of either the electric field or magnetic field is altered, which is also engaged in causing modifications in the projected strength of both gravity-field and electromagnetic-field motion-force(s), respectively.

Star Electro-magnetic Field-force, "the strong force" (between the Sun and the Planets) and "the weak force" (allowing planetary self-reflexive rotation, due to Newton's Third Law of Motion) are but differentiated expressions of the Electromagnetic Properties of Matter being "re-translated" as "Curvature pressure-force Energetics," complex interactions of which, engineering heliocentric motion patterns from which a Gravity-field emerges as a Motion-Force-property.

Matter will always embody "Motion potential" due to its electromagnetic constitution, as particles engage in "electro-motive-charge displacement-movement in Space." Centri-vectored motion Forms are composite-recombinant renditions of the dynamic engendered by centripetal and centrifugal force(s) impinging upon a particle, object or body-in-Space, impelling it to "negotiate" or "recalibrate" a Species-of-cycling-pattern that remains within the parameters, ranges, variables, co-determinants, and conditions of its specific-Frame-type.

For example, Planet Mercury must remain within the solar ellipsoid plane of Motion; thus, it "negotiates" changes in the rate of solar magnetic-field-force and gravity-field force Energetics, respective to its closest proximal distance from the Sun, by undergoing a perihelion shift in faithfulness to its frame cycling conditions that are determined by solar-interplanetary centers-of-mass, centers-of-field, and centers-of-gravity.

In the same vein, if the Earth stopped to emit a "gravimetric Field-of-influence" to sustain the Moon's rotary motion around its orbit, the Moon's trajectory would "transmute" into a "straight-line escape pattern," as "conservation-Forces" accountable for preservation of angular momentum would then act upon its lunar mass to "re-define" its "path of motion."

The same effect has been witnessed in particle accelerators whereby constant corrections must be imposed upon moving particles in order to keep them in sustained desired-preferred Curvature patterns.

In addition, Planet Mercury displays a perihelion shift because, even as it resists centri-vectored Motion, solar and interplanetary field momentum dynamics project such great Curvature energies upon its Frame, that it cannot "go into straight-line Motion."

As an action causes a commensurate reaction (Newton's Third Law of Motion), the "perihelion shift" compensates for the momentum mass-induced perturbations caused by all gravitational-field and electromagnetic-field dynamics, even as the Planet resists rotary motion forces. The shift "corrects" for perturbations by sustaining parameters contributing to rotary motion patterns, thus annulling Mercury's attempts at resisting Continuum Curvature tensor-stress-torque pressure-force Energetics (Mercury's counter-centric tendencies to carve an alternate path aimed at "escaping" the solar ellipsoid plane of revolution, "to fly off" into a straight-line Motion-pattern.)

Though particles and Space-bodies would tend to follow a straight line trajectory in the absence of gravitational-field Force and electromagnetic-field Force actions, it cannot be generalized that all motion is patterned after a "straight line path."

Thus, it will be observed that universal motion is not in straight line trajectories. For, the reality of Continuum Space-Time thrives in Curvature Forms of motion force. In the Universe, Curvature is a superintendent trajectory pattern that necessitates rotary Forms of motion-force responses.

Wherefrom would originate the external force that would supersede gravity –field and magnetic-field Curvature Forms of centri-vectored motion? In order for planets to follow a straight line trajectory, they would have to escape from the gravitational field Energetics of the Sun, or the Sun itself would have to be extinguished.

Earth surface radial Forms of motion though different in kind from field force rotary Forms of motion embody the same force-vectors that compound into relatively predictable motion patterns, because they are Gravity-induced. They lie in a Continuum because they are also centri-vectored or directed towards earth center-of-mass and center-of-gravity, even as external planetary rotary motion is heliocentric.

Complexities of conceptualization inhere in the fact that Curvature pressure-force Energetics engenders all Forms of Motion that are centri-vectored/pro-centric in pattern: rotary, rectilinear, linear, and curvilinear patterns, are "composite-recombinant trajectories" that are differentiated by Frame-specific Mass-conditions of thermodynamic Energy cycling, e.g., the Earth revolves around the Sun and rotates upon its own 23-degree tilt-axis; the Moon revolves around the Earth; a vehicle on Earth surface would follow a straight path or a curvilinear path along Earth line-of-radius, depending upon "obstacles" and road layouts in its path, towards Earth center-of-gravity and center-of-mass; electron trajectories embody complex nucleus-centric composite patterns of rotary, rectilinear, linear, and curvilinear motion, as engendered by micro-Gravity, electro-motive charge displacement-movement-in-Space, centripetal and centrifugal tensor-stress metrics that torque particles into spinning, and field force equivalents that engender "compressed time cycling Differentials."

(a)   "The Continuum Principle" implies that, as "Matter-Space-Time Complex," the Universe is an undivided and indivisible whole-Energy system;

(b)   "The Thermodynamic Principle" means that Mass-in-Motion-Frames cycle in temporal or measurable time, or, so-to-speak: "Thermodynamic Time;"

(c)   Electric field and magnetic field "Simultaneity Principle" infers that wherever an electric field exists, so will a magnetic field, and vice-versa;

(d)   "The Curvature Principle" means that Continuum Space-Time is "curved" and Mass-in-Motion Units — e.g., Planets, — must follow their pro-centric rotary trajectories within its pre-determined dimensions, as carved by its "Curvature displacement-in-Space Energetics."

(e)   "The Gravitational Principle" means that centers-of-mass "attract" centers-of-mass, as well as "bend" or "curve" the Space within which they are moving;

(f)   "The Input-Process-Output Mechanism" implies that all energy systems have "intake requirements" for frame-bounded energy cycling processes, as they are interconnected via common regions where cycling Differentials intersect "in dynamics of perturbation" for frame-specific equilibrium as impacted by centers-of-gravity, centers-of-field, centers-of-mass, and Curvature Energetics, for centri-vectored/pro-centric Forms of motion.

(g)   "The Relativity Principle" means that Mass-frames "move" within necessary interactive platforms interfacing via "bending" or "curving" of the Space within which their "range of Force influence" is respectively applied.

The Universe and Nature apparently display "the property of redundancy" or the characteristic of "synchronized Symmetry of Iterative or quasi-similar Forms," only because the Electromagnetic Properties of Matter necessitate Curvature patterns as Mass-in-Motion-Frames "carve" their respective rotary trajectories in Continuum Space-Time. As Frame-cycling Differentials and Force-Opposites interface, they "hide" certain "normalizing Constants" that serve as "catalytic reconcilers" to sustain whole energy system integration for unified-and-equilibrated Curvature pressure-force Motion in Continuum Space-Time.

Curvature Energetics must be continuously applied in order to sustain the rotary Forms of motion patterns as engendered by Gravity-force equivalents and the magnetic field force.

In short, in the absence of continuous solar gravitational Curvature binding energies, which would be the equivalent of "a change from without," all planets and the Moon would "go on their merry way" as they would escape from heliocentric motion forces that impose "rotary motion forms" upon their mass-Frames as engendered by conservation of angular momentum.

All Forms of Force in specifically-structured Energy transformation platforms made manifest as "thermodynamic cycles" being generated by Mass-in-Motion Frames embody "Curvature binding Motion" in "variant patterns" of "re-translated" electromagnetism, as different "converted states" of Gravity force energy. Why? Because of the Laws of Thermodynamics: "Energy is never created nor destroyed, but always transformed."

Consequently, when we take a look at structure-specific conditions that animate a respective Frame or Energy-state, we are always evaluating electromagnetic mass in "Curvature force motion" along with one of its "many forms of expression." All Forms of Force, e.g., electromagnetism, "the strong force," "the weak force," and gravity display "planes of interactive Curving energy" composed of motion-force dynamics, the complex sum of which, are framed as "tensor-stress-torque pressure-force energies" that cause Energy-Transformation-Mass Structures to possess routines modeling pro-centric Forms of Motion.

Given that Universe is composed of Matter (with Electromagnetic Properties engendering Curvature Motion-Force Energetics), then every measurable quality or quantity can be categorized as an "Entity" or "Class of Things" arising only from Matter, manifested as Mass-in-Motion Frames in Thermodynamic Energy Cycling.

The problem that remains to be solved, then, is to unify all these "converted states" of "transformed Curvature Motion-Force Energetics" into a mega-conceptual paradigm that in Form and application will integrate all Frames within Continuum Space-Time into a unified general theory of force, by "reconciling" all "re-translated Forms" of Curvature Motion-Force Energetics, e.g., Electromagnetism (with "the strong force," and "the weak force"), with Gravity in all its transmuted equivalents operating as complex structured Forms of "the mass-in-motion-energy-force complex," via alpha-numeric Constants that would "normalize" their thermodynamic cycling Differentials and Force-Opposites, while "conversion factors" would effect their "relative reconciliation," as "proportional equivalencies."

Curvature-pressure force-motion exists in every place where Mass-Energy temporally or thermodynamically cycles, in the pursuit of dynamic System equilibrium for Continuum unification.

Mathematical formulation of its many equivalent Forms, e.g., gravity-Field force, magnetic-Field force, mass-in-Motion momentum force, energy cycling force, the strong force, the weak force, electromagnetic charge displacement force, etc . . . will unify "all force forms," including those considered as "fundamental forces" into a coherent theory of Curvature pressure force motion that contains internally consistent incorporation of all patterns of the centripetal-centrifugal tensor-stress-torque complex: rotary, curvilinear, linear, rectilinear, and composite/recombinant trajectories, as "RELATIVE EQUIVALENCIES," regardless of the Frame within which they appear to occur, e.g., Quantum Mechanics, Newtonian mechanics, Relativity dynamics, solar system ellipsoid plane Energetics, solar core plasma condensate dynamics, molecular change mechanics, etc. . .

It is due to Mass-in-Motion-Frame cycling-Differentials "caught" in energy transformation for functional specificity as produced by operations of the Input-Process-Output Mechanism that Curvature pressure-force Energetics takes the Forms of Force-Opposites, such as Gravity, Electromagnetism, Magnetic field, "the strong force," "the weak force," displacement-in-Space as caused by the repulsion-attraction dynamic existing between Opposite Electrical Charges yielding centripetal-centrifugal Differentials, etc . . ., all of which, demonstrating Curvature-Motion as the fundamental characteristic of Matter-mass "trapped" in quasi-deterministic differentiated cycling patterns of "gravitation"— matter-mass Frames "attracted to each other" and "repelled by each other," at the same time, as "moved" by Curvature pressure force Energetics.

Differentiated mass-frame energy cycling structures are necessary because frame-specific functions determine fulfillment patterns for the Input-Process-Output mechanism, in order to allow for the greatest latitude in range variability so that in "negotiating" Functional Operational Entropy, dynamic System equilibrium is sustained for universal integration.

How does the Earth "process" gravitational inputs from the Sun as opposed to how Jupiter or Mercury would "process" the same? "Gravitational processing" is Mass-dependent! Earth constituent components and gravimetric variables for life-sustaining ecology determine cycling patterns in fulfillment of the Input-Process-Output Mechanism, as mass-dependent solar-activated ellipsoid plane field-strength dynamics engender heliocentric rotary motion Forms that sustains energy conservation.

Therefore, in examining all overlapping Frames-of-Force and the Curvature motion patterns they embody, e.g., ellipsoid plane with solar-inter-planetary Mass-dependent co-determinants of motion-force-energy equilibrium respective to each planetary Frame-of-reference, we are also encountering "composite gravi-magnetic field force energies" in various structured Forms that "re-translate" electromagnetic binding torque-tensing-stress power, respectively embodying specialized functional Frame-conditions responsible for operational dynamic System equilibrium, e.g., Though Mercury rotates, revolves, and displays a perihelion shift, it is still "a lifeless planet." Motion is a necessary condition for planetary functionality, but it is not a sufficient one for yielding life-support systems.

Dynamic equilibrium Differentials necessitate the linking of all overlapping structural forms via alpha-numeric Constants in order to "normalize" mathematical operations for relative Continuum uniformity — "Energy is never created nor destroyed but always transformed." For, from the frame of quantum mechanics where electrons revolve around the atomic nucleus, to the frame of Newtonian mechanics where planets revolve around the sun and rotate upon their own axes, differentiated but analogously structured rotary Forms of the same "transformed" Curvature Energetics motion Force, as a property of Matter-Mass-in-Motion, are electromagnetically "re-translated" due to frame-specific Differentials of temperature and pressure, Force-Opposites, and gravimetric Relativity-dynamics.

How are "redundant Frame embodiments" of the mass-in-Motion energy-force complex to be mathematically differentiated and then re-integrated for gravitational Continuum unification within Space-Time Curvature? By ferreting out the complex equivalencies integrated therein to "retranslate" Continuum Curvature Motion-Force in Space-Time into its many quasi-similar Forms! The pursuit of a "Unification Equation" for the Universe does imply the integrated operational entanglement of many hidden gravimetric equivalencies!

Frame-bounded but interconnected gravimetric cycling Differentials of thermodynamic energy transformation due to intrinsic rates-of-change emerging from Force-Opposites, do necessitate alpha-numeric Constants for "normalization" and "reconciliation."

Frame interface dynamics of Curvature as linked by magnetic fields and equivalents of the gravity force will display many Forms of centri-vectored motion "trapped" in pro-centric rotary, curvilinear, linear, and rectilinear patterns that allow overlapping regions to cycle simultaneously in dynamic System equilibrium within Continuum Space-Time while "correcting" for "areas yielding turbulence."

While they overlap and interconnect wherein are present "regions of turbulence," those "waves of perturbation" have to be "buffered" for Continuum dynamic universal System equilibrium via "normalizing" alpha-numeric Constants, e.g., Moon-engineered oceanic tide surges on the Earth; perihelion shift of planet Mercury.

In chemical reactions, elemental atomic integrity is "conserved" due to "reciprocally interactive strong-force nuclei," even as electrons bond at their outer orbits with electrons of other elemental atoms in forming greater Space-bodies-with-Mass.

Conservation of nucleic strong force amounts to preservation of the atomic Form in order that distinctive Elements can maintain their identity as solids, liquids, or gases, and/or composite/recombinant Forms thereof. For example, bronze is a composite metal alloy constituted of 20 percent tin and 80 percent copper; elemental atoms of tin and copper bond "in a molten soup" as temperature and pressure controls effect the reaction, after which the alloy is "cooled into solidity."

It is via conservation of the proton-neutron nucleic strong force that matter retains its distinctive characteristics for functional utility. Distinct particles interact via charge Motion-in-Space displacement Forces to Form elements. However, pressure and temperature Differentials are not so extreme as to transform the process into atomic fusion. Fusion and Fission are special Forms of Energy Transformation that require extremes of temperature and pressure not common or usual to Earth ecology for vitality and vibrancy of life-support systems.

Why do electrons bond at outer energy levels rather than at energy levels closer to the atomic nucleus? Magnetic-Field strength and Gravity-Field strength are more powerful nearer the nucleus, hence, the greater the energy required for "escape velocity."

The formation of a metal alloy does not consist of "metalized fusion" at "plasma level." If a certain Object made of bronze is re-melted, its tin and copper molecular constituents can be differentiated for re-separation.

In chemical reactions between metals for the formation of alloys, there is no "perfused merging" as in nucleated plasma condensate fusion. Particles maintain their identities and their "electro-motive charge binding energy distances" are sustained even when they interact for molecular change, e.g., There is a specific range-of-distance maintained between particular molecules of each Element. Binding-energy distances, and particulate identities pre-determine values of variables amenable to setting the density of a specific Form of Matter respective to its distinctive elemental structures, and also for such properties as electro-conductivity.

"The strong force" is the prevalent atomic Motion-force that sustains energy conservation due to its pro-centric vectored forms of "Curvature attraction" that simultaneously "repel" reacting electrons during molecular change so that atomic integrity, particle characteristics, and binding-distances are maintained for specific elemental properties.

The Planet's integrity is also "conserved" even as it receives solar gravitational inputs and interplanetary inputs from the interplay of shared gravimetric solar system and ellipsoid plane parameters. The integrity of the solar-Frame is "conserved" even as its inner-Curvature plasma condensate convection dynamics are "agitated" by polarized fields with multi-vectored and omni-directional counter-forces that compress nucleated reactions within instantaneous cycling-Time for energy transformation, complex operations of which, accounting for release of spectral radiations. Every 11 years, the Sun undergoes magnetic field polarity displacement and every 22 years, magnetic field polarity change. Solar system binding energies are "conserved" through ellipsoid plane operations even as Planets engage in "re-calibrating" their energy cycling parameters in frame-bounded gravimetric conditions of dynamic System equilibrium.

Per quantum mechanics, electrons revolve around the atomic nucleus; per Newtonian Mechanics, planets rotate on their own axes as they revolve around the Sun. "The strong force" keeps protons and neutrons together as "the weak force" accounts for electron orbital binding energies, released emissions, and molecular change dynamics. A magnetic field mimics the strong force in binding energy strength, as gravity force equivalents mimic the weak force. While a magnetic field displays electromagnetic projection characteristics, Gravity emerges from centri-vectored rotary Forms of motion elicited by magnetic field Energetics.

The Sun's magnetic field sustains planetary revolution and rotation, as planetary magnetic field engenders satellite orbital motion. Solar-induced electromagnetism acts as a force that binds particles, rays and waves into "cycling units of momentum radiation mass energy gains" for gravitational emissions that engender the magnetic field form of Curvature pressure force. There is Continuum between all these overlapping and interrelated unique structures of binding-energy Force that replicate differentiated forms of Curvature pressure-motion-Force Energetics for uniform universal integration, due to the Electromagnetic Properties of Matter, as bounded by the Speed of Light, which is the limiting factor in the Law of Energy Transformation.

Emission of spectral radiations by a Space-body with Mass-in-Motion necessitates "the shedding of Mass" in the Form of energy gains, and hence, why a Planet, as "congealed Energy," must "move" much more slowly than a Unit of Spectral Radiation energy, e.g., Gamma Rays.

Relativity interactions, thus engaged, climax into "binding energy-Force Opposites," yielding cycling-Differentials" that impinge on each Frame's "Curvature Force-Energy re-translation" of solar electromagnetic Field Energetics, into gravimetric binding energies.

"Re-translation of Curvature-Force Electromagnetism," is therefore, at the root of "reconciling" all "equivalent expressions of Field-force" for Frame-specific centri-vectored Forms of motion with functional Outputs that embody respective patterns of thermodynamic cycling.

Do not protons and neutrons "cling together" by a strong force compressing their masses upon their own axes? Do not electrons engage in nucleus-centric revolution? Curvature pressure force motion equivalents find relative expressions as "spheres-of-entanglement" constituted of gravity force, field force, electro-motive-magnetic charge displacement-in-Space force, "field-plane mass interactive force" and their common centers-of-influence.

All Matter is electromagnetic in natural essence or constitution: Even organic molecules like peptide chains constituting proteins are sensitive to polarized light, turning to the left or to the right, e.g., L-Lysine, L-Dopamine. As all Space-bodies-with-mass are constituted of atoms composed of positively charged protons, negatively charged electrons and charge-less neutrons, whole energy systems that are in motion possess an intrinsic affinity to respond to electro-motive magnetic field properties that engender gravitational Curvature motion.

The only Element, the lightest, Hydrogen, does not have a neutron in its nucleus, and hence its "affinity" for nuclear reactions during which, positrons are formed as neutrons are gained.

In biological organisms, DNA-base pairs are bonded by Hydrogen. Hydrogen chemically reacts with Oxygen to form water. The electromagnetic properties of Hydrogen gave it first place in the Table of Chemical Elements. It is a nuclear fuel in fusion reactions as well as a breathable atmospheric gas; it is a component of tissue fiber and protein chains; it can be transformed into combustible rocket fuel. Hydrogen's unique atomic structure, due to the absence of neutrons in its nucleus, offers electrons the greatest latitude in interfacing with positive protons during molecular change and nucleated chain reactions.

The Atom, — where electromagnetism — "the strong force" and "the weak force," — Gravity-field-force and Curvature pressure-force are joined together "in cycling entanglements," — is "suffused-and-perfused," with Relativity variables embodied in magnetic, electro-motive and gravity force equivalents.

This Frame possesses complex Forms of micro-Curvature binding Force-Energetics that:

1) Hold the nucleus together for common proton-neutron projected force-influence;

2) Hold Electrons in their respective orbital revolution paths around the nucleus;

3) Hold the proton and the neutron in a "tight embrace configuration" at the nucleic center to separate individual "electron fields of force" from each other, so that electrons don't "clump" as the proton and the neutron do; this "dynamic of force separation" contributing to the electron's affinity for molecular bonding without "short-circuiting";

4) Hold the atomic frame together as a whole energy system under structure-specific conditions of thermo-cycling dynamic System equilibrium;

5) Embody matter density frame Differentials that distinguish one element from another;

6) Encapsulate cause-and-effect mechanisms that pre-set chemical reactions between specific elements and not with others;

7) Sustain energy cycling Differentials that hold the atomic structure as a "congealed frame" as "the strong force" prevents neutron-escape and electron-release, under normal conditions;

8) Is responsible for Conservation dynamics in cycling Frames because the proton-neutron force is the strongest of all physical forces that counter Entropy processes facilitated by "the weak force;" during molecular change under "congealed conditions," electrons are not released but combine with others to form greater mass-bodies;

9) Is accountable for centri-vectored Forms of pro-centric rotary Motion that gives rise to Gravity equivalents and Field-force analogs, because "the strong force" prevails over "the weak force" and other Forms of motion-Force-energy.

Earth magnetic-field is of a Form different from the Sun's magnetic-field — the first is from a "congealed Space-body," whereas the second is from an "excited Space-body." The magnetic field of the Earth is a dipole holographic effect emerging from molten iron-magma-core electro-conductivity, in synergistic workings with atmospheric layers whose ionized properties cause the retention of static electrical charges, from complex interactions of which, arises "the dynamic property of planetary bi-polar magnetism."

The existence of an operating "dipole magnetic field" as detected by a metallic compass, is a complex derivative of relative interactions between the Sun's radioactive electro-emissions impacting the Planet, as processed by Earth ionosphere in "entangled interface" with Earth electro-conductive molten iron magma core.

Continuum Space-Time Curvature "bends" trajectories of Mass-in-Motion-Frames that exert relative binding energies upon each other's overlapping spatial environs, hence, causing centri-vectored or pro-centric Forms of Motion, such as Rotation and Revolution. Consequently, Gravity-field force, magnetic- field force, solar gravitational binding Energetics, "the strong force" at the atomic frame and "the weak force" allowing for molecular change and radio-emissive discharges, all of which, demonstrating differentiated pro-centric and counter-centric cycling patterns of motion-Force, are "transformed states" of Curvature torque-stress-tensor pressure-force Energetics emerging from the Electromagnetic Properties of Matter.

Gravity-field Motion-Force Energetics is multi-faceted: There is Gravity at the atomic Frame; Gravity towards the center of the Earth; Mass-dependent Gravity as in the Moon's geo-synchronous orbital motion; Gravity within solar core plasma radiation condensate processes; and Solar System Ellipsoid Plane Gravity as an emergent property of solar magnetic-field projections for Earth revolution and rotation!

What is Gravity? Wherefrom does it originate? Is there really a "Gravity-particle" called "graviton"?

Gravity is a Motion-causing emergent Force-property of Electromagnetic-field induced Curvature-activated centri-vectored/pro-centric Forms of Mass-dependent displacement-in-Space. Gravity, possessing both centripetal and centrifugal dynamics that engender radial patterns of vectored motion towards a Center-of-Mass, presents a great diversity of variegated mass-dependent Motion-vectors, e.g., Solar magnetic-Field-activated gravitational radiation projections yield Continuum Curvature Energetics effecting Earth magnetic-Field exertions whose induced Forms of heliocentric Motion climax into Revolution and Rotation; these rotary motion patterns acting upon the great mass of the Earth cause mass-dependent Core-centric and external-Force-induced counter-centric Forces, complex interactions of which, causing Objects "to move" along its line-of-radius, as equivalents of Gravity-field Force directed towards Earth center-of-Mass.

Just spin a record on a phonograph player; and beginning at the center of revolution, put a light plastic pebble near the center-hole of the disc, and observe what happens thereafter! When closer to the center of revolving record Mass, the light plastic pebble will appear to remain in static rotation. But after a while, due to vibrations and the cumulative effect of the outward-force or centrifugal force impinging upon the pebble, it will "fly off," away from the record's center of Mass to fall off the record player unto the floor. Thus, absent a centripetally motion-vectored Force to counter the centrifugal Force that compels the pebble towards the record's circumferential outer-edge — a centripetal Force-like effect towards the record's spinning center — the pebble will not remain on the record's disk to attain an "on-the-record spinning state" that allows it to remain on the disk in a specific location without "flying off!" Without the pro-centric or centripetal counter-Force to oppose the centrifugal Force "pushing the pebble" out of the disk's ecliptic, the pebble will never reach "dynamic System equilibrium."

Therefore, regarding on-Earth Gravity-field Force, both Revolution and Rotation are necessary "cycling Inputs" for putting into effect the great latitude-of-movement affording Objects the ability to "carve-out" a desired path or trajectory upon the Earth, apart or different from the rotary Forms of Motion demanded by Solar projections of Magnetic-field Curvature Motion-Force Energetics.

"This centrifugal effect" — causing the pebble to fall-off the rotating disk onto the floor below — would have controlled for all Space-bodies-with-Mass-in-Motion upon the Earth, if the Planet were to have engaged only in Revolution around the Sun but not in Rotation upon its own 23-degree-tilt axis; or if the Earth had smaller Mass or size that allowed it to rotate with greater velocity, it would be impossible for Gravity to remain at 1-G. But the Earth has great Mass; it takes 24 hours for only one rotational cycle upon its own axis.

The spinning disc revolving on the record player has no centripetal force to counter the centrifugal force causing it to "fly off unto the floor." But the Earth experiences motion in two rotary patterns: Revolution and Rotation that engender both centrifugal (counter-centric) and centripetal forces (pro-centric), whose "tensor-stress-torque pressure-force Energetics," contributes to Earth "wobbling," equatorial "bulging," and bi-polar/dipole oblation.

Thus, centripetal and centrifugal forces "balance each other out" so that there is dynamic planetary System gravimetric equilibrium — in that there are "Force-Differentials" as well as "Opposite-Force-Vectors" operating between them that allow us to be "attracted" towards Earth center-of-Gravity and center-of-Mass while we are able to move freely at the same time, to even counter pro-centric Force-vectors with centrifugal Motion patterns propelling us "away from the center of the Earth," e.g., A jet-propelled aircraft launched "straight-up" into Earth Atmosphere, away from Earth center-of-Gravity and center-of-Mass.

In addition, the counter-Force-effect of Earth Revolution around the Sun being greater than that of Rotation of Earth upon its own axis, climaxes into the centripetal force overcoming the centrifugal tendency to "fly off" away from Earth center-of-Mass. This gravimetric complex of "Force/counter-force Differentials" allows us to thrive under the Force of Gravity or "its equivalents," with freedom of movement and fluidity of motion.

There is a predisposition in equivalents of the gravity force to "pull all Objects" towards Earth center-of- rotation due to centripetal Core-centric forces overcoming centrifugal counter-centric forces.

The hypothetical "graviton" has not been detected during on-going operations of the Gravity-field force. However, a magnetic-Field force is an electro-Motive particulate phenomenon whereby opposite electric charges counter each other in Space-Displacement-Force, rotary Forms of Motion, and directional Vector; hence, a magnetic compass will point to the North Pole of the Earth.

It is not certain that there is a "gravity particle" in reality; but the theory is that "the graviton" is a "subatomic particle" surmised to be representative of Gravity in operating force action. Particles are constituent components of Atoms; but Atoms-qua-Atoms as "inert-Energy-units," do not "emit" particles.

Particles, rays, and waves are usually emitted by "excited Matter-mass Units of Energy," such as the Sun; or by radio-active elements like Uranium and Radium. For example, a microwave oven contains a light bulb that is compelled to emit non-nucleated radiation due to electrical current being driven to high frequency wave emissions by an electric transformer.

The Sun's "excited" nuclear plasma-condensate inferno projects the electromagnetic spectrum, ranging from X-rays to long wave radio. No so-called "graviton particle" has been detected within the electromagnetic spectrum. "Congealed bodies" like the Earth, do not emit a radiation spectrum which would include "the graviton." A non-burning stick of wood will not project light or heat as wave emissions; however, an ignited stick of wood will burn to release light and heat, both of which, belonging to the electromagnetic spectrum.

Gravity-field Force has no electromagnetic characteristic that would confirm the emission of any particle called "graviton." Gravity-field Force is not particles-based but is an emergent force-property of rotary patterns or Forms of motion, such as Revolution and Rotation, as activated by centri-vectored or pro-centric Forms of Continuum Curvature pressure-Force Energetics, e.g., Earth Revolution and Rotation "climax" into an "Interactive-Entangled Field-Complex" yielding "Gravity-force equivalents."

An electric field and a magnetic field embody Forms of "re-translated Electromagnetic Properties of Matter" that process "Curvature binding Energetics" in ways that cause centri-vectored or pro-centric Forms of Motion-force, e.g., Earth Revolution is clockwise heliocentric; Earth Rotation is counter-clockwise geo-centric. Both are induced by the Sun's extremely potent electromagnetic-Field. Both Earth Revolution and Earth Rotation account for the Gravity-field force equivalents.

Curvature pressure-force binding-Energetics involves "tensor-stress-torque-elastic pressure-Force-variability" — Force-range flexibility — in Frame-bounded proportions, as impacted by counter-vectored patterns or Forms constituting the "Entanglement Complex" engineered by attraction-and-repulsion Frames, contraction-and-expansion Frames, and compression-and-dilation Frames, relative interfaces of which, activating centripetal-forces and centrifugal-forces Energetics, and composite-recombinant equivalents thereof.

Gravity as "binding-energy," is different from an electric field force and from a magnetic field force, the complex sums of which, amounting to Electromagnetism. Gravity operates as a "Mass-dependent holographic attraction-Force" that "resonates outwards" from a center-of-Mass that is "trapped" in rotary Forms of Motion. Gravity permeates all Mass-in-Motion Frames that thermodynamically cycle in Energy Transformation. Gravity is Mass-in-Motion-dependent!

Atoms agglomerate into greater Space-bodies-with-Mass as particulate bonding climaxes into patterns effected by the complex interplay of nucleic and molecular forces.

Gravity, as an emergent property of Curvature-induced centri-vectored Motion-Forms, such as Revolution and Rotation, does not have a particle foundation or electro-motive basis analogous to that of an electromagnetic field force.

For example, the Moon displays a fraction of Earth gravity-quotient (1-G) because it is also revolving and rotating as it geo-synchronously orbits the Earth. A mere asteroid, even of great Mass, in a state of complete inertia in deep-void vacuum Space, would have no Gravity-field-Force whatsoever.

In addition, given that other Planets such as Mars and Venus are also in constant Continuum Curvature Motion, they also possess "gravity-quotients," as compared to 1-G or Earth gravity, in Mass-dependent Forms, respectively, as consistent with gravimetric variables whose relative proportions are determined by the respective cycling "Mass-in-motion energy-pressure-force complex" that animates their thermodynamic cycling planetary Frames.

Furthermore, there arise additional difficulties with accepting the supposition of "Graviton existence" in the absence of empirical reproducible experimental data substantiating its physically proven discovery, detection, emergent-Source, and operational dynamics.

As vacuum Space is continuously bathed in solar gravitational radiation, it would be logical to assume that "sub-atomic particles" also thrive there, including "Graviton," were it to exist in reality. Yet, even in orbit around the Earth, astro-scientists in the Space-Station do not detect any so-called "Graviton particle."

The whole electromagnetic spectrum of rays, waves, and particles is traveling through vacuum Space as it extends towards the Earth. How come there is no "Graviton particle" operating in Earth orbit when our astronauts are in the Space-Station or in the Space-Shuttle?

Where are the so-called "Gravitons" around Earth orbit, to generate a force analogous to Gravity upon the Earth? Why would "the Graviton particle" exist only on the Earth but not upon the Moon, which also has Mass? If the electromagnetic spectrum contained the "Graviton particle," it would be made manifest everywhere in the Solar System. Is not the solar system uniformly showered with radiation throughout the whole range-of-projection covered by the heliosphere?

The Space-Shuttle is revolving around Earth orbit at 117,000 miles per hour and mildly rotating upon itself in order to provide a modicum of gravity-like Force, but which is still referred to, as "micro-Gravity." Astronauts and scientists even conduct experiments in "micro-Gravity" in order to investigate occurrences that they assume would be different in Form and results, in comparison with those obtained on the Earth under normal 1-G Gravity conditions. Yet, the so-called "Graviton particle," unlike the proton, neutron and electron, has not been substantiated in a Form that is scientifically reproducible in laboratory or field experiments.

Dr. Isaac Newton was able to ferret out the law of universal gravitation and the laws of motion without the complexities of electromagnetism, spectral mass emission binding energies, gravitational Curvature Energetics, and tensor mathematics. He worked out equations for the laws of motion and the law of universal gravitation before the invention of the light bulb.

Gravity Force is an emergent property that inheres from specific patterns of centri-vectored or pro-centric Forms of rotary motion, as espoused by Revolution and Rotation. An example given to illustrate the operation of General Relativity is, that a man situated in a moving elevator, would not be able to tell the difference between Gravity towards Earth center-of-mass and the velocity-Force involved when the elevator is moving sideways, given that his feet would be pegged to the side opposite to the direction of Motion. It is analogous to the motion force experienced in a vehicle or airplane taking off suddenly, hence "jerking passengers backwards, the sum of which might result in nausea and cold sweats.

The problem that arises in this "thought experiment" is that Human Beings are pre-designed with "sensory metrics" that "tell them" or indicate to their organism that their position is "out of synch" with Gravity-centric indicators, e.g., Given that his body is not rigid and would require some kind of support for remaining in a "sideways position," the man in the elevator moving sideways would know for sure he is "going sideways," just as he would also know he is upside down whether in a vacuum or within an interactive sphere of normal Gravity-force influence. Astronauts in micro-Gravity do know when they are either right-side up or upside-down.

Centrifugal force generating machines utilized in astronaut training consistently engender analogous effects on the human body via specific forms of counter-centric-vectored Motion, while the centripetal force is held or presumed to remain at "a constant inertial state" while the Person is strapped to the seat of the centrifugally-moving machine. Restricting the body within the seat engenders a Force that substitutes for the absent centripetal force, which could be engineered by rotating the seat in the direction opposite to the centrifuge's Motion-vector — clockwise or anti-clockwise.

Therefore, because Gravity on the Earth emerges from the active influence of two counter-posit Forces, i.e., Revolution (clockwise) and Rotation (counter-clockwise), then, Gravity is intrinsically an emergent-Force property of oppositely operating centri-vectored or pro-centric Motion-Forms that counter-actively project "Curvature counter-posit Energetics" as controlled by co-determinants that factor in setting range, magnitude, and direction, e.g., Revolution and Rotation, as counter-vectored opposite Forms of pro-centric rotary Motion Energetics that account for Earth 1-G Gravity-force range-of-projection.

Because the impetus to unify the Universe via a general theory of force engages the mathematical formalization of electromagnetism, (that also includes "the strong force" and "the weak force") and Gravity into a common operational reference-Frame that accounts for Earth 1-G Gravity-force as well as for Mass-dependent Earth Revolution-and-Rotation momentum-moving-Force, then it is necessary to analyze how these Forces impinge upon Mass-in-Motion Frames — how they affect or impact the massive constituents of Planet Earth for inducing 1-G Gravity equivalents — for generating "Thermo-cycling Differentials" and "Motion-force Opposites," consistent with respective fulfillments of the universal Input-Process-Output Mechanism.

It is the Mass of moving-Planet-Earth that is the common co-determinant in establishing the range-of-projection for both Electromagnetism as a Motion-force and Gravity as a Motion-force. Thus, Electromagnetism accounts for both Revolution and Rotation, the complex interplays of which, accounting for pro-centric G-1 Earth-center-of-Mass-vectored Gravity-force. Revolution is moving the Mass-of-the-Earth in ways that are countered by those in which Rotation is moving the Mass-of-the-Earth, which they both share, in co-determining Earth G-1 Gravity-force.

****

QUANTUM MECHANICS AND SOLAR SYSTEM DYNAMICS RELATIVE TO EARTH 1-G MASS-DEPENDENT EARTH-CORE-PRO-CENTRIC GRAVITY-FORCE

When Atoms combine for molecular bonding reactions, particulate binding energies come into action between elemental nuclei and electrons. Protons and electrons are simultaneously attracted and repelled by each other as the neutron serves "to ground" particulate charge Motion-displacement Forces that interact in forming molecular bonds.

When molecules aggregate/coalesce to form greater Space-bodies-with-Mass, binding energies work via micro-exponential Forces to hold the Atomic-Frame as a whole Energy system in dynamic System equilibrium. Nuclei constituted of protons and neutrons also interact to relatively repel each other due to projection of similar electro-motive charge-Forces, which are mediated, however, by charge-less neutrons, the complex interactions of which, climaxing into elemental characteristics, molecular qualities, and dynamic properties accountable for density of Matter.

From Quantum Mechanics to Units of Moving-Mass formed by molecular change, and from intermediate Mass-frames to Newtonian Mechanics, all Mass-in-Motion-Frames participate in responding to the same Curvature binding pressure force Energetics projected by the Sun, as "negotiated" by proportions, variables, and thermodynamic mechanisms at work within such Mass-in-motion-force-energy Frames, respectively.

Each organized Frame, e.g., Planet Earth, presents specialized cycling variables consistent with fulfillment of the Input-Process-Output Mechanism towards its own particularly predetermined range-of-operational-projection for achieving dynamic System equilibrium.

Electro-Motive Charge-Displacement relationships accountable for Motion-in-Space, unfold under either "congealed conditions" or under "excited conditions" that embody analog or quasi-similar Forms of "re-translated Electromagnetism" from which emerge "Mass-dependent Gravity-equivalents," consistent with the predetermined purposive functions of each particular reference-Frame.

Atomic particles are still "actively engaged" within Mass-in-Motion Frames and within inertial Mass-frames but as configured by their "congealed-state thermodynamic boundaries," complex interplays of which, preventing the development of expression of their Energy-transformation "charge-displacement-exchanges" as Motion-in-Space pertinent to nucleated plasma chain reactions. Planet Earth remains in "a congealed Energy state" whereas the Sun remains in "an excited Matter state."

However, in the Sun, which is an "excited Mass-frame," particulate charges transmute and transubstantiate into "nucleated radiation-chains sequences" taking place in "compressed cycling run-Time," even as they "shed their mass energy gains" as spectral mass emissions that activate the tensor-stress pressure-Force Energetics animating the whole heliosphere within its "Solar System projection-range.".

The First Law of Thermodynamics, which is Conservation of Energy, states that: "Energy is never created nor destroyed but always transformed;" and its Second Law, being the Law of Entropy, is expressed as "the amount of energy unavailable for useful work in a system undergoing change."

Conservation and Entropy are inevitable thermodynamic processes or phenomena characterizing every Mass-frame in the Universe that undergo Energy Transformation, the complex interplays of which, giving rise to patterns of energy transformation that appear as "stochastic resource cycling" or "cycling periods of resource utilization," in differentiated fulfillment of the Input-Process-Output Mechanism by respective Mass-in-motion-Frames, or by presumed inertial Mass-frames.

Cycling Differentials between mass-Frames determine how they relatively interact within their overlapping regions, or areas, or theaters of intersection. Relative equivalency is established via "conversion factors" and alpha-numeric Constants that must "reconcile" cycling Differentials as "renormalized" with Motion-vector Opposites, into "a unified stream" of Continuum Space-Time Curvature Motion-Force operations.

Though there is relative equivalency between operational variables predominating in each mass-Frame, respectively, — given that the Universe is gravimetrically integrated for dynamic

System equilibrium in Oneness-Functioning — these variables account for differentiated cycling due to the Law of Energy Conservation that is "operationalized" within the boundary-ranges encompassed by proportional ratios of dissimilar and asymmetrical constituent variables.

As the same universal Energy is converted into its many Forms, — these Forms espousing structures consistent with their respective differentiated thermodynamic Energy-processing Mass-units or Mass-in-Motion-frames— the same Mass-dependent Gravity-force is also "replicated," as applied in its variously vectored equivalent motion-Force patterns, e.g., rotary, linear, rectilinear, and/or curvilinear, all of which depending upon each mass-Frame's cycling variables, parameters, co-determinants, and conditions.

Gravity-force operates in lifting a stone as it operates in launching a rocket into outer space. However, proportional ratios for each circumstance that effect this "equivalent exertion of the same Gravity force" would not be of similar quantities. However, there is uniform application for validly proven physical laws in relative functional equivalency, but within cycling-Frames where gravimetric conditions are similar, e.g., A valid physical Law for our Solar System would be scientifically applicable to other Solar Systems that reiterate or replicate our Solar System's gravimetric and Field conditions.

Because of cycling Differentials inherent in fulfillment of the Input-Process-Output Mechanism for Frame-specialized functionality, as Frames share overlapping regions of intersection, alpha-numeric Constants are always needed for "reconciled unification." There will be Differentials in mass, force, energy, temperature, pressure, and motion: the proton-neutron force is the greater atomic force called "the strong force," e.g., neutrons do not revolve around protons but "they cling to each other in strong bonding force."

On Planet Earth, centripetal motion-force towards Earth core-center of Mass is the greater mass-centric force as centrifugal motion-force is exerted in a "weaker magnitude," e.g., a man can walk or run with equal ease upon Earth surface; but he needs an aircraft to fly, as a bird needs wing-motion to reach a tree branch on which to perch. Conservation is the greater process in securing Continuum in the Universe, as Entropy involves operations of "the weak force" that allows for molecular change, atomic decay, and radiation emissions.

Dynamics engendered by conservation of energy, "the strong force," and centripetal motion are accountable for Curvature-Force in Continuum Space-Time, whose Energetics hold the Solar System together, hold the atom as a whole energy system, allows a person to walk upon the earth, drive an automobile, and fly an airplane. "The strong force" is at the center of the atom. It is encapsulated into proton-neutron bonds. The proton-neutron strong force causes nucleus-centric motion force Forms that impel electrons to revolve around the nucleus. Earth forms of radial motion are also core-centric towards its center-of-mass and center-of-gravity. In the same vein, the Sun's cumulative core-centric atomic strong force complex projecting the external magnetic field impels all Planets to revolve around solar system centers-of-mass, centers-of-gravity, and centers-of-field.

Curvature pressure force "spheres of interactive influence" are expressed as binding energies, taking the Forms of magnetic field force that engenders heliocentric rotary motion patterns such as Revolution and Rotation from which emerge equivalents of the Gravity-force accounting for geo-centric Moon orbital motion and Earth surface radial core-centric motion Forms.

The Mass-energy-in-motion-force complex accounts for mass-Frame constituents in differentiated cycling of the Input-Process-Output Mechanism as "the strong force" compels energy conservation for dynamic System equilibrium.

Centripetal forms of "motion-processing" having pre-eminence over centrifugal motion patterns cause all things "to gravitate towards the center." Tornadoes are formed in that manner as great mass-in-motion-force-energies are "potentiated" into a center of rotating "Curvature dynamo," as framed by pressure Differentials, oppositely vectored electro-motive charges, and temperature Differentials. Condensed traveling momentum mass energy gains, are then "transferred" to the environment, as destructive damage.

Electro-magnetic-Field-engineered mass-in-motion will generate a gravitational field from which will arise Curvature-Motion-Force properties that are expressed in equivalents of the Gravity force. Electromagnetism is responsible for Revolution and Rotation as standard Planet Earth external rotary Motion-Forms. Gravity-force equivalents emerge from these co-laboring rotary Motion-Forms. How such equivalents of the Gravity-force are expressed depends upon the Mass of the Object upon which they are prevailing. Where Curvature thrives, differentiated and opposite motion patterns will emerge, as conditioned by Mass-Frame cycling variables, parameters, and co-determinants, and operations of the Input-Process-Output Mechanism.

Electron-motion is in Continuum with planetary motion. Every Object and Space-body is in Continuum Motion elicited by relations between thermodynamically-cycling, opposite electro-motive charges.

An accelerated particle in a cyclotron is in "Continuum-Form-synchrony" with a particle emitted from the Sun. A compressed particle from an Earth-bounded nuclear fusion chain-reaction is in "Continuum-Form-Synchrony" with solar plasma condensate particulates. The difficulty arises in identifying, analyzing, quantifying, and inter-relating gravimetric Differentials that distinguish these various Frames wherein the electromagnetic properties of matter are being "re-translated."

For example, a traveling spaceship that is accelerated gains mass or has a mass increase due to increase in the rate of momentum energy. But that is of a kind different from a nuclear particle accelerated in a cyclotron that gains an increase in mass energy. The spaceship will not begin to emit cosmic radiation like the accelerated particle, but will disintegrate or "shed its mass" as it is approaching the Speed of light, which embodies a limiting factor for all Mass-in-Motion Frames in Continuum Space-Time Curvature. Specific-Frame variables-proportions in response to cycling Differentials establish the gravimetric force parameters operating in "re-translating" electromagnetic properties of matter as mass energy gains.

These "gravimetric Differentials" would apply also in analyzing the kinematic velocities of nuclear particles in a fusion reaction chamber to account for cosmic velocities of emitted radiation in a vacuum.

Acceleration of an Object by Gravity at Earth surface is estimated at 9.8 meters per second per second; the Speed of Light in a vacuum is 3.0×10^8 meters per second. How great is the force of Gravity in nucleated processes? Gravity-field Force is mass-dependent in conjunction with the Electromagnetic-field Force acting also upon the Object. The smaller the mass, the lesser a Force necessary to propel it into Motion, and the faster the Object can travel, relative to mass-dependent Gravity-equivalents acting thereupon.

How does the Speed of Light get "re-translated" in nuclear chain reactions? Given that in Stars, it takes nucleated chain-reaction radiation processes to generate Light-radiation-Energy, then, then no particle can exceed the Speed of Light while itself, is generating Light-radiation-Energy. Hence, Energy-radiating Objects, even as infinitesimal as a Photon, must have some amount or quantify of Mass, however, minute-micro in proportions and dimensions; and that's why the Photon is the lightest radiation-energy generating particle. Given that all things visible must possess the property of Mass, then the Photon must also possess an infinitesimal amount of Mass, which it "sheds" when accelerated to infinitely fast velocities that approach the Speed of Light. Hence, reasons why, once the original Source of radiation-Energy emissions is "shut-off," the Photon must "disappear," e.g., A flash-light switch is "turned-on" to project a stream of visible Light; once that switch is "turned-off," there is no longer a visible Light stream: It "shuts-off" as instantly as "turning-off" the switch.

Therefore, Light-radiation-Energy, in whatever Form, cannot propagate without an ever-present original "real-time on-Source," which means that it is not possible to "see" the Universe "as it was billions of years ago," by simply looking at starlight from the Heavens: Light-sources being perceived-visualized-seen are shining-and-radiating "in-the-Now," "in real-Time."

Given the Law of Entropy, Light-radiation-Energy cannot propagate indefinitely: It must have an ever-present real-time "on-Source" in order to remain as a stream of rays or particles from which derives the Light we perceive.

In the Star, such as our Sun,"the strong force," centripetal motion, and Energy Conservation cause Curvature Energetics, Gravity-field, Magnetic-field, and electro-motive-dipole-charge displacements-in-Space Forces, to cycle in "compressed-Time" or "shortened-Time," such that events occur in "nano-seconds of duration," as "all lines of Force" are "contracted" in the likeness of "Core-centric convolutions" and "Corona-vectored permutations," along the Sun's "line-of-radius."

Therefore, in order that Electromagnetism and Gravity "converge" or "commute" as "confluent Field-Forces," given that where Mass-in-Motion-Frames overlap there are different Forms of "Motion-force-energies" that "interface" as engendered by "Curvature Energetics," then this "transmutation-Complex" requires the interposition of alpha-numeric Constants for formalizing in Mathematical Theoretics, the "dynamics of unification," the realities of which, being already integrated in the Universe by "Momentum Curvature-pressure-tensor-stress-torque-Forces," e.g., Electrons revolve around Atomic Nuclei; Planets revolve around Stars; components of Galaxies, such as Solar Systems made-up of Star-Planets Constellations, revolve around the Galactic Center wherein radiates a super-massive Star, such as the Milky Way Galaxies; and Galaxies revolve around Galaxies, etc. . . . . .

The impact of a Spaceship upon another Spaceship will result in catastrophic structural damage to both Spacecrafts — this release of "Momentum-accelerated Mass-energy-gains" "expotentiates" the damage-causing collision with explosive shocks accompanied by shockwaves-Forces with attendant materials destruction and residual radiation.

"Momentum Accelerated Mass Energy Gains" are made manifest in different Forms, respective to "congealed Frames" and "excited Frames." When two Planets "collide," the residual impact is of a Form ("congealed") that is markedly differentiated from that resulting from "the collision" of two Stars ("excited"). Explosive shocks-and-shockwaves are also "transferred" to the "environment" in differentiated Forms, respective to void-vacuum-Space and Planetary or Earth-Atmospheric-Space.

The impact of a nuclear reaction chamber's explosion within a Spaceship entails radio-active emissions, immense shock, shockwaves, turbulence, heat, and light-and-particulate radiation-energy in a Form that is clearly differentiated from the impact of two "nuclearly-inert" Spaceships in terms of release of Momentum-Accelerated-Mass-Energy-Gains — Whereas the explosion of nuclear bomb or nucleated reaction chamber, whether as fusion or as fission, would be tantamount to the explosion of a Star.

Still, from "congealed" to "excited," there is "Continuum-of-Forms" embodied in both kinds of residual impact resulting from explosive phenomena, respectively, engendering therefore, "Frame-interactions" that "commute as relative equivalencies" when a "conversion-transmutation" between the two states occurs, in the sense that it is the same Energy that is being "transformed" into another state — "Energy is never created nor destroyed but always transformed;" hence, reasons why we observe quasi-redundant structures displaying "iterativeness of pseudo-similar Forms," e.g., Electrons revolve around a nucleic center-of-Mass; Planets revolve around a Solar-System center-of-Mass.

In a nuclear reaction, convergence or integration of "Relativity Differentials" is activated as Electrons are released and Neutrons escape from environmental and materials components, whereby nucleated atomic nuclei form Positrons as they process the released Energy. On the Earth, "star fuel" or "nucleated plasma-condensate" is created from the heretofore inert gases in the Atmosphere or in environmental-ecological constituents as residually impacted by the nuclear explosion detonation.

$E = mc^2$ is a specific application of the Law of Transformation of Energy, which when "processed" upon the Earth or within one of its constituent components such as the Atmosphere or the Hydrosphere, such as when an atomic bomb explosion is detonated, it results in a disintegrative reproduction of the universal Input-Process-Output Mechanism.

In such cases, electromagnetic properties of Matter are "re-translated" into an "excited Form" within a "congealed environmental Frame, "Cycling Differentials and Force-Opposites" of which, are already pre-bounded by the constant value of the Speed of Light as modulated by Earth Atmosphere, or respective to the Speed of Light in a planetary atmospheric medium. Thermo-nucleated and composite-recombinant chemical chain reactions are not self-contained, for the Earth is not an "artificial Sun." The whole Energy yield — kinetic or external Energy, the internal energy of nuclear particles and thermal radiation energy — are all dispersed in a catastrophic impact that deleteriously affects the whole earthly environment while there is no "escape mechanism" like outer vacuum Space into which explosive materials can disperse. The Earth has a breathable Atmosphere and a hydrosphere made-up of oceans, lakes and rivers, from which sweet water evaporation yields condensation for cloud formations, contingent with ionized radiation Energy generating "bouts-and-flashes of lightning," thus producing rain.

Due to the facts that there is no "container" or "encasement" for the nuclear interactions that form different atomic nuclei by redistributing protons and neutrons and positrons etc...in the atmosphere, land or oceanic hydrosphere, untold damage and destruction are inflicted upon Earth ecology, including the possible "rarefication" of the atmosphere, on the one hand, and the "nuclearization" of the oceanic hydrosphere, on the other.

Fission products will even fuse with particles of Earth; and in the upper atmosphere where air is already "rarefied," or is "less dense," which decreases shock wave dispersal but increases the release of thermo-nuclear radiation, there occur electromagnetic transmission disruptions that affect the formation of ionizing radiation for the onset of descending "cloud charge uptake" for lightning generation and other atmospheric processes of electro-static condensation.

Atmospheric gases are limited in volume, proportion, content, and composition, and when atoms are "stolen" by uncontained or unconstrained explosive atomic reactions, this implies a decrease in atmospheric Mass and therefore, also a decrease in atmospheric capacity to synthesize breathable variables for Human and animal life, and for photosynthetic requirements of plant growth.

The oceanic hydrosphere also contains Deuterium, a nuclear plasma-condensate fuel in the chain-reactions that comprise the continuous nucleation of Hydrogen into Helium, that exists in molecular arrangements structured with molecules of salted water and sweet water, the atomic nuclei and electrons of which, are again "stolen" by ocean-based nuclear explosions, which "rarefies" oceanic waters (alters their naturally created intrinsic functional composition for dynamic System equilibrium) with proportional disruptions of elemental constituents and molecular components that serve as variables, factors, parameters, and co-determinants of Qualitative Energy Conservation.

This complex disruption of dynamic System equilibrium engenders non-system variant intensities and catastrophic increases in pressure, temperature, radiation, and binding energies that destructively impact ecological processes calibrated to thrive under normal life-sustaining operational conditions.

Normally, only visible light, infrared heat, and ultraviolet rays penetrate the atmospheric layers to arrive at the Earth surface while the primary constituents of nucleated radiation such as Alpha and Beta particles, and Gamma rays engage the magnetosphere to energize the ionosphere for processing of electro-static charge displacements that give rise to lightning as an emergent property of "frictional interactions" between Earth iron core electro-motive-conductivity, solar radiation emissions, and ionized gases formation.

But in an artificial nuclear radiation condition propagated by an atomic bomb explosion, for example, large regions, quantities, or masses of air, land and water remain potentiated with radio-active contaminants which can cause genetic damage giving rise to birth defects and other DNA-affected abnormalities.

It is the "regular Hydrogen Atom," two of which, chemically react with "the regular Atom of Oxygen" in order to form sweet water for us to drink, and not the "Hydrogen position."

In the Hydrogen atom, it is not the electron that is missing, but the neutron, hence its affinity for positron formation in nucleated reactions when Hydrogen gains a Neutron. And that's why Hydrogen constitutes the greater amount of solar nuclear fuel, cycling from Deuterium, to Tritium, and to Helium.

Opposite charges attract. Positive protons are "attracted" to negative electrons within allowable distance ranges as calibrated by neutron mass and zero-charge characteristics. The "atomic interactive micro-sphere Energetics-Complex" arising from micro-Curvature, gravity, field force, centers-of-Micro-Mass, and particulate charge displacements-in-Space, account for "strong force and weak force Differentials," as engendered by Force-Opposites deriving from Electromagnetism and Gravity, in the same manner that the solar macro-gravity Energetics-Complex arising from polarized plasma-condensate-particulate Motions account for extreme solar magnetic field strength; hence, earth planetary Revolution around the Sun and Rotation upon its own 23-degree tilted-axis, Gravity-force equivalents on the Earth, and orbital motion-Force of the Moon around earth center-of-Mass, center-of-Field, and center-of-Gravity.

Atomic nucleus, Earth core, inner-Sun radiation core, and solar system centers-of-Mass, centers-of-Field, and centers-of-Gravity, respectively, while operating as "registration points" for differentiated Curvature-Energetics and Opposite-pressure-force-tensor-stress Motion-forces dynamics, are "sensitive-responsive" to energy changes that are effected by "ellipsoid plane tensing disruptions," hence, the Perihelion Shift of Planet Mercury and the "esoteric Form of ecliptic Revolution by outer-Solar-System Planet Pluto.

Ellipsoid plane tensing equilibrium accounts for planetary dynamic System equilibrium. However, planet Mercury displays a perihelion shift and the earth "wobbles," "bulges at the Equator, and is "slightly flattened" at both Poles. Solar ellipsoid plane tensing conditions derive from solar gravitational radiation and magnetic field, and solar-inter-planetary centers-of-field, centers-of-mass and centers-of-gravity.

How do we explain sustained periods of draught when we have the technological capacity for daily weather system forecasts? Could forecasting electronic technologies also "discover patterns" that predict evaporation-condensation cycles in conjunction with solar activity to climax into more accurate prediction of drought periods?

Due to "slight pre-determined within-ranges variations" in "center of ellipsoid plane tensor-stress-torque-pressure-Motion-force parameters," "relative to" Solar-System-Centers of Mass, Gravity, and Field as caused by rates-of-change in solar magnetic field strength, radiation emission rates, and inter-planetary Relativity dynamics embedded in rates of gravimetric radiation Energy absorption and processing, these Curvature-centers-of-Solar-System-Energetics embody the differentiated workings of all "thermo-cycling Ellipsoid-field plane-Energy-metrics" that effect quasi-iterative frame-specific Earth ecological patterns, e.g., Seasonal Cycles from Spring to Summer to Autumn to Winter and to/from Spring to Summer etc . . . . ad infinitum!

A simplified summary of all known Energy-Forms and-or of all known field-force types display only sub-divisions of: (1) Quantum Mechanics (the strong nucleic proton-neutron force, the weak force accounting for electron orbital binding energy, molecular change, particulate decay and radiative emissions); (2) Gravity (planetary revolution and rotation, Moon's orbit around the Earth, G-force equivalents at Earth surface); (3) Electromagnetism (particulate binding energy dynamics for molecular change, electromagnetic spectrum, electrical current, electric motor, radar, sonar, radio, television); and (4) solar gravitational nuclear plasma condensate-particulate Inner-Core-Dynamics of Field-Radiation-Energetics (atomic bomb, H-bomb).

Every "Form of Force Sub-Division," — "the strong force;" "the weak force;" Quantum Mechanics Forces; and Nucleated Solar Inner-Core Plasma-Condensate-Electro-Motive-Magnetic-Forces, — except Gravity-Field-Force, can be subsumed under Electromagnetism.

Thus, only two primary Forces need integration and unification for explaining "Continuum Momentum Curvature Motion Field Energetics:" Electromagnetic-field Force and Gravity-field Force.

FORCE, as a Physical 4-Dimensional Property of Motion within Continuum Space-Time Curvature, is expressed in terms of Mass-in-motion dynamics in which tensor-stress-torque-pressure-Energetics are "embedded within the devolution of Thermodynamic Time," due to "re-translation" of the electromagnetic properties of Matter.

This Mass-in-motion-force-Energy classification system as elaborated upon in the next Chapter, is not necessarily the one adopted by astro-evolutionist-physicists but is rather an "educated detailing" of major physical force-energy Forms-of-Expression into their "sub-component constituents" (or as expressed by "re-translated-Forms of Continuum Curvature Motion-Force).

Rather, in order that they might interface with each other in Continuum, in accordance with "attached-motion patterns" peculiar to each Form; — ("Energy is never created nor destroyed but always transformed") — they share "regions of overlap," respectively.

And in order that they (Forms-of-Force-Expression) interface with "attached-motion patterns" that each "Force-Form-of-Expression," or a particular sub-counterpart Form thereof, appears to have "an affinity," for accomplishing or fulfilling the predetermined purpose of climaxing into an exposition of a general theory of Force, as prescribed by "A Unified Motion-Force Theory of Universal Space-Time Curvature Continuum," that scientifically explains the workings and operations of "Continuum Momentum Curvature-Motion Field Energetics," the complex Frame-relations of which, as prescribed by conjoined Electromagnetic and Gravitational Fields, the sum of which, being accountable for Universal Whole-System Dynamic Equilibrium!

"The strong force," "the weak force," Gravity and Electromagnetism are called Fundamental Forces by astro-evolutionists; however, the "strong force" and "the weak force" are "re-translated expressions" of Electromagnetism, which accounts for them as "Curvature Motion-force sub-counterparts or sub-types."

But all Fundamental Forces impact the way in which energy cycling Differentials are distributed throughout interactive Mass-in-Motion Frames, as "modulated" by Motion-force

Opposites that impact cycling variables for the dynamic operation of "Relativity determinants of Continuum Equilibrium."

In reality, the two major Force-categories or mathematically measurable quantities involved in conceptualizing a general theory of force, engage Electromagnetism and Gravity, as differentiated, first in the Atom, then in "congealed Frames" such as Planets, and finally in "excited structures" such as the Star.

"The strong force" and "the weak force" already possess properties of Electromagnetism in the Form of opposite-particulate charge displacement-in-Space motion-force dynamics.

Gravity, however, has no particulate characteristics that would base or frame its Form or type-of-Force into "relative equivalency" with "re-translated Forms of electromagnetism," such as "the strong force" and "the weak force." For, while Electromagnetism has "an affinity" for displacement-in-Space Motion-force Energetics via interactive electro-motive magnetic charges, — as engendered by relative interactions of opposite electro-motive charges, — in contrast, Gravity has "an affinity" for Mass-in-Motion variables-and-parameters that account for its emergence as a Mass-dependent Force, as derived from Motion already caused or predetermined by Electromagnetism, respective to particular amounts or quantifies of Mass.

Iterative redundancy of Forms animates overlapping-interacting Mass-in-Motion Frames. However, due to cycling Differentials indirectly and directly caused by Motion-force Opposites that are naturally geared for Frame-specific functions, alpha-numeric Constants are necessary in order to link or interface Relativity variables into Curvature Momentum Force Continuum Energetics, while "conversion factors" are also engaged in establishing "relative equivalency of functional co-determinants" between analogous Forms of the same universal Energy that effect Motion-Force Patterns peculiar to each framed Mass-in-Motion expression of Energy Transformation in Thermodynamic Time!

Not only, then, is there "Continuum Space-Time," but there is also "Continuum Momentum Mass-in-Motion Curvature Force," even when each Mass-Frame, respectively, is fulfilling the universal Input-Process-Output-Mechanism, via "stochastic variables or co-determinants" that together work for maintaining Whole-Universe dynamic System equilibrium with integrated, coordinated, consistent, and constant Motion.

Thus, not only are there apparent iterative redundancy in Energy Forms but there appears to be also apparent iterative redundancy of Forms in "processing structures" and in Motion-patterns, as organized for their overlapping functional operations to climax into "collectively co-laboring for Thermodynamic Continuum."

Because of the Law of Energy Conservation, it is expected that expressions of the Fundamental Forces will find "Mass-Frame equivalents" in "converted states" that embody the same Form of Momentum Curvature Motion-Force Energetics Continuum, as accounted for by the differentiated centri-vectored rotary Motion patterns encapsulated in relative interactions between equivalents of the Gravity-field force and analogs of the Magnetic-field force.

"All Force categories" are integrated for contiguous operations that predetermine how the universal Input-Process-Output-Mechanism will be fulfilled for each Mass-in-Motion-Frame, respectively.

Magnetic-field force, "the strong force" and "the weak force" are all specialized embodiments of force-Forms that encapsulate "re-translations" of the electromagnetic properties of Matter, the complex sums of which, cycling within operational ranges as bounded by various differentiated Frames, or as peculiarly organized and ordered, for Energy Transformation, respective to the kind or type of Frame they are.

Thus, it is Electromagnetism, —expressed as Magnetic-field-force — in all its differentiated cycling Forms, that must find a Common Frame-of-Integration for unified operational Continuum with equivalent Forms of the Gravity-force — expressed as Gravity-field-force.

******

INTEGRATING ALL FORMS OF FIELD-FORCE IN A UNIFIED CONTINUUM OF OVERLAPPING MOTION-FRAMES!

The following rendition is not patterned merely after Force-categories such as Gravity, Electromagnetism, "the strong force" and "the weak force." It could be laid out as a "Force matrix" whereby Force interactions could be exposed for overlapping regions where cycling Differentials are operating in response to Motion-force Opposites.

As the Input-Process-Output Mechanism is being fulfilled via each Mass-in-motion-energy-force Frame Complex, respectively, differentiated Input-Process-Output patterns in fulfillment of the thermodynamic mechanism of functional operations, are accounting for the way in which each Frame maintains its integrity, respectively, especially in the presence of "disruptions" in Continuum Space-Time-Motion-Force-Field(s) Energetics, e.g., Planet Mercury's dynamic System equilibrium is "restored" after each Perihelion Shift "in response to" field-Force activated, range-specific variations, that "trigger recalibrations" of relative Gravity-field strength, (Perihelion Shift still "keeps" Mercury, within the Solar System); angular momentum (Mercury's ecliptic respective to the ellipsoid plane is still preserved); and Motion-Form trajectory is conserved (returns to pro-centric Rotary Motion).

Likewise, because of these range-specific boundaries that allow "thermodynamic cycling overlapping" while sustaining gravimetrics of wholeness-integrity, the Atom can "meta-transmute" from within a "congealed Frame" to another structured "excited Form."

Type of Force: Electromagnetic "congealed"

Indicators: Particle/Wave; Polarity; Attraction/Repulsion; Proton-neutron strong force; Orbital electron force

Effects/Applications/Action: Audio-Visual – Telegraph/Sonar/Radar/Radio/Television/Electric motor; electrical current; high frequency electricity

Counter-Indications: Possible electrocution; tumors if intensity is extreme at close proximity

Type of Force: Sound (Atmospheric Distance Traveled as waves); non-electromagnetic

Indicators: Vibration/Sound Waves/Air Turbulence

Effects/Applications/Action: Shock; Voice/Noise/Sound/Echo/Speakers/Ultra Sound Technology

Counter-Indications: No "known" dangers to human health, except ear damage in case of extreme loudness

Type of Force: Molecular Bonding Energy

Indicators: Atomic electrons bond at outer orbital levels to form compounds

Effects/Applications/Action: "congealed," when combined, form liquids/solids/gases; "excited," when combined, form nuclear radiation plasma condensate as in solar inferno

Counter-Indications: can form poisons, deadly drugs, conventional explosive compounds; when "fused," atoms form Hydrogen bomb; fission creates an atomic bomb

Type of Force: Gravity; non-electromagnetic; non-sound; special sphere/field-of-force-influence; Rotary-Motion induced Category that is Mass-in-Motion-dependent; derived from Electromagnetism-caused pro-centric Forms of motion such as Revolution and Rotation

Indicators: None - "Invisible/undetectable Attraction" (centripetal); also "invisible/undetectable repulsion" when triggered by an additional external force (centrifugal); (not a wave, not a particle, not a vibration)

Effects/Applications/Action: Rest/Inertial mass; Centri-Vectored Motion along line-of-radius, with presence of Applied Multiples of G-Force; accelerated mass, "congealed," destructive physical impact

Counter-Indications: Velocity-and momentum sensitive, due to Mass-variables causing discomfort in humans

Type of Force: Atomic/Nuclear; Electromagnetic "excited"

Indicators: Nuclear Quantum Mechanics – Attraction/Repulsion; Proton-neutron strong force; Orbital Motion of Electrons; particulate charge transubstantiation; electromagnetic spectrum

Effects/Applications/Action: "Congealed:" Molecular Bonding/Change; "Excited": Atomic Bomb/Hydrogen Bomb/Sun; X-ray machines

Counter-Indications: Accelerated "excited": radiation burns, "instant vaporization," radiation sickness, cancers, eventual death

Type of Force: (assumed) Sub-atomic Realm/Frame

Indicators: None

Effects/Applications/Action: None

Counter-Indications: None or unknown

Type of Force: Curvature Pressure Force Motion Binding Energies

Indicators: Solar radiation emissions as Magnetic field-and-gravitational Energetics

Effects/Applications/Action: Centri-vectored Forms of Motion, e.g., revolution, rotation, orbital trajectory, e.g. electron, earth, the Moon; radius-lined motion pattern (e.g., apple's rectilinear fall), curvilinear motion pattern (e.g., aircraft flight); solar system ellipsoid plane

Counter-Indications: Perturbations, turbulence, (e.g., Mercury's perihelion shift; earth "wobbling;" "bulging" at the Equator; and "bi-polar oblation")

Typical classification of the above force-types as differentiated Forms of thermodynamic Energy cycling is usually expressed as: "the strong force," "the weak force," Gravity, and Electromagnetism. However, all Force-Forms are Mass-in-Motion-dependent! Attempt was made to include "converted sub-forms" that are expressions of the same Fundamental Forces in "differentiated bounded-states" as embodied in Frame-specific thermodynamic cycling conditions. These Force-types appear in all Frames constituting the Mass-in-motion-Curvature-Force-Energetics Complex, in differentiated analog Forms that account for their specialized functional cycling patterns, in fulfillment of the Input-Process-Output Mechanism. All Force-Forms are reiterative or "quasi-replicating renditions" of Electromagnetism and Gravity, but within the environmental context of each kind of Mass-in-Motion-Frame, respectively, where corresponding thermodynamic conditions prevail for fulfillment of dynamic System equilibrium, e.g., The Earth as a "congealed Frame" operates like a Planet should; The Sun as "an excited Frame" displays a repertoire of processes and phenomena akin to those of a Star.

How is Gravity to be differentiated by other types of Force? Gravity as a Force does not "share" Electromagnetic properties of moving Mass as induced by particulate-electro-motive-charge characteristics of Matter. Gravity emerges from rotary Motion-Forms that are directly or indirectly caused by Electromagnetism. Thus, Gravity, as a Field-Force, is "particles-free!"

As an a priori principle, Force does not give rise to particles; particles give rise to Force, due to the Electromagnetic properties of Matter, because Matter is constituted of "counter-charged particles" (positively charged Protons; negatively charged Electrons; Charge-free Neutrons), complex interactions of which, engendering Force projections that are "re-translated" or "meta-transmuted" as Frames that are "operationally-functionally enabled" by Mass-in-Motion-Curvature-Force-Energetics."

Gravity is a different "type of force" in that its "absent indicators" do not correspond to any identified particles and/or indices, as displayed by the "other Force-types,' e.g., wave, particle, vibration, oscillation, ray, beam, radiation, etc… One is not aware of Gravity until a Unit-of-Mass "moves" in a particular way, in a particular direction, and in a specific vectored path or trajectory.

When one needs to see in the dark, one shines a light; when one wants to detect Gravity, one interjects an external Force "in the picture," causing an already Moving Unit-of-Mass or an Inertial Unit-of-Mass to change its operational repertoire or routines; and then, Gravity's presence is marked by its effects upon said Inertial Unit-of-Mass or upon said Moving Unit-of-Mass, as indicated by observable and measurable "rates-of-change" in the values of variables, parameters, and co-determinants characterizing its operational repertoire or routines, respectively, e.g., What happens when a non-moving rock (at inertia or at rest) is "pushed into motion" by an external Force such as a Person's foot; or What happens when an already moving rock such as one going downhill is "accelerated" at even a greater speed downhill, thus causing it "to crash into" or be "stopped" by a planted tree resting at the bottom of the hill!

Gravity is not a subsonic wave/particle, given that no "Graviton" was discovered, via experimental operations during which, Scientists had lowered the frequency and lengthened the wavelength of Sound, in order to determine whether "Gravitons" were somehow "related" or "connected" to sound waves in the lower decibels spectrum. "Graviton" has not been detected "above" or "below" the "Sound barrier."

In the same vein, given that the method of "echo-location" utilized by certain animals such as bats or dolphins is well-known and can be electronically detected by pertinent machines, if "Gravitons" really existed as real particles, so too, they should have been detectable via comparably or correspondingly designed electromagnetic-spectrum-based machines. But no such particles as "Gravitons," have ever been detected, whether on Earth or in void-vacuum Space.

Gravity has no known "particles," and no detectable "waves." Gravity is of a different "order," or "type of force." It is a Force that emerges from centri-vectored or pro-centric rotary forms of Motion in the presence of Electromagnetism, via application of Continuum Magnetic Field Curvature Energetics, or other "types of power" externally imposed upon "Units-of-Mass" that are either already moving or at inertia; and Gravity is operationally explained by its effects upon such Mass-Frames, as determined by factors characteristic of rates-of-change, such as, in magnitude and/or direction.

Thus, Gravity is Mass-in-Motion-dependent. And in the absence of Mass as the Object of observation upon which Gravity must be imposed, — or if no "Units-of-Mass" are interjected within the active Magnetic Field concurrently causing rotary Forms of Motion in the Frame under examination, — Curvature pressure-force binding-Energetics takes the form of a Magnetic Field-force expression only, e.g., In the absence of Gravity, there would have been only Revolution, but not Rotation. For, given that Gravity is Mass-in-Motion-dependent, — "Inertia" is a "mathematical convenience" that gives Scientists "a starting point" from which to "start a measurement" in the rates-of-change undergone by cycling variables. Gravity effects cannot be observed within an operational Frame if no Mass-Units are present to display imposition of its Motion-Force thereupon.

In order to demonstrate that Mass "curves or bends Space," something, such as a beam of Light, has to be projected within "its sphere of influence," so that the "Curvature expression" can be detected.

For example, why is there no Gravity-Field-Force within the Space-Station that is orbiting the Earth? In void-vacuum Space where the Space-Station is orbiting the Earth, it is not close enough to Earth Core-center-of-Mass "to experience" Gravity as a Field-Force in the same manner that an aircraft within Earth Atmosphere would "experience" Gravity.

Per Newton's Law of Universal Gravitation, the force of Gravity is directly proportional to the product of the two masses and inversely proportional to the squared distance separating them, as "reconciled" by the gravitational Constant, as in $F_g = G\, m_1 m_2 / r^2$.

Thus, the farther away from Earth Core-center of Mass and center-of-Gravity, that is, the greater the distance between the Space-Station and Earth-Core center-of-Gravity, the less Gravity-field-Force it will "experience."

The Space-Station, far away from Earth centers-of-Gravity and centers-of-Mass, if not contained within a moving, greater Object-with-Mass like the Earth within which Gravity-field-force is being exerted, there is no great Object with a "center-of-mass" towards which the Space-Station could "gravitate" as on Earth surface or in Earth atmosphere; thus, due to the great distance between Earth Core-centers of Gravity and centers-of-Mass, the Space-Station must "become like the Moon," but as an artificial or Human-created satellite.

A stone at rest, — which is in equivalent relationship with the relative G-force exerted upon its frame in conjunction with the pressure its Mass is exerting upon Earth surface vis-à-vis Earth center-of-Mass and center-of-Gravity, — can be pushed out of place: Inertial or at-rest Mass becomes the equivalent of 1-G Gravity-acceleration (at 9.8 m/s/s, as it also represents the equivalent of pressure-force being exerted by the Mass of the stone upon the surface of the Earth in pounds/square inch) — Thus, necessitating that an external Motion-initiating Force must be introduced for pushing the stone out of place.

The internal combustive violence of a rocket equipped with ignited volatile gases like Hydrogen and Oxygen can be thrust out of its place from inertial rest, away from Earth surface Gravity, propulsion-forces of which, must overcome the equivalents of Earth center-of-Mass pro-centric gravitational Force 1-G — the rocket's propulsion thrust must be greater than 1-G Earth surface Gravity-force at inertial Mass.

But no "particle" or "wave" is present in order that inertial-Mass "gravitates" towards the center-of-the-Earth, nor in order to engender an acceleration of Mass that is already-in-Motion — no "particle" or "wave" is needed, but only some Form of externally applied Force greater than G-1 Gravity force and/or greater than the momentum Mass-pressure-force being exerted by the rocket upon Earth surface, as necessary to initiate movement away from Earth core centri-vectored Motion.

The rate of acceleration of an Object by Gravity on the surface of the Earth is already known as, g = 9.8 meters/s/s; and pressure-force exerted by that same Object, is expressed as pounds per square inch or as lbs/inch^2. That acceleration-rate by Gravity is the equivalent-Force-exertion of the inertial pressure-Force "pushing" that same Object against the surface upon which it is "at rest."

Gravity-field-force is "entangled" with Magnetic-field-force projections at the solar-planetary frame of Newtonian Mechanics, hence the great difficulty in differentiating it from another field-force possessing inherent electromagnetic properties, i.e., Magnetic-field-force. However, a Magnetic-field-force embodies tensing-stressing-torque mechanisms of Curvature pressure-force-motion Energetics, as engendered by opposite electro-motive-charges that are "transubstantiating" the electromagnetic properties of matter; whereas Gravity, as a field-force, is but an emergent property from Mass-in-Motion that is directly caused by Magnetic-field-pressure-tensor-stress-torque-force.

A Magnetic-field-force necessarily projects Torque-motion-force; but Gravity-field-force is only a projection of pro-centric Motion-force towards a center-of-Mass, not necessarily accompanied by Torque-force-action (centripetal Motion); or away from a center-of-Mass (when an external force is applied to engender centrifugal Motion).

An Object "responding" to Gravity, will undergo either centripetal or centrifugal Motion; but an Object "caught" in a Magnetic Field, will undergo Torque-Motion-force-action that then "triggers" either Revolution (clockwise) or Rotation (anti-clockwise). Therefore, Torque-Motion-force-action is peculiar only to a Magnetic-Field, and thus, differentiates a Magnetic-field-force from a Gravity-field-force.

As Earth Magnetic Field externally operates to engender Moon geo-centric-synchronous Motion patterns, Gravity-Force upon Objects "moving" within the geo-sphere or upon the Earth, is not expressed fundamentally as a tensing-stressing-Torque Force-Form of Curvature Energetics; but rather, geo-sphere-bounded Gravity-Force is expressed as pro-Earth-Core-centric vectored Motion along Earth line-of-radius.

It must be remembered that General Relativity was conceptualized in conjunction with Special Relativity, and not in consonance with Newtonian Law of Universal Gravitation, because the way in which a Gravitational Field engenders Motion-force is differentiated from the way in which an Electromagnetic Field engenders rotary Forms of pro-centric Motion-force.

While Gravity can take many vector-Forms — rectilinear, linear, curvilinear, and/or rotary Motion-Forms, — Electromagnetism engenders only one specific vector-Form of Motion-Force: Pro-centric Rotary, or namely, as "Curvature Motion-Force Energetics," manifested either as Revolution (clockwise) or as Rotation (anticlockwise), e.g., Such as displayed by the Earth!

Electromagnetism will cause rotary vector-Forms of Motion-force such as Revolution and Rotation while Gravity-field-Force is "attractive" or centripetal towards a center-of-mass, or can also be centrifugal or away from a center-of-mass, but vectored along the Object's line-of-radius.

Though Electromagnetism and Gravity are in "entangled Field-exertion dynamics," lines-of-force engendered by Electromagnetism engage only tensing-stressing-Torque "Motion-action variables," or "dimensions" that must directly cause only rotary patterns of centri-vectored or pro-centric Motion-force-action.

Gravity gives rise to Motion-force Energetics along "lines-of-attraction" towards a center-of Mass, or along "lines-of-repulsion" away-from a center-of-Mass ,while "recombining" or "re-translating" differentiated vector-Forms comprising rotary, curvilinear, linear, and rectilinear patterns of Motion-force, either as naturally elicited in the Universe by inherent sets of predetermined Field-algorithmic principles, e.g., the Moon's geo-synchronous orbit; or as "artificially triggered" by Human Beings who exercise deliberate functional guidance-control over mechanisms or instruments that steer the Object in the direction of chosen Force-vectors, e.g., A jet-aircraft pilot can "move" the aircraft in any direction made possible by its purposively designed avionics.

It has also been suggested that the "Gravity particle" might exist below sound waves and frequencies that are inaudible to Human ears, audible range of which spanning from 12 Hz to 20 kHz, — the "hertz," named after Heinrich Hertz (1857-1894 AD), being the unit of frequency, or the equivalent to the number of "cycles per second" of a periodic process, event, or phenomenon.

The "Gravity particle" often referred to as the hypothetical "Graviton," has been inferred to exist as a "Sound property," due to the more extreme physical vibrations produced by lower-spectrum Sound waves that cause "oscillations of pressure" in the medium within which Sound waves are traveling, as indicated by propagation. Sound is measured as propagating waves composed of frequencies that cycle as "pressure oscillations" traveling through a medium such as air molecules, sweet water, ocean waters, or metal.

However, Motion-pressure-Force-action, as exerted by Gravity-Field, does not possess a "traveling index" such as wavelength or frequency that embodies a physical indicator belonging to the Gravity-Force itself.

The Force of Gravity has no physically detectable indicators such as wavelength, vibration, shock, frequency, particles, or cycles, as it is not marked by any physical properties, safe those displayed by moving Matter-Mass-in-Motion (or Matter-Mass at inertial-state) upon which the Force of Gravity is being exerted.

Gravity might cause rates-of-change in "pressure oscillations" typical to Sound waves traveling at specific frequencies or cycles, because it is a "given Force" permeating all physical events, processes, and phenomena. However, "pressure oscillations" that occur due to "alternating Differentials" between the pressure within the traveling Sound wave and the pressure within the medium, e.g., Earth Atmosphere, that is external to the traveling Sound wave, are not intrinsic properties of Gravity-Force itself.

The "inherent oscillation properties of Sound" do not cause the emergence of "equivalents" of Gravity-Force. But rather, it is Gravity that allows or makes possible, these "pressure oscillations" that occur in order that Sound can be heard and/or be instrumentally "recorded," and/or "re-translated" by Human ear-drums or tampanic membranes, as electrical impulses that travel to the Human brain, to be "registered," after conversion to cognitive understanding in the Human Mind or intellect, as a real out-there external physical phenomenon.

For, in vacuum void-outer Space where there also is "Gravitational Energetics" but in the absence of atmospheric molecules as a medium within which Sound waves could travel, and thus the absence of oscillation properties, Sound cannot propagate so as to be instrumentally "recorded," audibly "registered," or understood by the Human Mind.

The Sun is a Space-Object, an entity-in-Space, or a body-with-Mass constituting a "steady-state System" in dynamic equilibrium, nucleated processes of which, activated by "excited Electromagnetic Mass," the totality of which, forming or creating "binding-and-curving Energetics" as "planes-of-attraction," e.g., The ellipsoid revolutionary plane.

Thus, it is necessary to formulate "a theory of Continuum unification," in a manner that "reconciles" gravimetric Differentials between "congealed Electromagnetic states" and "excited Electromagnetic states" made manifest in Units of Mass-in-Motion, respectively, as articulated by Electromagnetism-and-Gravity Fields, in integration with "Curvature Motion-Force Energetics."

Consequently, the force of Gravity as a derivative of centri-vectored motion Forms (from Mass-in-Motion undergoing Revolution and Rotation) would be "normalized" with Gravity as "displacement of Mass-in-Space" (Motion initiated via a Force that counters "Inertia-Force" or 1-G Force at 9.8 meters/s/s rate-of-acceleration, as "naturally exerted" upon all "within-Earth" Objects at rest or moving upon Earth surface).

Devising a gravimetric accelerated momentum Field-Curvature tensor-stress pressure-force-Torque Constant, would also "reconcile" all "cycling differentials" and Motion-force Opposites that exist between all cycling Mass-in-Motion Frames.

****

HOW ELECTROMAGNETISM IS DIFFERENTIATED THROUGH DISTINCT BUT QUASI-SIMILAR FRAMES-OF-REFERENCE!

Let us remember that the Universe is constituted primarily of Continuum Space-Time and sporadically distributed or within predetermined distances from each other as Frames or Units of Moving Mass, the sum of which, "trapped" or "caught" within Continuum Field Curvature Tensor-Stress-Pressure-Torque Motion-Force-Action, as engendered first by Electromagnetism, and then, from which derives, Gravity.

In Effects of Nuclear Weapons, edited by Samuel Gladstone (1964), the author explains the difference in radiation energy dynamics thriving in "a unit volume of Matter" in temperature equilibrium. In a "conventional chemical explosion, the radiation energy density" is less when compared with the material energy, where the radiation energy is a small fraction of the total energy; whereas in a nuclear explosion, with extreme temperatures prevailing, "radiation energy density" is far greater than the material energy — estimations are that "in a nuclear explosion some 80 percent of the total energy may be present as radiation energy." (p. 24). In the sun, density of matter is a co-determinant variable in effecting pressure and temperature Differentials that "explosively process" the ratio of radiation energy to material energy — as in the roles played by lighter and heavier elements in plasma condensate dynamics, hence, these various "fuel cycles" to compensate for internal core and corona-vectored turbulence caused by moving nucleated-mass-energy Differentials. Solid Mass is a greater material Energy determinant in "congealed states" whereas condensate or nucleated plasma Mass controls as the primary radiation Energy determinant in "excited states." Corresponding "effect processes" due to ratio of radiation energy to material energy, follow a path patterned after the type of energy system involved, e.g., conventional TNT device as opposed to a nuclear device. No electromagnetic-nucleated plasma condensate radiation Mass is generated by a conventional explosive device.

Not only does the Sun possess the most tremendous Magnetic field force that inwardly compresses all fused nuclei, but each reactant mass therein possesses polarized magnetic field quanta that repel each other as a counter-force to core-centric compression. This dynamic engenders corona-directed convection Energetics — as internal "polarized force fields" that are pluri-potent in dispersive action and omni-directional in magnitude trans-morph in "counter-force entanglements" to transform the Sun into a convolution-vectored raging, restless, turbulent nucleated plasma condensate Mass inferno-dynamo.

In "congealed relationships" peculiar to certain Mass-Frames, such as the Earth-moon equilibrium steady-state, the interactive Forces exerted are connected to directly observable effects that do not require technologically attuned electromagnetic devices for detection and study.

Dr. Isaac Newton discovered the Law of Universal Gravitation before "the age of the light bulb." It is the discovery of electric current that gave the impetus to Scientists for research-and-experimentation in Electromagnetism, nuclear energy, mechanical engineering, fluid dynamics, quantum mechanics, and modern astro-physics.

Now, then, "as light is chasing light," or "as radiation Energy is chasing radiation Energy," are there limitations to such techno-electro-mechanical processes for purposes of data collection and operational analysis?

How then to test the validity of Mathematical Theoretics if the technology utilized by the researcher factors the researcher's "conscious apparatus" into the equation?

Planet Earth with all its attendant gravimetric co-determinants affects the experimental enterprise in conjunction with electronically-based measurement systems that add external energy to the quantifiable Frame. Accelerated particles moving at only a fraction of the Speed of Light will begin to emit cosmic radiation as a Form of "mass-energy-gains shedding" when either the electric field or the magnetic field is re-calibrated in order to correct for deviations from Curvature trajectories.

Electronic instrumentation, the Human brain, Human vision, research measurement systems, and particulate Motion-repertoires or Movement-routines, are all activated by differentiated Forms of Electromagnetic Spectrum radiations. "Energy's quantum" or "unit of expression" is carried by the Photon, the accepted particle of radiation. Forms of radiations comprised in the electromagnetic spectrum are also animated by "electro-motive field force motion" as engendered by oscillating electric charges and their associated magnetic fields. A television antenna utilizes the same principle in a Form amenable to audio-visual data and information broadcasting.

The Photon is the particle of common visible Light. All Forms of spectral radiations are expressed in terms of wavelength (in centimeters), frequency (in cycles per second or hertz units), and Photon energy (in electron volt units). The standard velocity for all Forms of radiation motion is the velocity of Light at 186,000 miles per second (or, 3.00×10^8 meters per second).

The Speed of Light implies traveling Motion and distance covered by Light as it moves. Wavelength as expressed in centimeters characterizes "length-distance" of the Unit-of-Spectral-Radiation-Energy, the sum of which, being a dimensional property of Matter-mass.

Frequency, in cycles per second, expresses "traveling motion" momentum-velocity Mass-pressure-force as "chunks" or "streams-of-quanta," dispatched/emitted/transmitted as iterative periods in "Thermodynamic Time."

Certain energies of electromagnetic radiations, in electron volt units, embody electro-motive-magnetic-charges that can burn and cause cancer, e.g., X-rays.

Space, Time and Gravity present no intrinsic physical dimensions, e.g., length, width, or height for isolation as "closed systems," except when their indicators are combined with or provided by Matter-Mass dimensions that physically cycle in Thermodynamic Time, hence the CGS system (centimeter, gram, second) and the MKS system (meter, kilogram, second).

Field equations utilizing tensor and stress mechanisms in attempts to interface Electromagnetism (field) with Gravity (field) engendering Curvature pressure-Force-Motion-Action, should also include Torque-Action Energetics. Torque-Rotary-Action-Energetics constitutes the essential Fundamentals of Electromagnetism! However, Gravity is exerted upon Objects-in-Space that are either at-rest or in Motion!

The Sun projects gravimetric electro-motive magnetic radiation emissions that climax into Curvature motion Torque pressure force thus engendering the ellipsoid plane. The solar System ellipsoid plane displays tensor, stress, torque, pressure, power, force, and Curvature elasticity parameters that re-combine to effect "interactive Fields of motion pressure force" within which Planets revolve and rotate as their overlapping regions of turbulence "re-translate" perturbations as "degrees of deviation" from rotary forms of centri-vectored patterns of motion, e.g., planet Mercury's perihelion shift.

"The Ellipsoid Plane Complex" constituted of centri-vectored rotary pro-centric Froms-of-Motion is not rigid but tensile in working operations. Interactive and recombinant factors, variables, parameters, and co-determinants of Momentum-Gravitational electromagnetic Mass-energy-radiation-Motion-Force form "a Holistic Curvature-Field of Gravitraction," that engenders "Torque pressure-stress-tensor-binding-Force Energetics," imbued by "elastic qualities of tensile strength" —thus, allowing for Mercury's Perihelion Shift; Earth "wobbling," dipole oblation, and equatorial "bulging."

These "elastic tensor-stress-Torque-Force metrics" acting upon the Planets cause them to revolve and rotate "in response to" counter-vectored dynamics engineered by centripetal and centrifugal Motion-Force-interactions. 'Relativity Effects' are induced by Motion-Force-

Opposites animating solar-inter-planetary centers-of-Mass, centers-of-Gravity, and centers-of-Field. Their "bending-curving-effects" drive Planets to project their "Momentum Curvature Gravitraction Field Energetics" in Force-exertion-Patterns that are geared for "merging as counter-clashing-overlapping intersections." Where their overlapping intersection-regions interface, climax "Counter-Vectoring Motion-Force Sets" that alternate for fulfilling the universal thermodynamic Mechanism (Input-Process-Output) while sustaining dynamic System equilibrium: between pushing and pulling, contracting and expanding, compressing and dilating, compacting and distending, stressing and relaxing, tensing and releasing. Hence, the Solar System Ellipsoid Plane of Revolution is analogous to "a stretched rubber-band" — bounded by intrinsic "Curvature-Field Gravitration Energetics" — that is prevented from collapsing inwardly, on the one hand; and on the other hand, that is inhibited from outwardly planetary dispersal! Fine-tuned integrated Dynamic System Ellipsoid-Plane Equilibrium reigns, while also accommodating with "tensile flexibility," slight planetary deviations from routine repertoires that "hold the Solar System together."

Though rays, waves, and particulates are detectable, the invisible force they elicit, must be indirectly ascertained by the presence of moving-Mass(es) that interface with intersecting, overlapping, or interactive spheres of relative influence.

An interactive sphere-of-influence can be referred to as a "Field," in the same manner that Gravity-induced "attraction-repulsion dynamics" between centers-of-Mass that also "bend each other's Space," can be termed as the "Bounded Tensile-Force" that causes Mass(es) to "coalesce together" via "counter-vectored Gravitraction Energetics."

Both "Magnetic-Field pro-centric rotary Force" and "Gravitraction Field-Counter-Force" cause things to "gravitate" towards each other, but in various Forms that display differentiated patterns of Motion. And that's why, in order to test the theory that massive Space-bodies "bend" or "curve" a beam of Light passing near them, or in their vicinity, there has to be an actual experiment that involves thrusting a Light-beam into their Mass-contoured Space so as to have it interface with their respective, range-specific, "Fields-of-influence."

However, the same Relativity-effect is activated by electronic instrumentation that injects "external Energy force" into the experimental Frame at the same time that the routines or repertoires of accelerated particles are being measured for mathematical quantification, hence the difficulty in simultaneously determining their position and momentum.

Within the quantum mechanical Frame, a quasi-similar effect is observed as micro-Curvature, micro-Gravity-field, micro-Electromagnetic-field, and forces induced by Opposite-particulate-charges that cause displacement-in-Space, in conjunction with experimental instruments electronics, together, engender particulate repertoires interpreted as "waves of probability" due to particulate Mass-energy-gains that are "re-translated" into compensatory trajectory dynamics that are then re-interpreted as unpredictable-chaotic "random Motion-Forms."

But in the Universe and in Nature, nothing happens by random chance accident. The concept of "probability" serves to assuage uncertainty born of Human failure or inability to fathom or decipher all the physical cause-and-effect relationships accountable for the particulate repertoires being instrumentally observed.

A particle can be technologically detected directly, as Photon energy embodied in visible Light is seen by Human eyes. But a Field requires the presence of cycling Mass-in-Motion-Frames to demonstrate its operationally active presence.

The magnetic Compass will seek Earth North Pole, as it performs a uni-polar pseudo-electro-magnetic operation of "invisible attraction," thereby providing by means of indirect detection, a way of knowing that the bi-polar or dipole Magnetic Field really exists.

Thus, we see visible Light with our biological eyes but must technologically and mechanically detect a Magnetic Field.

Curvature pressure-Motion-Torque is caused by a Continuum Force. Continuum Curvature Energetics (Continuum Torque-Force) is the Fundamental Force that is originally activated upon Matter-Mass by Electromagnetism, from which ensue, all other Forces.

Continuum Space-Time Curvature follows from the interfacing of gravitation with electromagnetism, the sum of which, referred to as "Gravitraction Field Energetics." And Quantum Mechanics follows from electromagnetic counter-charges, causing displacement-in-Space or inducing Rotary Pro-centric Motion-Force upon Matter-Mass, as framed by Relativity-dynamics between Continuum Curvature and Cycling Energetics, e.g., As the Earth rotates and revolves, ecological processes and hydro-geo-meteorological events are taking place, all of which, in fulfillment of the universal Thermodynamic Mechanism (Input-Process-Output), while sustaining dynamic System equilibrium.

Newtonian Mechanics — in conjunction with Electromagnetism, Relativity, and Curvature — and Quantum Mechanics, represent Frames of Understanding within which Force and Motion as induced by the dual-action of Electromagnetic Fields and Gravity-Fields or their equivalents, simultaneously cycle in centri-vectored rotary patterns, geared for unifying the Universe as a Whole-Energy System in Thermodynamic Equilibrium.

The mechanics of the Quantum Frame presents complexities, when in isolation, that gave rise to the discovery of "the Uncertainty Principle," as propounded by Werner Heisenberg (1901-1976 AD), whereby the position and momentum of a particle cannot be measured or ascertained simultaneously, the causes of which, being also uncertain. Causes of "Uncertainty" could have arisen from "experimenter or observer effects," the measurement process, electromagnetic-spectrum-based experimental devices or instruments utilized in research experiments, and variability-in-range properties inherent in universal events, cosmic phenomena, and natural processes.

Atomic-ionized nuclear plasma-condensate fuel is a radioactive complex in which electro-motive particles interchange charges in compressed core-centric polarized electro-magnetic chain-reactions engendered via extremely hot thermonuclear temperatures and "Gravitation-Magnetic Fields pressure-Torque-forces, creating convection dynamics that engineer "solar tornadoes" that expel plasma rains, flare, winds, and coronal mass ejections.

In the same manner that Earth ecology is powered by pressure and temperature Differentials and Oppositely-vectored-ionized-air-currents-Motion-Forces that climax into "cycling tornadoes" or "aero-vortices," the Sun itself undergoes rates-of-change in its Cycling variables and determinants that attest to its own prerogatives for "negotiating" Energy

Conservation and for "navigating through" Entropy Processes, while it must also fulfill the thermo-nuclear requirements of the Input-Process-Output Mechanism for sustenance of dynamic Helio-System equilibrium.

In a so-called "free fall state," gravimetric parameters, variables, and co-determinants that commonly operate to regulate natural phenomena and events, are taken as "Constants," e.g., a ripe apple falling from a tree is said to be "in a free fall;" a coin tossed into the air is said to go into a state of "free fall" once released by the Person throwing it; a rocket ship expelled from Earth-orbital-bands-of-rotary-Motion are said to then enter a state of "free fall," so to speak, as soon as it enters outer-void cosmic Space.

However, "free fall" is a notion utilized to denote the absence of additional external Forces or Man-made Forces that would have then contributed to the routines or repertoires of the apple, coin, or rocket ship. For, "free fall observations" ensue from pre-determinant variables that crafted the resultant routines and repertoires.

Gravity, wind turbulence, surface topographical features, throwing-force with which the Person released the coin, angular momentum at Earth-surface-contact for yielding final resting position, etc . . . operate to determine its "post-fall resting position" consequent to Motion-Force variables, co-determinants, and parameters.

Thus, in the Universe, end-results cannot be in a state of "free-fall" and are not "open-ended."

For, every process, phenomenon, or event, — even though appearing to be "random" or "stochastic," accidental or due to chance, and hence, said to be subject to "the laws of probability,"— is pre-determined by fixed laws of Physics that possess applicable ranges, variations in execution, and Mass-dependent diversity-of-Effects, as well as "peculiar specificity" in the structured-Forms they assume, e.g., Planetary structured-Form; Star structured-Form, yet the same Energy-Source, because "Energy is never created nor destroyed but always TRANSFORMED."

But in essence, by created Design, there is holistic synergism and symbiotic inter-relations between the operational determinants of all these "Thermodynamic Energy-Cycling Structured-Forms," conducive to "Universal-System-Unification," for maximum organized complexity and whole-System integrated Co-Ordination, via application of all the Principles that inhere in the Laws of Physics, to processes, phenomena, and events.

Regardless of how "theoretically fragmented," "randomly disjointed," or "conceptually compartmentalized" the Universe appears to Mathematicians-Physicists who rely upon electromagnetic-Spectrum-based instrumentalities for "data collection and analysis," in factual Reality, the Universe is One Entity that exists in "Perfect Unification for Continuum Thermodynamic System Equilibrium" in "Effective Fulfillment of the Input-Process-Output Mechanism of Thermodynamic Integration."

It is in this context that the Laws of Thermodynamics cannot engender "anarchy," "disorder," "chaos," "randomness," "chance," "accident," or "free fall."

Integrated-Thermodynamics fosters "structured functional organization" of steady-state Frames that are together cycling as Whole Energy Systems in "shared-overlapping-interfacing-intersecting" Continuum interactive operations, for the sustenance of created Earth Life-Support Systems that secure "Phenotypic Geneticization" for each particular Species, e.g., In adherence to the First Law of Bio-Genesis, Life perpetuates from pre-existing Life of the Same Kind: A Sperm-Whale will not give birth to a Spider or a Scorpio!

Conservation Principles predominate over Entropy Processes, as "the strong force" prevails for pro-centric or centripetal Motion-Forms over tendencies of "the weak force" to engender dispersive, counter-centric, or "centrifugal equivalents."

"Things gravitate towards each other" even as fulfillment of the Input-Process-Output Mechanism in differentiated Patterns-of-Thermodynamic Operations engenders Frame-cycling Differentials from Motion-Force Opposites.

In short, in accordance with above-expounded scientific analysis from applicable Principles that inhere in known-valid Laws of Physics, it is conclusive then, that no other end-result but the current Universe could have emerged from those intrinsic God-designed physical Laws. Newton's Prime Mover being God, then God's prime instrument is: "Thermodynamic Electromagnetism" as an intrinsic Property of Matter-Mass from which ensue all other physical processes, events, phenomena, variables, parameters, and co-determinants, especially Motion-Force and Gravity-Equivalents for "vivifying," "triggering," or "quickening" Continuum Space-Time Curvature Energetics.

"Probability" and "Infinity" are mathematical constructs-of-convenience that define immeasurable quantifies and proportions conceptualized to continue to "never-ending limits" or to be "pushed to their maximum expansion." However real in functions performed, they are "anthropocentric formulations" of "sensed-apprehended-intuited-felt Uncertainty" being inherent in Biological-Spiritual-Intellectual Human Nature! However gifted we are intellectually, Kant's "Der Ding an sich" remains an imperturbable and impregnable "Form of Resistance-Force" to our ever reaching complete fullness of Knowing!

We are sentient, emotional, moral, and "psychological" Beings with spiritual knowledge and understanding that facilitate Judgment-Choices in Decision-Making.

Though "Uncertainty" is mathematically configured as "Probability" and "Infinity," we ought to spiritually and morally "handle Uncertainty" in ways that foster God-inspired Hope, in that, confidence in God's love for us empowers us with Faith — "Faith is the substance of things hoped for and the evidence of things not seen." (Hebrews 11:1, KJV, Holy Bible).

Thus, God-inspired Humanism is not "secular," for "Love your neighbor as yourself" is God-founded. Christ, the Son of God also became Human, the Son of Man. "Theistic Humanism," having had a Biblical foundation, then, is therefore, a healthy concept that sustains our love for God and our love for each other. \

Human Beings are worthy, for God cares for us and continuously enacts his Will for our advantageous benefit as we rely upon his loving divine Grace, Mercy, and Providence. When framed in this perspective, "Theistic Humanism" climaxes into healthy self-concept and does not result in "the worship of Nature," the "deification of Man," "the fetishism of Ethnicity," "the reification of Self," or "the idolatry of Materialism."

In "the tossing of a coin," results are determined by actual physical Laws that process co-determinant variables for certain outcomes. The concept of "Probability" as mathematically formalized, allows for "statistical projections" of outcomes, but cannot predict the actual occurrence of "single events."

Why did Halley's Comet follow the same path, visible from Earth, every 75 years, without ever "losing steam"? It does not "remain parked" in another corner of the Universe while waiting for its Solar System appearance – Halley's Comet had followed "a circuitous Orbit."

Why does the Sun remain as a Star, without ever changing in essential nature?

Why does each specific Planet remain as it is, for thousands of years, without ever changing in essential "Whole System Energy state?"

As each cosmic Space-body-mass cycles via articulation of Conservation Principles and Entropy Processes that climax into "Natural Resource Thermodynamics," be it a Planet or a Star, its specifically designed Energy state is sensitively dependent upon initial conditions of Creation that pre-determined its fulfillment patterns for the Input-Process-Output Mechanism, from which it cannot deviate, except "within allowable ranges of variability," after which, it must return to dynamic System equilibrium, e.g., the Earth "wobbles" and is slightly "flattened" at each Pole; Planet Mercury undergoes a Perihelion Shift; Planet Jupiter radiates more Energy than it absorbs.

There is no "Law of Transformation of Time;" there is no "Law of Transformation of Space." For, only "what happens to Matter-Mass in Continuum Space-Time" determines "qualifications" or "quantities" for Time-and-Space.

There is only the Law of Transformation of Energy. Because Continuum Time-and-Space has no "physical indicators" or "thermodynamic markers," only Energy can be consistently scientifically manipulated: Energy is Matter-Mass-dependent! Time-and-Space is not!

A Force is exerted in terms of "Angular Momentum Motion-Energetics" as prescribed by gravitational Curvature.

"A Field-Force" should be expressed as a cycling unit of "Mass-in-Motion-Curvature-Torque-Energetics!"

As above analyzed, universal phenomena cannot transpire in a "free-fall," nor can they be as "indeterminate" as events hypothesized in the "Cat thought experiment" as proposed by Erwin Schrodinger (1887-1961 AD), in the sense that physical phenomena have "catalytic cause-and-effect mechanisms" that work independently of "the Frame-of-Mind of the Observer."

But it is required that the Observer-Experimenter takes precautions that account for all active variables that would compromise the validity of his or her findings, e.g., accounting for his own relative displacement-in-Space if he or she is operating "from a moving Frame," from the vantage point of which, he or she is not able to derive validly obtained data that are conducive to authentic scientific analysis for true-to-Reality results.

The Schrodinger "thought experiment" illustrates a "problem of Mind" rather than one of data evaluation. All universal processes, cosmic events, natural phenomena, and "Earth ecological repertoires," are governed by specific physical Laws whose "pre-set Frames-of-causality" within "ranges-of-application" cannot give rise to "random chance," chaos, or "un-causality"— For there are always a "cause-and-effect Mechanisms" to effect observable events and experienced phenomena, e.g., An earthquake does not "happen by accident;" nor does a tornado occur "due to random chance."

Gravitational Field and Magnetic Field cause-and-effect relationships that give rises to "the strong force," Energy Conservation, and pro-centric centripetal Motion-Forms climaxing into predominance of Continuum within Space-Time Curvature, are responsible for the Speed of Light having a fixed value of 186,000 miles per second in a vacuum, the sum of which, being "pegged" to "Angular Momentum Motion-Force parameters" that are pre-determined by Electromagnetism and Gravity.

And thus, it is not "due to random chance" that no process, phenomenon, or event within the Thermodynamic Frame of Continuum Space-Time Curvature Energetics can "travel" faster than the Speed of Light. For, given that "the fastest of all Thermodynamic Entities" in the Cosmos is Light-radiation-Energy, the Speed of which, cannot be exceeded by any other Object or unit of Matter-Mass in the Universe — Light-radiation-Energy which itself must "slow down" in a "medium" such as Water, then, it is incontrovertibly factual-and-true that every other Object or Entity within that same "Universe-Frame" must "travel" slower than the Speed of Light. Hence, cause-and-effect relationships effecting scientific "sequence-of-events" resulting in a Photon being "the lightest particle" or as "nearly mass-less as possible;" and thus, why isolated-particles accelerated in a Cyclotron "beyond their routine-normal-regular "Thermodynamic Cycling Velocities" as they are "approaching the Speed of Light," must "shed their Mass-Energy Gains" as cosmic radiation units of Energy. For, in the Cosmos, in the same manner that there is "Specific Gravity," then, "specific-Object-Mass" (e.g., the Specific-Mass of a particular Particle — an Electron, for example) is directly "tied-to" its intrinsic pre-determined "Specific Thermodynamic Cycling Velocity," e.g., A Planet cannot "travel as fast" as an Electron; nor can a Rocket-Ship ever "travel faster" than an Alpha Particle!

Consequently, a so-called "Tachyon World" so prevalent in Science-Fiction movie-productions, and in some "theoretical Physics quarters," is an immutable impossibility! There cannot ever be any such things as "Tachyons!"

Every process, phenomenon, or event, even though appearing to be "random," and hence, subject to "the Laws of Probability," is already pre-determined by fixed Laws of Physics, Principles of which, "offering ranges-of-effectiveness," such as "the limits-in-strength" of a Magnetic-Field; variability in co-determinants and parameters, such as numerical values, ratios, and proportions characteristic in their rates-of-change while thermodynamically cycling for fulfillment of Frame-specific operational functions; and diversity, in the Thermodynamic-Forms they assume, given that "Energy is never created nor destroyed but always TRANSFORMED," e.g., Electrons revolve around their Atomic Nucleus (Quantum Mechanics): Planets revolve around their Solar System Star (Newtonian Mechanics); A blade-stem assembly with various blades designed "for moving-the-air" by means of an Electric Fan, revolves around the copper-coil producing the Magnetic-Field for inducing a rotary Form-of-Motion (the Fan's electric motor); — The same Thermo-Electro-Motive-Dynamic Principles apply to propeller aircrafts, helicopters with rotor-blades assemblies; jet-propelled aircrafts via a "jet-engine" having "micro-mini-blades assemblies on a centrally-located revolving Shaft," the sum of which, achieving the same results from taking advantage of "air-pressure-Differentials" for "producing propulsion and/or thrust."

For example, the "tossing of a coin" is "not a random-chance-probabilistic event" in the sense that it offers a "range-of-possibilities" within which a specific number of "throws" will result in "heads," while the remainder of the "throws" will result in "tails." But "each throw-event" presents its own Relativity variables whose proportions are physically measurable and quantifiable, in terms such as "inertial throw angle," initial Curvature throw-force, angular momentum at launch, momentum-Force during "throw-flight," traveling velocity, air resistance or drag, magnitude of contact-impact with the ground-surface, throw ground-surface-topological-features, angular-momentum Conservation, density of the coin, weight of the coin, thickness of the coin, faithfulness of its manufacturing production to Properties of a Geometrical "Circle-with-thickness," the coin's kinetically-induced final resting position/angle, etc, . . .

Therefore, these physical variables belonging to particular Laws of Physics "primarily touching" Gravity, Mass, Inertia, and Force have specific numerical values that "suffer rates-of-change" during "throw-flight," the complex interplays of which, conducive to mathematical computations, even though the complexity involved in measuring them, precludes the formalized conceptualization of a mathematical formula for experimental reproducibility by independent Human Experimenters that would embody said experiment's operational, procedural, and environmental dynamics.

Gravity, Space, and Time present analogous perplexities of mathematical formulation when Mass-in-Motion Force Energetics yield Cycling Differentials and Oppositely-vectored-Forces to be factored into "A General All-Encompassing Totalizing Theory of Force" that accounts for Electromagnetism as the foundational-fundamental Property of Matter-Mass par excellence.

Could Time be "curved" or "suffer bending" as "embedded" within the Continuum Space-Time Complex while "Field Curvature Gravitraction Energetics" (could be either "attraction" or "repulsion") are being applied to Mass-in-Motion-Frames that are cycling in "temporal Time" or "Thermodynamic Time?" No! Not thermodynamically possible!

("Field Curvature Gravitraction Energetics:" When an Electron "loses a Photon," it "descends to a lower Orbital" — for "pro-centric-Nucleus centripetal attraction-action;" when an Electron "gains a Photon" it "ascends to a higher Orbital" — for "centrifugal counter-centric-Nucleus repulsion-action!)"

Time is a measurable quantity encapsulating a "quantum-state-distance," — such as "from Point A to Point B" — "found-to-be-trapped" in "re-translating," "apprehending," "catching," or "registering" iterative periodic rates-of-change in "cycling Mass variables," into "Thermodynamic Motion-Processes Embodiments" as "durations" of Energy Transformation, e.g., From starting-point A to destination-point B, or from "a congealed state" to "an excited state."

Cycling Frames pursue dynamic System equilibrium while "progressing" or "transmuting," "from Energy-state to Energy-state" even as Continuum Space-Time is occupied by Thermodynamically-Operating Electromagnetic Matter-Mass that is primarily constituted of "Gravitational Particulate Volume" (with "Curvature-Within-Substances" that are "caught" in three-dimensional Voluminal Density,)" hence, the existence of Continuum Space-Time, the sum of which, having no authentic attributes, no intrinsic qualities, and no inherent characteristics of-their-own: It is Matter-Mass "that gives dimensional-substance" to Continuum Space-Time due to its Electromagnetic Properties from which emerge Field, Curvature, Gravity, Motion, and Force as it "continuously occupies" Space-Time!

Time is "embedded" in thermodynamic operations of Mass-in-Motion-Force Energetics as these Frames fulfill the Input-Process-Output Mechanism in apparently "stochastic cycling patterns" that inhere in Continuum Energy Transformation to which must also apply Entropy Processes, thus giving rise to Gravimetric Differentials, i.e., Every process, phenomena, or event "must thermodynamically cycle," from a Beginning-Point to an Ending-Point.

Thus, Continuum Space-Time Dimensions are Matter-Mass Dimensions as Matter-Mass "occupies" it. Time Dimensions are Mass-in-Motion Dimensions while Matter-Mass is engaged in Energy Cycling Transformation under thermodynamic conditions of Curvature Pressure-Tensor-Stress-Torque-Force Binding Energetics.

A cosmic Frame operating as a "whole Mass-Cycling System" cannot "consume" Time and Space and then proceed to "replenish" its "supplies of Time and Space;" it can only produce, utilize, and replenish "conditions of Energy Transformation."

Time-and-Space, even as a Continuum, is not "a consumable category, entity, or quantity;" — Nor does it partake of the thermodynamic-electromagnetic physical-material properties of Matter-Mass.

Energy is a specific property of Mass — Electromagnetic Mass. Given that Matter-Mass is finite in both Substances and Dimensions, even within the vast infinite-void of the limitless Cosmos, therefore, the Total Energy of the Universe-as-a-Whole is also indeed finite: This means that he Sun will eventually "suffer Terminal Entropy".

In the full context of Thermodynamics, whole Energy systems seek dynamic equilibrium via uniquely specified cycling ratios, properties, conditions, Laws, and proportions.

Earth natural processes work synergistically and symbiotically for ecological efficiency and dynamic gravimetric equilibrium. Since every process climaxes into a "global" or "wholistic by-product" by engaging the Whole Energy System in its symbiotic totality, then "functional entropic outlets" must be devised for such by-products, e.g., We, Human Beings, do exhale Carbon Dioxide that is utilized by plants and trees in photosynthesis for manufacture of chlorophyll; trees and plants produce Oxygen which we, Human Beings, utilize in respiration.

But not all "functional Entropy by-products" are amenable to transformable utility, e.g., dioxin is a deadly poison.

Then, what about nuclear waste? Radioactive Decay or "Half-Life" is not fast-enough for ultimate protection of Life — Human Beings as a Biological Species have a life-span that is much shorter than "the Half-Life" of nucleated waste-products! Individual Human Beings do die, though the contemporary Generation continues!

Can Nuclear Waste be assembled-compacted in a rocket's payload bay and propelled towards the Sun at the greatest velocities possible?— Would not this be the safest method of nuclear waste disposal, given that the Sun is already a nucleated whole-Energy system and void-cosmic Space is already filled with "excited radiation Energy?"

The Universe displays an intrinsic affinity for what we've come to know as "information processing," in quantifiable proportions, ratios and equivalencies that are consistent with its "Created programming" for Energy Transformation whose cycling patterns resonate with sensitive dependence upon initial conditions of Thermodynamic Conservation.

Hydrogen Protons in Human brain neurons return to their original polarity after being "de-polarized" while the Person undergoes a Magnetic Resonance Imaging (MRI) medical procedure. Operational patterning parameters are already pre-configured in every whole Energy system that is "pre-programmed" to achieve dynamic System equilibrium as a "steady-state entity" specifically designed for purposeful functions.

Temperature, pressure, density, and other gravimetric Differentials that affect the rate at which restoration of dynamic System equilibrium is accomplished, are "re-normalized" via "hidden determinants" (or alpha-numeric Constants), the ranges of which, can be ascertained from "processing routines, or repertoires" and/or "operational activities" that cause a counter-Entropic reaction in a System undergoing drastic changes, geared at sustaining Qualitative Conservation, e.g., When strenuous activities cause the Human Organism to exceed homeostatic equilibrium temperature, this dynamic "triggers" sweating that then proceeds to "cool the Human body" in restoration of parameters controlling for dynamic System equilibrium.

The Solar System is in gravimetric-Field dynamic System equilibrium; the Human body is in homeostatic equilibrium; the Earth is in ecological Systems equilibrium.

All complex Macro-Systems reflect stable equilibrium patterns of Mass-in-Motion-Energy-Force Cycling, even though they are built upon apparently much-smaller, yet as complex, dynamic Micro-Structures marked by "intricate Motion-Force Information Processing Mechanisms" — The physical world, erected upon the Atom; and the Biological Organism, "built" upon DNA.

This is the Universal Organizing Principle: The organization of universal Reality rests upon the Principle that every Macro-Structure is erected upon a Micro-Structure-Complex unrevealed to the Human eye, at first, and then discovered by "remote sensing" microscopes that help unravel its composition, structure, content, organization, processes, parameters, functions, and cycling patterns, demonstrating at every stage, level, or step, "increasing levels of complexity," e.g., DNA's double helix structure within which are embedded Hydrogen-bonded base pairs with a Phosphate-Sugar backbone.

Are there any non-hypothetical so-called "sub-atomic particles"? Is the sub-atomic structure itself built upon a level lower or more invisible level than the Atomic Frame? Micro-structures that give rise to Macro-Frames yielding "increasing orders of complexity" possess qualitative characteristics that typify them as "building-blocks" of greater structures; however, these specific qualitative characteristics usually have "a limiting signature type" that channels all their pre-imprinted "building -block programming" into one special kind of end-product, e.g., for Matter, it is the atom constituted of Protons, Neutrons, and Electrons; for biological Life, it is DNA, made-up of Genes that are themselves composed of Chromosomes.

Human Beings, as a Biological Species, have predispositions or capacities amenable to a Species-Specific Genetic Foundation or Basis, i.e., Human Beings (genotype) qua Human Beings, due to Species-Specific DNA-Genetic Manifestations, (genotype) possess certain attributes, qualities, capacities, predispositions, and characteristics uncommon to other Forms of terrestrial Life. However, it is important to note that, though DNA is Species-Specific in accordance with the Laws of Bio-Genesis, DNA is not deterministic of the contents and direction of abstract or transcendent personal skills and knowledge and social skills and knowledge yielding activities such as touching Human Spirituality, Religious Morality, Systems of Belief, Creative Expression, and/or Exercise of Faith (phenotype).

But how these innate capacities or predispositions are developed, fulfilled, and actualized depends upon multi-causal factors that are often referred to as "consequences of Nurture" as circumscribed by "environmental or social conditions, or "educational conditioning," as synchronized or attuned in conjunction with God-endowed Free Will that allows Human Beings to make deliberate choices, albeit, in the presence of occasional circumstances that can prove to be beyond the reach or control of the Individual Person, e.g., a birth defect; a vehicular accident; a natural disaster; an economic depression, a Stock-Market Crash, etc...

Due to boundaries imposed by the prevalence of the Speed of Light, "the strong force," pro-centric Motion-Forms, and Energy Conservation as universal standards determining "cycling Frame routines;" due to Motion-Force Opposites that yield Energy-Cycling Differentials; and due to the role of Entropy in the thermodynamic operations of Cycling Mass-Frames, possibilities are not endless for discovering physical Laws that govern "unified, interactive, overlapping spheres-of-Force(s)" animating Continuum Space-Time Curvature Energetics. Even there, limits also apply!

Given that "the congealed Energy state" is thermodynamically differentiated from "the excited Matter state," all transactional Input-Process-Output operations — erected upon Quantum Mechanics as framed by "the Properties of Electromagnetism" (for "congealed Energy states"), e.g., the Planet; and upon Quantum Mechanics as framed by "the Properties of Electromagnetism" for Thermo-Nucleated Chain-Reactions Chambers (for "excited Matter/Energy states") e.g., the Star, (wherein are also manifested: "the strong force" "the weak force," electro-motive charges, magnetic fields) — are driven by gravitational relationships of thermodynamic cycling, climaxing into Relatively-Overlapping-Frame-Differentials that "interface" on account of Field-Curvature-Motion-Force-Energetics, as impacted by "positional locus" or "place of operation" within a greater whole system, the complex sum of which, being pre-configured by Gravity, Relativity, Conservation, and Entropy within Continuum environmental conditions or "Continuum thermodynamic theater interactions," e.g., the Earth is the third Planet from the Sun as it revolves around the Sun and rotates upon its own 23-degree tilted axis within the Solar System ellipsoid-tensile Binding Energetics Plane.

The Atom, the Molecule, the Planet, the Sun — all lie on a "Thermodynamic Continuum" that makes them "responsive" to each other's "theater of influence interactions," respectively.

We usually refer to such a "theater of influence" as a "Field." Any time that there is "a theater-of-thermodynamic-influence" interfacing with a Mass-in-Motion Cycling Frame, it can be mathematically-considered as "a Field." Hence, why we speak of "Gravity-Field(-force)," "Magnetic-Field-(force)," "Electric-Field," "Curvature-Field-(Energetics)."

Curvature-Field Energetics encompasses all the Fields in all their relative interactions that interface for sustaining Motion-Force-Continuum within the Space-Time-Continuum! Given that when we say "Field" we are referring to "an area of manifestation" or "a theater of influence," then, we might say that there is also "a Conservation-Field-routine" (Spring to Summer, each characterized by specific ecological interactions routines, respectively), as well as "an Entropy-Field-repertoire" (Autumn to Winter, each marked by specific ecological interactions repertoires, respectively).

However, at our present stage of technology, not all enumerated or discovered Fields or "theaters of interactions" are mathematically quantifiable as "relative equivalencies" that can be embodied or represented by equations reflecting measurable numerical computable values.

Given that in the Field of Physics, Energy is defined as "ability to do work," then, even the "Universal Property of Continuum" embodies "Forms of Energy" heretofore unaccounted for; and hence, our referencing it as: "CONTINUUM CURVATURE MOTION-FORCE ENERGETICS," a term accounting for "all Forms of Energy" that "do work" within the Universe, generally manifested as "Motion-Force," regardless of the kind or type of "work" effected by such "Motion-Force(s), respectively, e.g., whether Revolution, Rotation, Radiation, — or "Gravitraction:" whereby both Electromagnetic Fields and Gravity Fields and their respective equivalents, are operating to effect observed processes that are "found to be" in "Entanglement-counter-dynamics" — "Gravitraction" (centripetal-repulsive / and centrifugal-attractive as in the Infernal Solar Nuclear Dynamo wherein Nucleated-Atomic-Motion-Force-Opposites and rates-of-change in Plasma-Condensate-Energy Cycling Differentials are "kept-entangled" by tremendous, extreme, and immensely powerful "Solar Magnetic Field Energetics," the whole complex of which, operating "as a Form of exploded implosion," or "explosively imploded nucleated violence dynamics" ("an explosive implosion"), or "as a Form of imploded explosion," or as "compressed explosive nucleated violence dynamics," ("an imploded explosion") whereby inwardly-vectored contraction and outwardly-vectored dilation "oscillate" between inner-Core-centripetal Motion-Force Energetics AND outer-Corona-centrifugal Motion-Force Energetics.

Because of "Continuum Curvature Motion-Force Energetics" activated by Mass-in-Motion-Frames undergoing "Continuum Thermodynamic Energy Transformation Cycling" as "embedded" within "The Universal Continuum Space-Time," we live in "an on-going ever-present Infinite Physical-Material Universe," albeit with "a life-span" or with "a half-life" extending to billions and billions and billions of years into the Future!

Consequently, all Whole System Energy-states — or converted Forms of Energy — are interconnected as overlapping-interfacing-intersecting "spheres-of-influence" via these relatively entangled "comprehensive Curvature interactions" as made-manifest within the Frame of the same universal Curvature pressure-force tensor-stress Motion-Force Property, to which is assigned or attributed: "Continuum Energetics," in which the whole Universe partakes, and for which, logical, cause-and-effect mechanisms of operation, in the Forms of mathematical expressions, must be discovered.

There is a "Spectral Continuum" for the Curvature pressure-Force Motion-Energetics Property, as embodied in all Frames, respectively, as "they overlap in relatively entangled effects" within Continuum Space-Time, which is analogous to the way in which Electromagnetic Radiation Units belong to a Spectrum that range from X-rays to radio waves, e.g., All of the following events occur in accordance with: Continuum Curvature Motion-Force Energetics: An apple falls from its branch to Earth Core-centric surface; an Electron is "attracted" by a positively charged Proton; the Moon orbits the Earth or is "attracted" by the Earth; Planet Mercury has a precession shift in response to "rates-of-change in gravitational (Field) Energetics" at Perihelion.

Curvature as condensed within the Atomic Frame; Curvature as articulated within the Planetary Frame; Curvature as embodied in molecular change dynamics; Curvature as pressurized-compressed within the solar plasma condensate frame; Curvature as expressed within the ellipsoid tensing plane: All of these expressions of Curvature are Forms of gravitationally projected binding energies, finding equivalents in Magnetic-field-force, Gravity-field-force, Space-displacement-force from opposite particulate-charges, "the strong force," and "the weak force," all of which, made manifest via "complex agitations" from centripetal-centrifugal composite-recombinant Forces having rotary, rectilinear, linear, and curvilinear Patterns-of-Motion that are centri-vectored or pro-centric in magnitude, strength, vector, and direction.

Given that Matter as moving-Mass has an electromagnetic constitution — arising-emerging from positive Protons, negative Electrons, and charge-less Neutrons, then, all Frames thrive in a Continuum comprised of overlapping Mass-in-Motion-energy-Force structures that thermodynamically cycle in "temporal Space-Time thermodynamics," as they articulate complex, yet relatively equivalent Forms, of "re-translated" or "re-transmuted" Electromagnetism.

The Universe was created as a "unified Mega-theater, Mega-sphere or Mega-Field of cycling Frames" constituted of Units of Mass-in-Motion as activated by Continuum Space-Time Curvature Energetics as integrated via physical Forces taking Forms that model "spheres or regions of interactive influence," at times referred to as "Gravity-force," at other times referred to as "Field-force."

For all practical and mathematical purposes, Gravity can be considered a "Field" in the same manner that Electromagnetism is considered a "Field," to merely represent the recognition that a "Field" constitutes "an area of manifestation' or "theater of interactive influence."

"The strong force" is but another Form of "sphere of interactive influence" between the positive-Proton and the charge-less Neutron, whereas "the weak force" is another Form of "sphere of interactive influence" between the Atomic Nucleus and the Electron, as well as "an area of Force manifestation" between interactive nuclei undergoing reactive changes for molecular formations, and/or that cause atomic decay and radiation emissions.

In short, overlapping cycling-Mass-Frames "gravitate towards each other" as they interact in transforming Energy in accordance with the Input-Process-Output Mechanism, whose specialized functional conditions, engender Energy-cycling Differentials and Motion-force Opposites that distinguish one Frame from another Frame, e.g., the Earth is a Life-planet that "wobbles" a little bit, and displays slight bi-polar oblation; Mercury undergoes a perihelion shift; Jupiter "keeps" 67 Moons.

Each Planet's uniqueness in "negotiating" Cycling-Differentials and Force-Opposites, consistent with its overlapping functional operations while thriving to maintain dynamic System equilibrium, constitutes the characteristics of its "Gravimetric Signature."

Continuum Space-Time is said to be "curved" or "bent" due to cycling mass-Energy-motion-Force properties that are measurable and quantifiable. Such quantifiable properties, for purposes of mathematical convenience and efficiency, are commonly calculated in terms of the CGS System (centimeter, gram, second) or the MKS System (meter, kilogram, second). There is also the (miles, pounds, second) MPS System, such that the Speed of Light can be expressed as 186,000 miles/second rather than as 3.00×10^8 meters/s or as 299,792,458 meters/second; or such that pressure can be expressed as pound-force/inch^2.

Matter-mass-in-Motion "embodies" many forms of Energy, either as "congealed Energy" (non-nucleated Energy,) or as "excited Matter" (nucleated Energy). However, because Matter-Mass Energy is in Continuum-Motion while being "embedded" within Continuum Space-Time for thermodynamic Conservation dynamics and Entropy-processes, "Time-cycling," — as a "period-distance of Temporality traveled from point A to point B," e.g., from Sunrise to Sunset, manifested in quantifying processes inherent in Mass-in-Motion-Energy-force Frames, i.e., Revolution and Rotation — cannot be factored as a "curved parameter," even when being practically treated as a "mathematical-physical dimension" that is "embedded" within gravimetric-and-Field-properties of Mass, variables and parameters of which, undergoing rates-of-change due to thermodynamic Energy cycling.

For, then, "Time" has become "Thermodynamic Time" or "Temporality periods" having to do with measuring "rates-of-change" within the Energy-Transformation Properties of Mass-in-Motion as "put-into-effect" by the Gravity-field and the Electromagnetic-field.

"Time" is no longer "an abstract conceptual variable" but gets transmuted or transubstantiated into a Matter-Mass-in-Motion bounded physical-dimension, akin to distance, Energy, Force, and Momentum. Rates-of-change measured in Time as "temporal thermodynamics," also qualify Time as a physical dimension serving as a signpost or landmark for Energy Transformation Processes that must yield functional outputs consistent with universal Space-Time Curvature Continuum.

Time-patterns of Mass-Energy cycling partaking in Motion-Force dynamics, are representative of Curvature Energetics of Electromagnetism, as "re-translated" unto differentiated Forms, via equivalents of Gravity-field-force and Magnetic-field-force parameters, "condensed" or "emulsified" within "Matter-Mass-in-Motion-frame interactions," e.g., But it is the "trajectory path" of Rotation (counter-clockwise or anticlockwise) and Revolution (clockwise or anticlockwise) that is being "curved" as rotary Forms of centri-vectored Motion-force as followed by Mass moving thereunto, and not Time itself which is only measuring the duration of the Rotational Period or Revolutionary Cycle. Space-curving or Space-bending does not occur without Matter-Mass "acting thereupon;" and again, Matter-Mass moves in "a curved trajectory" or in "a bent path," and in the case of Rotation, it is the Earth that is rotating and not the 24-hour duration-period of Rotation Time; likewise, in the case of Revolution, it is the Earth that is revolving around the Sun in an ellipsoid-curved trajectory, and not the 365 ¼ day-duration-period of Revolution Time.

As previously stated above, Time has no material-physical substance and no particulate constitution that can cause it to be "curved" or "bent" — For, only "things physical" can "curve" or "bend," such as a piece of metallic tubing made-up of pliable Aluminum or soft Tin.

But Time, is an abstract-intangible yet still-real "Thermodynamic Operant" in measuring the phenomenon of Energy-Transformation, to be operationally factored as "a catalyst" for fulfillment of the universal Input-Process-Output Mechanism as Mass-in-Motion-Frames are performing their pre-designed or pre-determined gravimetric functions in maintaining Earth Life-support Systems.

Thus, "Thermodynamic Time," marking the duration of natural phenomena or of cosmic events, signifies, in other words: "Temporal Thermodynamics" (Thermodynamic rates-of-change in Time as framed by gravimetric parameters that determine the magnitude and vector of Energy Transformation processes).

"Thermodynamic Time" involves two major categories of measurable parameters: Processes and Operations of Mass-Energy-Frames "as they are moving" — from point A to point B or from one thermodynamic-Energy-state to another — through Continuum Space-Time.

"Thermodynamic Time," as such, is indirectly measured in terms of numerical values assigned to operational co-determinant variables of Mass-Energy-Frames that have practical portent for technologies produced in industrial manufacturing, because such variables belong to Motion-Frames of Matter-Mass undergoing Thermodynamic Energy Transformation Changes, rates of which, being measurable as Timed-processes, Timed-events, or Timed-phenomena, operations of which, "stochastically-streaming-elapsing in measurable duration periods."

In such cases, moving-Matter-Mass possesses "physical dimensionalities" or "material properties" that can be "objectively handled" for measurements that are tangible, definite, and precise, as "concretized" when made manifest "moving Space-Objects' that are operationally marking "distance-in-Time" (duration) or "distance-in-Space" (displacement-movement), specific functions or end-results of which, made quantifiable by concise scientific measurements of parameters and variables, dimensionalities and properties, and/or qualities and "operationalizations," as "practicalities" having to do with thermodynamic cause-and-effect transactions or relationships yielding consequences that impact either Human Beings in society and/or things in the environment, e.g., The Sun's "coronal mass ejections, "solar flares," or "solar winds," lasting several minutes or even several hours, can disrupt telecommunications devices and broadcasting instruments and the services they provide, due to their thermodynamic impacts upon microwave towers and revolving Earth-orbiting artificial satellites, functional operations of which, relying-dependent upon the Electromagnetic Spectrum as "a transporter-carrier" or "traveling-encoder" of "messages-and-instructions" that "transfer" audio-visual and/or graphics Data and Information from one place to another, from one Person to another, from one institution to another, or from one computer to another.

In addition, extreme or increased Solar activity can effect rates-of-change in geo-atmospheric-meteorological variables that might result in draught conditions lasting for months, in certain regions of the Earth, the sum of which, impacting agricultural production and harvest, and thus, having deleterious consequences for Human Life, as well as for oceanic animals, and for landmass fauna and flora on our Planet.

Physical things "moving" or "at inertia" in the Universe commonly possess three material dimensions, from which emerge other physical parameters yielding operational processes, elapsing in Time as a "Thermodynamic Operant" measuring duration of Energy-Transformation processes.

Thus, Time is "indirectly re-transmuted" into a scientifically observable and experientially factored measurement of a quantifiable parameter, as "spent-duration-period" or as "lapsed-duration-observed," as "Matter-Mass is in Motion," going from point A to point B," either such as a vehicle in Space, or such as a chemical reaction-process undergoing rates-of-chance in temperature and pressure, e.g., "Thermodynamic Time" as perceived/measured on a clock-device per "movements of its hands" such that mechanical or digital clock-operations are utilized in measuring "processes-with-duration," as proceeding from functionally operating Matter-Mass "occupying" Continuum Space-Time, e.g., An artificial satellite, such as the Space-Station, relative to its Mass-in-Motion around Earth-orbit, can have a Orbital Velocity of 117,000 miles/hour.

Given that Time is a Thermodynamic Operant, then, as Space is "occupied" by Matter-Mass that is constituted of "material-physical weight," — physical dimensions of which, being length, width, and depth or height, while "moving" in Continuum Space-Time, — "Space-dimensional characteristics," such as "borrowed" from Matter-mass that is "occupying" Space "within the thermodynamic periodicity of Time," can be delineated in terms of other measurements as "bounded" by Matter-mass — such as perimeter, circumference, surface, volume, speed, momentum, force, etc . . . , — pertaining to Mass-in-Motion-Frames that "cycle in periods of temporal Time" or "Periods of Thermodynamic Time," in addition to which, tensor metrics, torque factors, and stress Dynamics can be added due to "Curvature-pressure-Force-Energetics" being "clothed" as "Gravity-field-force and Magnetic-field-force equivalents" that account for centri-vectored or pro-centric Motion-Forms, i.e., Revolution and Rotation, from which emerge "Gravity-Force analogs," such as the "equivalent Forces" that can be generated by an aircraft engaging in acrobatic flight movements, or by an aircraft built to fly at speeds exceeding the Speed of Sound, otherwise referred to as "supersonic aircrafts."

The qualitative characteristic of "Continuum Curvature-Motion-Force Energetics" attached to Continuum Space-Time comes from application of "Accelerated Field Momentum-Force" as exerted by "Electromagnetism in-entanglements with Gravity" upon Mass-in-Motion-Frames that possess physical-objective / material-value-laden "Space-Time Thermodynamic-Energy-Cycling dimensionalities," i.e., characteristics, qualities, and attributes that can be quantifiably observed, experienced, detected, analyzed, defined, operationalized, measured and practically applied.

Thus, there cannot be any "Curvature Quality" attached to "stand-alone Time" or to "Time-qua-Time per se," such that its qualities could be comparable to "Curvature characteristics" that are "attached to Space" — given that Space, is a real entity that possesses "an external self-evident indicator" manifested as "the container of Matter-Mass" that "occupies" or "indwells it," from dimensions of which, Space indirectly obtains its "material dimensionalities," e.g., Volume of Space as "contained" or "enclosed" within a cubic cardboard box!

But, in contradistinction, Time has no "physical properties" or "material dimensions" in-and-of-itself. However, as a Thermodynamic Operant, Time "makes itself felt" as the operational catalyst within which Energy Transformation processes take place as "duration periods."

Matter-Mass is "contained" in Continuum Space-Time such that Space is "occupied" by Matter-Mass, and hence, cause-and-effect relationships giving rise to "measurable dimensionalities" imparted to Space by physical dimensions of Matter-Mass. However, Time marks the duration of a specific process that yields rates-of-change in variables and parameters of Matter-Mass having numerical values that can be quantified while undergoing thermodynamic cycling operations — But in such cases, Matter-Mass cannot impart any physical-material dimensionalities to Time-qua-Time.

But, even before the advent of Thermodynamics becoming an additional component that effects "Operationalization of Space-Time," (as Mass-in-Motion Field-Curvature Energetics), Time-qua-Time is in "an Inertial-Continuum-state with Infinite Space," such that it undergoes "a trans-mutational dynamic" that unifies it with Space as "Continuum Space-Time," up to the minute that Matter-Mass "enters into the greater cosmological picture."

In short, Space can exist in or with Time, without the presence of Matter-Mass "indwelling it;" which means that Space-and-Time can exist together as a Continuum without the presence of Matter.

But, Matter-Mass-in-Motion must have Continuum Space-Time for "an indwelling" so that its cause-and-effect thermodynamic processes, events, and phenomena can be operationalized for quantifiable measurements in accordance with Equilibrium Principles that inhere in afore-designed or pre-established Physical Laws, applications of which, practically proceeding during periods of Time (Thermodynamic Time).

"Time-Conservation embedding" into Infinite Space, is mathematically inescapable, for, "Curvature gravitraction" or "Thermodynamic Field(s) Energetics" requires "quasi-Self-iterative operations or proceedings," e.g., such as Sunrise and Sunset, — to be attached to geographical-spherical dimensionalities or "geo-spherical localities," — complex thermodynamic transactions of which, being precursors of geo-hydro-meteorological-atmospheric conditions predicated upon specific amounts of Solar radiation upon Earth-surface from contact-angles by Solar radiation, and dependent upon spectral intensities of "Solar Field-Gravitraction radiation-Energy received" by the Earth, which only "Continuum Curvature can transform" for numerical quantization, or to which only "Continuum Motion-Force Curvature Energetics" can impart "mathematically quantifiable modalities."

Gravity-field and Magnetic-field momentum Motion-Force Energetics engender Continuum Curvature Motion, i.e., Revolution and Rotation, "within the dimensionalities of Continuum Space-Time," the complex interplays of which, to "befall" every Object "within reach of" or "within the specific-Range of" its "projected spheres-of-influence" or "theaters-of-impact."

Hence, cause-and-effect interactions predicting rotary Motion-forms "clothing" artificial satellites, launched destination of which, being within Earth rotational-revolutionary Space-of-"acquired" centri-vectored Motion, lying between the Moon and Earth outer-atmospheric layers, both of which, constituting the "boundaries of Earth orbital bandwidth," relative to the Mass of the "rotating Object" and the "orbital distance" at which said Object is "positioned-located."

How could Time be "curved" in the same way that Space is ascertained to be "physically curved" via matter-Mass properties of material dimensionality?

How is the "physical dimensionality of Time" to be "mathematically operationalized?"

Is not physical-Space-Curvature dimensionality "operationalized" via matter-Mass momentum motion-Force Energetics having quantifiably measurable variables, parameters, and co-determinant factors? What makes this possible?

"Continuum Curvature," is thought to require special mathematical treatment via "Riemann-Space Geometry," as proposed by Bernhard Riemann (1826-1866 AD), encompassing multi-dimensional gravimetric parameters that engage a comprehensive "total-system matrix" for how "the dimensionalities of Matter-mass" — such as friction, material strength, matter density, crashing, pushing, pulling, crushing, field tensing dynamics, "gravimetric stressing and special folds," etc …, — affect Continuum Space-Time, as made possible only by the interposition of Time as "the unification variable."

Though "Time dimensionality" — (whether Time itself is "linear," "non-linear," or is also "physically curved-bent") — remains a contested subject in current-day theoretical Physics, as demonstrated above, Time has no physical parameters or variables that can be numerically measured apart from operationalized processes undergone by Matter-Mass-in-Motion-Frames.

Given the crucial role its "operationalized thermodynamic cycling" would play in devising a mega-unification Theory of Continuum Curvature Motion-Force Energetics for the Universe that would finally integrate/unify Gravity with Electromagnetism, Quantum Mechanics with Newtonian Mechanics; and Relativity Dynamics with Field-Force Motion Energetics, (as well as with "the strong force" and "the weak force,") — Then, the conceptualization of Time as "a Thermodynamic-Operant-of-Measurement" and the mathematical formalization of Time as a "gravimetric Factor," are therefore paramount to new, ground-breaking advances, in Physics Theoretics.

"Time-cycling," or the measurement of physical indicators of Thermodynamic Transformation processes in a "moving Object," requires "material dimensionality," as attached to properties of Mass-Energy within the "spheres of influence" of Continuum Curvature Motion-Force Energetics.

"Continuum Field-Curvature Motion-Force" — induced by "entangled" Magnetic-field and Gravity-field equivalents — implies physical dimensionality. "Gravity Curvature motion-Force" — induced also by Magnetic-field and Gravity-field equivalents — implies physical dimensionality. Thus, rotary motion-Forms imply physical dimensionality. Gravity-pressure-tensor-stress-Force implies physical dimensionality.

In sum, given that all categories of "things-that-are" persist through cause-and-effect "indicators of influence," then, no quantifiably measurable Property exists in this material Universe without "attachment" to "Mass-in-Motion-Frames' physicality-dimensions" that prove its existence as a functional Frame-of-Operations. In consequence, nothing happens "by random-chance accident probability." Every event, process, or phenomenon is amenable to cause-and-effect modalities, mechanisms, principles, transactions, and/or relationships.

The "things-that-are," functionally operate in accordance with specific, succinct, physical Laws possessing "spheres of influence," with quantifiable characteristics having "ranges-of-operation" that every Object having Mass-in-Motion-Frame must observe or adhere to.

Mass-in-motion-Frames, cycle in Time, within "Fields of physical-Curvature-tensing-Force."

It is by the utilization of alpha-numeric Constants that "renormalize" or "reconcile" all "overlapping operations" of all "moving Frames" as bounded by "thermodynamic ranges-of-effectiveness,' within the universal Time-cycling of Continuum-Energy, as "an integrated Whole" with dynamic System equilibrium!

For example, "Thermodynamic Time" can be "operationalized" or "concretized for functioning" as "distance traveled from point-A to point-B;" or as an "unfolding" process, event, phenomenon, or mechanism, that is being quantized: From place to place, from state to state, from Form to Form, from recurring periodic cycling to recurring periodic cycling, from process to process, e.g., Rotation, "as a Motion-Force" indicates that Planet Earth has "moved in Space" self-reflexively upon its own 23.8-degree tilted axis, within a 24-hour period of Time, also constituting a Time-period of One Day.

In this context, all Force-categories and all Space-Time parameters could be theoretically construed and "mathematically operationalized," as "physical-dimensionality variables" of "Continuum-overlapping moving-Frames, all of which, possessing "momentum Mass-in-motion-Energy."

To which "category of dimensionality" of "Things-in-Physics," belong height or depth, length, and width? To Matter as a Unit of Mass-in-Motion! This "category of Things-in-Physics" is called: "The Three Dimensions of Space," to which Time is added Time as "the Fourth Dimension." But Time is rather a Thermodynamic Operant during periods of which proceed Energy Transformation operations.

The "three-dimension quality of Space" or Space as "3-D Space," is therefore defined by physical variables/parameters and co-determinants that belong to the "Matter-category."

In addition to these three physical dimensions, Matter has "inertial-pressure-Force" called "weight," and "momentum moving- heaviness" otherwise referred to as "motion-Force;" thus, weight ("inertial moving-Force") and Mass (inertial pressure-Force) are also "equivalent Properties of Matter."

However, Weight and Mass are often used interchangeably, because, when originally, inertial-Weight-and-Mass become "kinetic Matter" as Mass-that-is-in-Motion, or as "kinetic moving-Mass," (no-longer inertial) then, they exponentially configure together into the motion-pressure-tensor-Force Energetics with which Matter-mass-weight is characterized for encompassing and embodying "Gravity-force equivalents" or "G-force equivalents," e.g., An acrobatic aircraft executing maneuvers, the complex interplays of which, equating "G-force equivalents" that can be "greater than Earth 1-G" or using engine-thrust exerting a "force-equivalent" that registers at "5-Gs," as experienced, sensed, or felt by the biological body of the Human pilot.

When Matter-Mass (inertial Weight-force and inertial Pressure-mass) is in Motion that is then accelerated: Momentum-Force and Motion-Force also become "Properties of Matter" ($F =$ ma). Because of Continuum-Curvature-Tensor-Force Energetics acting upon Matter-mass-in-Motion, the force of Gravity is defined as "attraction" between masses ($F_g = G \, m_1 m_2 / r^2$ — discovered by Newton) — such that, Matter is referred to as "transformable congealed Energy."

But because of these same Continuum-Curvature cause-and-effect interactions, Matter is also transformed into nucleated Energy that acquires a "momentum-Force strength-and-intensity" (as in Relativity Dynamics — discovered by Einstein) — causing Energy to be referred to as "excited Matter" — ($E = mc^2$). And in the same manner that in the Classical Physics Frame, acceleration of an Object by Gravity on the Earth embodied in the constant $g =$ 9.88 meters\second\second or 9.88 m\s^2 also "compounds" velocity into Accelerated Momentum Motion-Impact-Pressure-Force via the equation $F = m \times a$, due to "seconds" (Time) operating at the squared-Exponential factor, e.g., A two-vehicle collision will inflict severe damages on both vehicles, including triggering cause-and-effect consequences resulting in death of drivers and passengers; then, in the Relativity Theory Frame, because c, the Speed of Light is also elevated unto the Speed of Light Squared, in $E = mc^2$, these exponential operations also elevate "congealed Energy" to "excited Matter" as "nucleated Mass" is re-transmuted unto "compounded" Radiation-Momentum-Motion-Impact-Pressure-Force, e.g., Nucleated plasma condensate radiation burns and kills!

Momentum-pressure-tensor-stress-Force from non-nucleated Mass-in-Motion (e.g., the Atom, as comprised within a "congealed Energy-form," such as Planet Earth), is "transubstantiated" or "re-transmuted" into "excited momentum-pressure-tensor-stress-Force" arising from nucleated Mass-in-Motion — originating from its transformation into "nuclear Plasma-force Energetics." ($E = mc^2$: Matter as "congealed Energy" is converted into "Excited Matter" or "nucleated plasma condensate Energetics" via a Form of "physical transubstantiation" that "re-transmutes Excited Matter" into a "Gravitraction-Radiation-Field-Force." In this context, E is "the relative equivalent of "Radiation-Force.")

Continuum Momentum Curvature stress-tensor-pressure-Force Energetics is also a property of "moving-Centers-of-Mass as projected as accelerated-projected from operating-transacting Centers-of-Field (Magnetic-Field(s) and Centers-of-Gravity (Gravity-Field(s) or their "equivalents."

Energy is also a property of Matter. Matter, being constituted of atomic particles that have "Electro-motive displacement-in-Space-causing Charges" in that are active in molecular changes or elemental combinations, essentially has a fundamentally "electromagnetic constitution."

Thus, Electromagnetism is also a property of Matter. Gravity-field, Magnetic-field, and Curvature-Energetics expressed as characteristics that are "transformational-Force states" engendered by electromagnetic properties of matter, then, Force is also a property of Matter.

Moving Matter-Mass under operational dynamics of Field-Force also displays rotary Motion patterns: thus, Rotary motion patterns are also properties of Matter-Mass. Thus, the "property of Force" itself that causes rotary motion patterns is also a characteristic of Electromagnetic Matter-mass-in-Motion within Continuum Space-Time Curvature.

In sum, (1) Given that the Universe only comprises two distinctly differentiated "categories of things-that-are," that is, Space-Time and Matter; (2) Given that Space-Time has no intrinsic physical-material attributes or dimensions of its own; and (3) Given that it is "Mass-in-Motion Thermodynamics" that gives Space-Time "the Property of Continuum" due to objective results from proceedings that fulfill requirements of the universal Input-Process-Output Mechanism, then, (4) ALL measurable variables, factors, parameters, attributes, characteristics, properties, and qualities, such as Electromagnetism, Gravity, Field, Force etc… that are mathematically expressed in numerical values or as quantified equations, originate ONLY from Matter!

Thus, given that Matter is tangible in Form, content, composition, constitution, and structure as expressed or "as embedded" within physical dimensions and material properties that are mathematically measurable and computable, then, it is therefore very ascertainable that both Electromagnetism and Gravity having one and only origin: Matter, can be integrated via a "grand unification theory of Motion-Force Energetics!"

So, is there any "measurable quality or property category" in the Universe that cannot be quantified as an intrinsic characteristic of Matter-Mass thermodynamically cycling in "Temporal Time" as Motion-Force Energetics?

All "quantifiable categories" are measurable properties of Matter-Mass thermodynamically cycling in "Temporal Time" as Motion-Force Energetics, hence, the CGS system (centimeter, gram, second) and the MKS system (meter, gram, second.)

What is accomplished by the operational thermodynamics of "the exponential-mass function mc^2" that is "the equivalent of Energy," needs to be "detailed" in a manner that maximizes understanding of Solar Gravi-Magnetic Field Radiation Emissions, complex projections and interplays of which, effecting all Force-Forms active" (a) Within the ellipsoid plane of revolution, (b) Within the Earth, (c) Within intermediate Mass-Frames, and (c) Within the Atom, amenable to distinctly differentiating "congealed Energy" (Classical Physics Frame) from "Excited Matter" (Relativity Frame)?

Determining their relativistic interactions within the Frame of transubstantiated exponential functional units-of-Mass "trapped" or "caught" in Continuum Curvature tensor-stress-pressure-Force-Motion could elucidate how to mathematically factor their "Gravi-Magnetic Field Transactions" as "Force Equivalencies," expressed via a General Grand-Theory of Universal Force Unification.

For example, acceleration of an earthly Object by Gravity-field-force is expressed as 9.8 meters per second per second; and the Speed of Light approximates 3.0×10^8 meters per second. But in the exponential-Mass function mc^2, the Speed of Light to the second power is 're-transubstantiated" into "meters squared per second squared," thus, also denoting "the operation of something even greater than a Force," complex practical applications of which, also including "Continuum Curvature Tensor-Stress-Pressure Motion-Force Momentum-Mass Energetics," as made operationally manifest by exertions of Magnetic-fields-of-force and Gravity-fields-of-force that effect rotary Forms of pro-centric Motion (As "Gravitraction," which includes both centripetal and centrifugal configurations, especially within our earthly environment, e.g., An aircraft goes forward, backward, sideways, up, down, or in "recombinant-vectoring directions thereof "— as in a 360-degree fulcrum of Motion-Force Energetics!")

Therefore, Electromagnetism (Magnetic field, "the strong force," and "the weak force") and Gravity as universal "measurable Field categories" can also be quantified as Time-cycling properties of Matter-Mass-in-Motion-Force Energetics, "re-translated" or "re-transubstantiated" into "tensor-stress-Torque characteristics intrinsic to Continuum Curvature Pressure-Force Motion-Energetics.

An equation implies the establishment of relative equivalencies between its intrinsic variables. Thus, A General Theory of Force that is to be embodied in "The Unified Field Theory of Continuum Curvature Motion Torque-Force" would have to "re-translate" all the "stochastic cycling manifestations" of Electromagnetism and Gravity as Fields-of-Force, "the strong force," "the weak force," and all the Forms espoused by "their equivalents," in conjunction with Quantum Mechanics and Newtonian Mechanics, Relativity interconnections dynamics and Curvature interactions Energetics, into "a mathematical calculus" that discovers "relative equivalencies" between Tensing-force, Pressure-force, Stress-force, and Torque-force" for re-transubstantiation into "A Continuum Motion-Transform Force-Dynamic" accounting for all Vectored-Forms of Motion-Force emerging from "Curvature Rotary Motion Torque-Force Energetics," i.e., Revolution and Rotation.

This "mathematical calculus" would integrate all cycling Matter-mass Frames into a universal whole Energy system Curvature complex, as the climax of all extant Forces that are fulfilling the "Input-Process-Output Mechanism of Energy Transformation," as "cycling Differentials" and "Motion-Force Opposites" combine to configure the routines and repertoires of Mass-in-Motion-Frames — the Complex Relative Summation of which, being "reconciled" via alpha-numeric Constants, conjoined with "conversion factors" that "re-normalize" all variables, co-determinants, and parameters, unto "proportional equivalencies" (direct or indirect).

These specifically structured Mass-in-Motion-Units, as "framed by Thermodynamics," "overlapping-intersecting regions of which," are being operationally linked-for-interfacing, via interactive alpha-numeric Constants that account for their overlapping, contiguous, and continuous structures that are processing Differentials in composition, content, organization, and patterns of thermodynamic cycling as well as Motion-Force-Opposites animating fulfillment of the Input-Process-Output Mechanism that is so intrinsic to Thermodynamics.

There are also sub-patterns of Motion, deriving from processes internal to whole-system Energy steady-states, as they "negotiate" Functional Operational Entropy — for example, on Earth, a tree might get dislodged by flood waters from heavy rains to follow a "curvilinear Motion pattern" towards Earth center-of-Mass, which it "negotiates" in terms of "compensatory dislocation-Motion patterns" due to obstacles in its path — even Earth surface constitutes "a limiting obstacle" preventing all Objects from "gravitating" towards its Core.

This "push-Form of caused Motion" by flood waters, though different in Form from the electromagnetic-Field-induced rotary form of Motion, also lies in a Continuum with Gravity-field-caused rotary Motion-Forms, because all Motion-Forms lead to Earth-Core center-of-Mass — Such Motion-Forms are also "traveling" along "attraction lines-of-force" that "seek-along" Earth-radius. These Motion-Forms constitute a Form of Core-centric Motion along Earth line-of-radius, as "negotiated" via paths or trajectories that account for apparent "surface and/or other obstacles."

Electromagnetic-field-forces cause centri-vectored or pro-centric Motion; Gravity-field-forces cause centri-vectored motion along Earth-lines-of-Radius. Rotary, rectilinear, linear, and curvilinear Motion-patterns are "re-negotiated Forms" of centri-vectored or pro-centric Motion Forms — while within "inner-Earth," Gravity-field-Force has "differentiated expressions" from "external Gravity-field-Force: the Earth has a surface, an atmosphere and an ocean. But because Curvature pressure-Torque-Force Energetics represents a priori Forms of universal Motion-patterns, it also causes or generates composite/recombinant Forms of Motion as Objects and processes "re-translate" the Electromagnetic Properties of Matter that are embodied or embedded within Mass-in-Motion-Frames.

It is the "Electromagnetic Property" of Matter-Mass as embedded in Frame-specific "re-translated Forms" that causes thermodynamically dependent Time-cycling processes to be an "Operant" or "co-determinant" of operational Energy-Transformation dynamics, i.e., Due to Thermodynamics controlling Curvature Motion-Patterns engendered by Electromagnetism and Gravity, the Earth is a "congealed Frame," whereas the Sun is an "excited Frame."

In the cosmographic vacuum Space environment where Stars and Planets operate, including in Quantum Mechanics and molecular-change bonding-dynamics, all Motion is caused by Curvature pressure-Torque-Force binding Energetics taking the Forms of "spheres of interactive influence" commonly referred to as "Fields," or as "places/environments where things happen," as expressed via equivalents of Electromagnetism, Gravity, "the strong force," and "the weak force," engaged in "triggered Opposite-Motion-patterns" wherein are "embedded" Frame-specific electro-motive thermo-dynamic Energy Transformation cycling-Differentials — (Or Frame-specific electro-motive thermo-nuclear Energy Transformation cycling-Differentials in the case of Stars.)

The Universe displays an affinity for "Forces of cohesion" that cause all Frames to "coalesce into Oneness" as energy transformation cycling Systems that consistently pursue dynamic System equilibrium. Given that Curvature propagates as "differentiated-momentum

Torque-exertions" upon variegated reference-Frames via "binding ties of Energy" between Space-bodies as Mass-in-Motion-Units, "Curvature pressure-Torque-Force" takes "cycling Forms" that create the need for a "gravi-electro-motive-magnetic Constant-metric" (incorporating Motion, Radiation, Mass, and Field(s) as subsets of Curvature-Torque-Force-Energetics) that will "reconcile" all Frame-specific Motion-Opposites and Energy-cycling Differentials into a coherent, internally consistent, "normalized" Universe that is "trapped" or "caught" within Continuum Space-Time, while it is unified-integrated by "Continuum Curvature Energetics."

The evidence for A General Theory of Force can only engage "comprehensive Continuum Curvature Energetics" that "triggers" rotary motion-Forms, with a dynamic that incorporates/includes all the numerous variegated Forms of demonstrated cohesiveness uniting-integrating Cycling-Mass Frames.

From the smallest or micro-level (atomic) through the intermediate level (molecular), up to the largest or macro-level (planetary), all universal Forces need to be integrated as one "Mega-Curvature pressure-Torque Field-Force binding Energetics" that engenders "Relative-Interactive Planes-of-Motion," complex interfaces of which, holding Matter and Energy "in Continuum Curvature Transactions."

As centri-vectored or pro-centric "Forms of inter-exchange," e.g. Solar system-Milky Way Galaxy relationship) Curvature Torque-Force Energetics is expressed via specific Motion-patterns akin to each respective Frame-of-Reference, e.g., rotary, rectilinear, linear, curvilinear and composite-recombinant patterns that ultimately pursue centri-vectored or pro-centric Motion.

Per cycling-Frame Differentials and Motion-Force-Opposites hat articulate function-specific processes of Energy Transformation, Curvature-Motion Forms are: Atomic nucleus-centric, Earth geo-centric, and Solar-System heliocentric.

It is due to prevalence of the atomic "strong force" that Curvature-Torque-Force Energetics, as expressed via Gravity-field and Electromagnetic-field effects, is oriented towards nucleus-centric, geo-centric and heliocentric Forms of Motion that comprehensively yield Energy Conservation. As far as Scientists can assess, there is no such thing as "Proton-decay."

Field-vectored rotary Motion, equivalents of Gravity-directed recombinant patterns, and Curvature pressure-Torque-force Motion-states, as well as "the strong force," "the weak force," are all intricately interwoven "retranslations" of the Electromagnetic Properties of Matter, as "entangled pro-centric motion-force exertion Forms," in order to sustain thermodynamic cycling-Systems equilibrium in Continuum Space-Time, via Pressure-tensor-stress-torque Dynamics that climax into "Gravitational ellipsoid plane elasticity Energetics."

The Unified Field Theory of Continuum Curvature Pressure Torque-Motion Force, would integrate Gravity and Electromagnetism, Quantum Mechanics and Newtonian Mechanics, Relativity dynamics and Field mechanics, within overlapping Frames that operate in Continuum Space-Time, as "entangled planes of Field motion-force Interactivity."

A General Theory of Force comprises all projected Force-forms that climax into Curvature pressure Torque-Force binding Energetics, ranging from Electromagnetism (magnetic field, "strong force," and "weak force") to Gravity, complex interfaces of which, effecting rotary, curvilinear, linear, and rectilinear Motion-Force-patterns/paths/trajectories, in all their "recombinant/composite dynamics," as they engender pro-centric Motion-Force-Forms, designed for Energy Conservation.

Each Frame, e.g., atomic, Earth surface, planetary, Solar System, Galaxy-wide, etc… encapsulates its respective prevailing Forms of pro-centric rotary Motion Torque-force. Due to "entanglements" of Electromagnetism and Gravity climaxing into Curvature Energetics, the atomic Frame displays Electron rotary motion –Forms that generally conserve angular momentum; the Solar System embodies planetary rotary motion-Forms such as Revolution and Rotation.

Within the Earth, equivalents of the Gravity-force climax into Core-centric motion-Forms along Earth line-of-radius encompassing recombinant/composite rotary, rectilinear, linear, and curvilinear patterns/paths/trajectories of Motion-Force.

In all these Frames, atomic and planetary, Solar System and Galactic, Curvature pressure Torque-force expressed as "pressure-tensor-stress-Torque motion-Force Energetics" (atomic Electron revolutionary orbital-levels plane and solar system planetary ellipsoid plane) must be accounted for, as effecting Electron trajectories incorrectly interpreted as "waves of probability;" as well as the perihelion shift of Planet Mercury; and Earth "wobbling" and "bi-polar flattening;" while also gravitationally differentiating between such "thermodynamic deviations" and internal Earth-surface, Atmosphere, and waterways-oceanic Motion-Forms, Earth Revolution and Rotation, all of which, being expressed as Core-centric motion-Force as "radially-vectored" trajectories, "along Earth-Core lines-of-Attraction" — hence necessitating alpha-numeric Constants to "renormalize" Electromagnetism (magnetic field, "strong force" and "weak force") and Gravity, and its variegated G-force equivalents, for "relative equivalency," even as each Frame, as Mass-in-Motion-Units, operates within its own Frame-specific "cadre-of-thermodynamic-environment conditions."

In each Frame, e.g., atomic, Earth surface, planetary, Solar System, Galaxy, etc. . . , Curvature binding pressure-tensor-stress Torque-Force Energetics in the Form of "Curvature Energetics," is applied respectively, in response to the "sphere/theater" of "Field-plane interactions," as synchronized with motion-Force patterns corresponding to their particular cycling mass-Frames, e.g., Electrons revolve; Planets rotate and revolve; Earth-surface Objects move in Core-centric motions-Forms along Earth line-of-radius which can take curvilinear, linear, rectilinear, or circular trajectory-Patterns.

As quickened by Motion-Force-Opposites yielding thermodynamic Energy-cycling Differentials in processes of Energy Conservation, these overlapping Frames intersect via "transformative regions of perturbation" (conjoined with Entropy-processes) that are "linked," "normalized," and "reconciled" via alpha-numeric Constants, (such as "the gravi-electromagnetic-Curvature Constant-metric," "complex entanglements" of which, engaging motion, force, gravity, radiation and field), that unify the whole Continuum Space-Time Complex, into a Mega-Curvature pressure-tensor-stress Torque-Force Energetics whose "Fields of interactive influence" bind interdependently cycling Mass-Frames, thus causing them into "gravitating towards each other for Universe-Oneness."

In A General Theory of Force, the key is to synchronize the metrics of "Curvature pressure-tensor-stress-Torque-Force Energetics," for "relative equivalency," with "all cycling Mass-in-motion-Frames" as they are processing Thermodynamic Energy Transformation, respectively, as effected by "conversion factors" and as "normalized" via alpha-numeric Constants that "reconcile" all-quantities, proportions, and values utilized in "interfacing" "Conservation-Differentials" and "Entropy-Differentials" emerging from Motion-Force-Opposites, for equilibrium universe-wide Curvature-Field-Torque Motion-Force Energetics-System Unification.

****